ISBN 978-1-5278-3757-7
PIBN 10892085

This book is a reproduction of an important historical work. Forgotten Books uses
state-of-the-art technology to digitally reconstruct the work, preserving the original format
whilst repairing imperfections present in the aged copy. In rare cases, an imperfection in
the original, such as a blemish or missing page, may be replicated in our edition. We do,
however, repair the vast majority of imperfections successfully; any imperfections that
remain are intentionally left to preserve the state of such historical works.

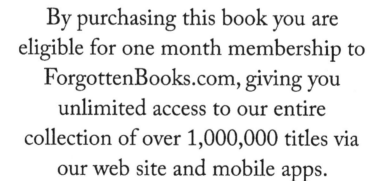

English
Français
Deutsche
Italiano
Español
Português

www.forgottenbooks.com

Mythology Photography **Fiction**
Fishing Christianity **Art** Cooking
Essays Buddhism Freemasonry
Medicine **Biology** Music **Ancient
Egypt** Evolution Carpentry Physics
Dance Geology **Mathematics** Fitness
Shakespeare **Folklore** Yoga Marketing
Confidence Immortality Biographies
Poetry **Psychology** Witchcraft
Electronics Chemistry History **Law**
Accounting **Philosophy** Anthropology
Alchemy Drama Quantum Mechanics
Atheism Sexual Health **Ancient History**
Entrepreneurship Languages Sport
Paleontology Needlework Islam
Metaphysics Investment Archaeology
Parenting Statistics Criminology
Motivational

INTRODUCTION TO
QUANTITATIVE GENETICS

INTRODUCTION TO
QUANTITATIVE GENETICS

D. S. FALCONER

Agricultural Research Council's Unit of Animal Genetics
University of Edinburgh

THE RONALD PRESS COMPANY · NEW YORK

FIRST PUBLISHED IN GREAT BRITAIN 1960

Printed in Great Britain by
Robert MacLehose and Company Limited, Glasgow

PREFACE

My aim in writing this book has been to provide an introductory text-book of quantitative genetics, with the emphasis on general principles rather than on practical application, and one moreover that can be understood by biologists of no more than ordinary mathematical ability. In pursuit of this latter aim I have set out the mathematics in the form that I, being little of a mathematician, find most comprehensible, hoping that the consequent lack of rigour and elegance will be compensated for by a wider accessibility. The reader is not, however, asked to accept conclusions without proof. Though only the simplest algebra is used, all the mathematical deductions essential to the exposition of the subject are demonstrated in full. Some knowledge of statistics, however, is assumed, particularly of the analysis of variance and of correlation and regression. Elementary knowledge of Mendelian genetics is also assumed.

I have had no particular class of reader exclusively in mind, but have tried to make the book useful to as wide a range of readers as possible. In consequence some will find less detail than they require and others more. Those who intend to become specialists in this branch of genetics or in its application to animal or plant breeding will find all they require of the general principles, but will find little guidance in the techniques of experimentation or of breeding practice. Those for whom the subject forms part of a course of general genetics will find a good deal more detail than they require. The section headings, however, should facilitate the selection of what is relevant, and any of the following chapters could be omitted without serious loss of continuity : Chapters 4, 5, 10 (after p. 168), 12, 13, and 15–20.

The choice of symbols presented some difficulties because there are several different systems in current use, and it proved impossible to build up a self-consistent system entirely from these. I have accordingly adopted what seemed to me the most appropriate of the symbols in current use, but have not hesitated to introduce new symbols where consistency or clarity seemed to require them. I hope that my system will not be found unduly confusing to those accustomed to a different one. There is a list of symbols at the end, where some of the equivalents in other systems are given.

Acknowledgements

Many people have helped me in various ways, to all of whom I should like to express my thanks. I am greatly indebted to Professor C. H. Waddington for his encouragement and for the facilities that I have enjoyed in his laboratory. It is no exaggeration to say that without Dr Alan Robertson's help this book could not have been written. Not only has his reading of the manuscript led to the elimination of many errors, but I have been greatly assisted in my understanding of the subject, particularly its more mathematical aspects, by frequent discussions with him. Dr R. C. Roberts read the whole manuscript with great care and his valuable suggestions led to many improvements being made. Parts of the manuscript were read also by Dr N. Bateman, Dr J. C. Bowman, Dr D. G. Gilmour, Dr J. H. Sang, and my wife, to all of whom I am grateful for advice. I owe much also to the Honours and Diploma students of Animal Genetics in Edinburgh between 1951 and 1957, whose questions led to improvements of presentation at many points. Despite all the help I have received, many imperfections remain and there can hardly fail to be some errors that have escaped detection : the responsibility for all of these is entirely mine. To Mr E. D. Roberts I am indebted for drawing all the graphs and diagrams, and I greatly appreciate the care and skill with which he has drawn them. I am indebted also to the Director and Staff of the Commonwealth Bureau of Animal Breeding for assistance with the preparation of the list of references.

D. S. FALCONER

Institute of Animal Genetics, Edinburgh
December, 1958

CONTENTS

INTRODUCTION

Quantitative genetics is concerned with the inheritance of those differences between individuals that are of degree rather than of kind, quantitative rather than qualitative. These are the individual differences which, as Darwin wrote, "afford materials for natural selection to act on and accumulate, in the same manner as man accumulates in any given direction individual differences in his domestic productions." An understanding of the inheritance of these differences is thus of fundamental significance in the study of evolution and in the application of genetics to animal and plant breeding; and it is from these two fields of enquiry that the subject has received the chief impetus to its growth.

Virtually every organ and function of any species shows individual differences of this nature, the differences of size among ourselves or our domestic animals being an example familiar to all. Individuals form a continuously graded series from one extreme to the other and do not fall naturally into sharply demarcated types. Qualitative differences, in contrast, divide individuals into distinct types with little or no connexion by intermediates. Examples are the differences between blue-eyed and brown-eyed individuals, between the blood groups, or between normally coloured and albino individuals. The distinction between quantitative and qualitative differences marks, in respect of the phenomena studied, the distinction between quantitative genetics and the parent stem of "Mendelian" genetics. In respect of the mechanism of inheritance the distinction is between differences caused by many or by few genes. The familiar Mendelian ratios, which display the fundamental mechanism of inheritance, can be seen only when a gene difference at a single locus gives rise to a readily detectable difference in some property of the organism. Quantitative differences, in so far as they are inherited, depend on gene differences at many loci, the effects of which are not individually distinguishable. Consequently the Mendelian ratios are not exhibited by quantitative differences, and the methods of Mendelian analysis are inappropriate.

It is, nevertheless, a basic premiss of quantitative genetics that the inheritance of quantitative differences depends on genes subject to the same laws of transmission and having the same general properties as the genes whose transmission and properties are displayed by qualitative differences. Quantitative genetics is therefore an extension of Mendelian genetics, resting squarely on Mendelian principles as its foundation.

The methods of study in quantitative genetics differ from those employed in Mendelian genetics in two respects. In the first place, since ratios cannot be observed, single progenies are uninformative, and the unit of study must be extended to "populations," that is larger groups of individuals comprising many progenies. And, in the second place, the nature of the quantitative differences to be studied requires the measurement, and not just the classification, of the individuals. The extension of Mendelian genetics into quantitative genetics may thus be made in two stages, the first introducing new concepts connected with the genetic properties of "populations" and the second introducing concepts connected with the inheritance of measurements. This is how the subject is presented in this book. In the first part, which occupies Chapters 1 to 5, the genetic properties of populations are described by reference to genes causing easily identifiable, and therefore qualitative, differences. Quantitative differences are not discussed until the second part, which starts in Chapter 6. These two parts of the subject are often distinguished by different names, the first being referred to as "Population Genetics" and the second as "Biometrical Genetics" or "Quantitative Genetics." Some writers, however, use "Population Genetics" to refer to the whole. The terminology of this distinction is therefore ambiguous. The use of "Quantitative Genetics" to refer to the whole subject may be justified on the grounds that the genetics of populations is not just a preliminary to the genetics of quantitative differences, but an integral part of it.

The theoretical basis of quantitative genetics was established round about 1920 by the work of Fisher (1918), Haldane (1924–32, summarised 1932) and Wright (1921). The development of the subject over the succeeding years, by these and many other geneticists and statisticians, has been mainly by elaboration, clarification, and the filling in of details, so that today we have a substantial body of theory accepted by the majority as valid. As in any healthily growing science, there are differences of opinion, but these are chiefly matters of emphasis, about the relative importance of this or that aspect.

The theory consists of the deduction of the consequences of Mendelian inheritance when extended to the properties of populations and to the simultaneous segregation of genes at many loci. The premiss from which the deductions are made is that the inheritance of quantitative differences is by means of genes, and that these genes are subject to the Mendelian laws of transmission and may have any of the properties known from Mendelian genetics. The property of "variable expression" assumes great importance and might be raised to the status of another premiss: that the expression of the genotype in the phenotype is modifiable by non-genetic causes. Other properties whose consequences are to be taken into account include dominance, epistasis, pleiotropy, linkage, and mutation.

These theoretical deductions enable us to state what will be the genetic properties of a population if the genes have the properties postulated, and to predict what will be the consequences of applying any specified plan of breeding. In principle we should then be able to make observations of the genetic properties of natural or experimental populations, and of the outcome of special breeding methods, and deduce from these observations what are the properties of the genes concerned. The experimental side of quantitative genetics, however, has lagged behind the theoretical in its development, and it is still some way from fulfilling this complementary function. The reason for this is the difficulty of devising diagnostic experiments which will unambiguously discriminate between the many possible situations envisaged by the theory. Consequently the experimental side has developed in a somewhat empirical manner, building general conclusions out of the experience of many particular cases. Nevertheless there is now a sufficient body of experimental data to substantiate the theory in its main outlines; to allow a number of generalisations to be made about the inheritance of quantitative differences; and to enable us to predict with some confidence the outcome of certain breeding methods. Discussion of all the difficulties would be inappropriate in an introductory treatment. The aim here is to describe all that is reasonably firmly established and, for the sake of clarity, to simplify as far as is possible without being misleading. Consequently the emphasis is on the theoretical side. Though conclusions will often be drawn directly from experimental data, the experimental side of the subject is presented chiefly in the form of examples, chosen with the purpose of illustrating the theoretical conclusions. These examples, however, cannot always be taken as substantiating the postulates that underlie the conclusions they illustrate. Too often the results of experiments are open to more than one interpretation.

No attempt has been made to give exhaustive references to published work in any part of the subject; or to indicate the origins, or trace the history, of the ideas. To have done this would have required a much longer book, and a considerable sacrifice of clarity. The chief sources, from which most of the material of the book is derived, are listed below. These sources are not regularly cited in the text. References are given in the text when any conclusion is stated without full explanation of its derivation. These references are not always to the original papers, but rather to the more recent papers where the reader will find a convenient point of entry to the topic under discussion. References are also given to the sources of experimental data, but these, for reasons already explained, cover only a small part of the experimental side of the subject. In particular, a great deal more work has been done on plants and on farm animals than would appear from its representation among the experimental work cited.

CHIEF SOURCES

(For details see List of References)

FISHER, R. A. (1930), *The Genetical Theory of Natural Selection.*
HALDANE, J. B. S. (1932), *The Causes of Evolution.*
KEMPTHORNE, O. (1957), *An Introduction to Genetic Statistics.*
LERNER, I. M. (1950), *Population Genetics and Animal Improvement.*
LI, C. C. (1955), *Population Genetics.*
LUSH, J. L. (1945), *Animal Breeding Plans.*
MALÉCOT, G. (1948), *Les Mathématiques de l'Hérédité.*
MATHER, K. (1949), *Biometrical Genetics.*
WRIGHT, S. (1921), Systems of Mating. *Genetics* 6: 111–178.
—— (1931), Evolution in Mendelian Populations. *Genetics* 16: 97–159.

CHAPTER 1

GENETIC CONSTITUTION OF A POPULATION

FREQUENCIES OF GENES AND GENOTYPES

To describe the genetic constitution of a group of individuals we should have to specify their genotypes and say how many of each genotype there were. This would be a complete description, provided the nature of the phenotypic differences between the genotypes did not concern us. Suppose for simplicity that we were concerned with a certain autosomal locus, A, and that two different alleles at this locus, A_1 and A_2, were present among the individuals. Then there would be three possible genotypes, A_1A_1, A_1A_2, and A_2A_2. (We are concerned here, as throughout the book, exclusively with diploid organisms.) The genetic constitution of the group would be fully described by the proportion, or percentage, of individuals that belonged to each genotype, or in other words by the frequencies of the three genotypes among the individuals. These proportions or frequencies are called *genotype frequencies*, the frequency of a particular genotype being its proportion or percentage among the individuals. If, for example, we found one quarter of the individuals in the group to be A_1A_1, the frequency of this genotype would be 0·25, or 25 per cent. Naturally the frequencies of all the genotypes together must add up to unity, or 100 per cent.

EXAMPLE 1.1. The M-N blood groups in man are determined by two alleles at a locus, and the three genotypes correspond with the three blood groups, M, MN, and N. The following figures, taken from the tabulation of Mourant (1954), show the blood group frequencies among Eskimoes of East Greenland and among Icelanders as follows:

		Blood group			Number of individuals
		M	MN	N	
Frequency, %.	Greenland	83·5	15·6	0·9	569
	Iceland	31·2	51·5	17·3	747

Clearly the two populations differ in these genotype frequencies, the N blood group being rare in Greenland and relatively common in Iceland. Not only is this locus a source of variation within each of the two populations, but it is also a source of genetic difference between the populations.

A population, in the genetic sense, is not just a group of individuals, but a breeding group; and the genetics of a population is concerned not only with the genetic constitution of the individuals but also with the transmission of the genes from one generation to the next. In the transmission the genotypes of the parents are broken down and a new set of genotypes is constituted in the progeny, from the genes transmitted in the gametes. The genes carried by the population thus have continuity from generation to generation, but the genotypes in which they appear do not. The genetic constitution of a population, referring to the genes it carries, is described by the array of *gene frequencies*; that is by specification of the alleles present at every locus and the numbers or proportions of the different alleles at each locus. If, for example, A_1 is an allele at the A locus, then the frequency of A_1 genes, or the gene frequency of A_1, is the proportion or percentage of all genes at this locus that are the A_1 allele. The frequencies of all the alleles at any one locus must add up to unity, or 100 per cent.

The gene frequencies at a particular locus among a group of individuals can be determined from a knowledge of the genotype frequencies. To take a hypothetical example, suppose there are two alleles, A_1 and A_2, and we classify 100 individuals and count the numbers in each genotype as follows:

	A_1A_1	A_1A_2	A_2A_2	*Total*
Number of individuals	30	60	10	100
Number of genes $\begin{cases} A_1 \\ A_2 \end{cases}$	60 0	60 60	0 20	120 80 }200

Each individual contains two genes, so we have counted 200 representatives of the genes at this locus. Each A_1A_1 individual contains two A_1 genes and each A_1A_2 contains one A_1 gene. So there are 120 A_1 genes in the sample, and 80 A_2 genes. The frequency of A_1 is therefore 60 per cent or 0·6, and the frequency of A_2 is 40 per cent or 0·4. To express the relationship in a more general form, let the frequencies of genes and of genotypes be as follows:

	Genes		*Genotypes*		
	A_1 A_2		A_1A_1	A_1A_2	A_2A_2
Frequencies	p q		P	H	Q

so that $p+q=1$, and $P+H+Q=1$. Since each individual contains two genes, the frequency of A_1 genes is $\frac{1}{2}(2P+H)$, and the relationship between gene frequency and genotype frequency among the individuals counted is as follows:

$$\left.\begin{array}{l} p=P+\frac{1}{2}H \\ q=Q+\frac{1}{2}H \end{array}\right\}\ \dots\dots(I.I)$$

EXAMPLE 1.2. To illustrate the calculation of gene frequencies from genotype frequencies we may take the M-N blood group frequencies given in Example 1.1. The M and N blood groups represent the two homozygous genotypes and the MN group the heterozygote. The frequency of the M gene in Greenland is, from equation *I.I*, $0.835 + \frac{1}{2}(0.156) = 0.913$, and the frequency of the N gene is $0.009 + \frac{1}{2}(0.156) = 0.087$, the sum of the frequencies being 1.000 as it should be. Doing the same for the Iceland sample we find the following gene frequencies in the two populations, expressed now as percentages:

	Gene	
	M	N
Greenland	91·3	8·7
Iceland	57·0	43·0

Thus the two populations differ in gene frequency as well as in genotype frequencies.

The genetic properties of a population are influenced in the process of transmission of genes from one generation to the next by a number of agencies. These form the chief subject-matter of the next four chapters, but we may briefly review them here in order to have some idea of what factors are being left out of consideration in this chapter. The agencies through which the genetic properties of a population may be changed are these:

Population size. The genes passed from one generation to the next are a sample of the genes in the parent generation. Therefore the gene frequencies are subject to sampling variation between successive generations, and the smaller the number of parents the greater is the sampling variation. The effects of sampling variation will be considered in Chapters 3–5, and meantime we shall exclude it from the discussion by supposing always that we are dealing with a "large population," which means simply one in which sampling variation is so small as to be negligible. For practical purposes a "large population" is one in which the number of adult individuals is in the hundreds rather than in the tens.

Differences of fertility and viability. Though we are not at present concerned with the phenotypic effects of the genes under discussion, we cannot ignore their effects on fertility and viability, because these influence the genetic constitution of the succeeding generation. The different genotypes among the parents may have different fertilities, and if they do they will contribute unequally to the gametes out of which the next generation is formed. In this way the gene frequency may be changed in the transmission. Further, the genotypes among the newly formed zygotes may have different survival rates, and so the gene frequencies in the new generation may be changed by the time the individuals are adult and themselves become parents. These processes are called selection, and will be described in Chapter 2. Meanwhile we shall suppose they are not operating. It is difficult to find examples of genes not subject to selection. For the purpose of illustration, however, we may take the human blood-group genes since the selective forces acting on these are probably not very strong. Genes that produce a mutant phenotype which is abnormal in comparison with the wild-type are, in contrast, usually subject to much more severe selection.

Migration and mutation. The gene frequencies in the population may also be changed by immigration of individuals from another population, and by gene mutation. These processes will be described in Chapter 2, and at this stage will also be supposed not to operate.

Mating system. The genotypes in the progeny are determined by the union of the gametes in pairs to form zygotes, and the union of gametes is influenced by the mating of the parents. So the genotype frequencies in the offspring generation are influenced by the genotypes of the pairs that mate in the parent generation. We shall at first suppose that mating is at random with respect to the genotypes under discussion. *Random mating,* or *panmixia,* means that any individual has an equal chance of mating with any other individual in the population. The important points are that there should be no special tendency for mated individuals to be alike in genotype, or to be related to each other by ancestry. If a population covers a large geographic area individuals inhabiting the same locality are more likely to mate than individuals inhabiting different localities, and so the mated pairs tend to be related by ancestry. A widely spread population is therefore likely to be subdivided into local groups and mating is random only within the groups. The properties of sub-divided populations depend on the size of the local groups, and will be described under the effects of population size in Chapters 3–5.

HARDY-WEINBERG EQUILIBRIUM

In a large random-mating population both gene frequencies and genotype frequencies are constant from generation to generation, in the absence of migration, mutation and selection; and the genotype frequencies are determined by the gene frequencies. These properties of a population were first demonstrated by Hardy and by Weinberg independently in 1908, and are generally known as the *Hardy-Weinberg Law*. (See Stern, 1943, where a translation of the relevant part of Weinberg's paper will be found.) Such a population is said to be in Hardy-Weinberg equilibrium. Deduction of the Hardy-Weinberg Law involves three steps: (1) from the parents to the gametes they produce; (2) from the union of the gametes to the geno-types in the zygotes produced; and (3) from the genotypes of the zygotes to the gene frequency in the progeny generation. These steps, in detail, are as follows:

    1. Let the parent generation have gene and genotype frequencies as follows:

$$
\begin{array}{cc@{\qquad}ccc}
A_1 & A_2 & A_1A_1 & A_1A_2 & A_2A_2 \\
p & q & P & H & Q
\end{array}
$$

Two sorts of gametes are produced, those bearing A_1 and those bear-ing A_2. The frequencies of these gametic types are the same as the gene frequencies, p and q, in the generation producing them, for this reason: A_1A_1 individuals produce only A_1 gametes, and A_1A_2 indi-viduals produce equal numbers of A_1 and A_2 gametes (provided, of course, there is no anomaly of segregation). So the frequency of A_1 gametes produced by the whole population is $P + \frac{1}{2}H$, which by equation *1.1* is the gene frequency of A_1.

    2. Random mating between individuals is equivalent to random union among their gametes. We can think of a pool of gametes to which all the individuals contribute equally; zygotes are formed by

random union between pairs of gametes from the pool. The genotype frequencies among the zygotes are then the products of the frequencies of the gametic types that unite to produce them. The genotype frequencies among the progeny produced by random mating can therefore be determined simply by multiplying the frequencies of the gametic types as shown in the following table:

		Female gametes and their frequencies	
		A_1	A_2
		p	q
Male gametes and their frequencies	A_1 p	A_1A_1 p^2	A_1A_2 pq
	A_2 q	A_1A_2 pq	A_2A_2 q^2

We need not distinguish the union of A_1 eggs with A_2 sperms from that of A_2 eggs with A_1 sperms; so the genotype frequencies of the zygotes are

$$\left. \begin{array}{ccc} A_1A_1 & A_1A_2 & A_2A_2 \\ p^2 & 2pq & q^2 \end{array} \right\} \ldots\ldots(1.2)$$

Note that these genotype frequencies depend only on the gene frequency in the parents, and not on the parental genotype frequencies, provided the parents mate at random.

3. Finally we use these genotype frequencies to determine the gene frequency in the offspring generation. Applying equation *1.1* we find the gene frequency of A_1 is $p^2 + \frac{1}{2}(2pq) = p(p+q) = p$, which is the same as in the parent generation.

The properties of a population with respect to a single locus, expressed in the Hardy-Weinberg law and demonstrated above, are these:

(1) A large random-mating population, in the absence of migration, mutation, and selection, is stable with respect to both gene and genotype frequencies: there is no inherent tendency for its genetic properties to change from generation to generation.

(2) The genotype frequencies in the progeny produced by random mating among the parents are determined solely by the gene frequencies among the parents. Consequently:

(a) a population in Hardy-Weinberg equilibrium has the relationship expressed in equation 1.2 between the gene and genotype frequencies in any one generation. And,

(b) these Hardy-Weinberg genotype frequencies are established by one generation of random mating, irrespective of the genotype frequencies among the parents.

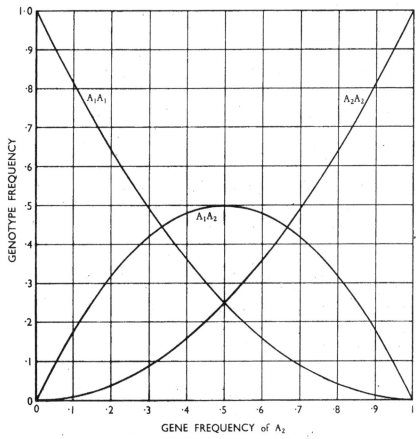

FIG. 1.1. Relationship between genotype frequencies and gene frequency for two alleles in a population in Hardy-Weinberg equilibrium.

We shall later give another proof of the Hardy-Weinberg law by a different method. Let us now first illustrate the properties of a population in Hardy-Weinberg equilibrium, and then show to what uses these properties can be put. The relationship between gene frequency and genotype frequencies expressed in equation 1.2 is

illustrated graphically in Fig. 1.1, which shows how the frequencies
of the three genotypes for a locus with two alleles depend on the gene
frequency. As an example of the Hardy-Weinberg genotype frequencies we shall take again the M-N blood groups in man.

EXAMPLE 1.3. Race and Sanger (1954) quote the following frequencies
(%) of the M-N blood groups in a sample of 1,279 English people. From
the observed genotype (i.e. blood group) frequencies we can calculate the
gene frequencies by equation 1.1. These gene frequencies are shown on
the right.

	Blood group			Gene	
	M	MN	N	M	N
Observed	28·38	49·57	22·05	53·165	46·835
Expected	28·265	49·800	21·935		

Now from the gene frequencies we can calculate the expected Hardy-
Weinberg genotype frequencies by equation 1.2, and we find that the
observed frequencies agree very closely with those expected for a population in Hardy-Weinberg equilibrium.

Comparison of observed with expected genotype frequencies may
be regarded as a test of the fulfilment of the conditions on which the
Hardy-Weinberg equilibrium depends. These conditions are:
random mating among the parents of the individuals observed, equal
fertility of the different genotypes among the parents, and equal
viability of the different genotypes among the offspring from fertilisa-
tion up to the time of observation. In addition, the classification of
individuals as to genotype must have been correctly made. The
blood group frequencies in Example 1.3 give no cause to doubt the
fulfilment of these conditions. It should be noted, however, that a
difference of fertility or of viability between the genotypes, though it
can be detected, cannot be measured from a comparison of observed
with expected frequencies (Wallace, 1958). The expected frequencies
are based on the observed gene frequencies after the differences of fertility or viability have had their effect. In order to measure these effects
we should have to know the original gene or genotype frequencies.
At the beginning of the chapter we saw, in equation 1.1, how the
gene frequencies among a group of individuals can be determined
from their genotype frequencies; but for this it was necessary to know
the frequencies of all three genotypes. Consequently the relationship
in equation 1.1 cannot be applied to the case of a recessive allele,

when the heterozygote is indistinguishable from the dominant homozygote. Consideration of the population as a breeding unit, however, shows that when the conditions for Hardy-Weinberg equilibrium hold, only the frequency of one of the homozygous genotypes is needed to determine the gene frequency, and the difficulty of recessive genes is thus overcome. Let A_2, for example, be a recessive gene with frequency q; then the frequency of A_2A_2 homozygotes is q^2. In other words the gene frequency is the square root of the homozygote frequency. Thus we can determine the gene frequency of recessive abnormalities, provided that selective mortality of the homozygote can be discounted or allowed for. But we can go further, and this is often the more important point: we can also determine the frequency of heterozygotes, or "carriers," of recessive abnormalities, which is $2q(1-q)$. It comes as a surprise to most people to discover how common heterozygotes of a rare recessive abnormality are.

EXAMPLE 1.4. Albinism in man is probably determined by a single recessive autosomal gene, and the frequency of albinos is about $1/20,000$ in human populations (see Stern, 1949). If q is the frequency of the albino gene, then $q^2 = 1/20,000$, and $q = 1/141$, if selective mortality is disregarded. The frequency of heterozygotes is then $2q(1-q)$, which works out to about $1/70$. So about one person in seventy is a heterozygote for albinism, though only one in twenty thousand is a homozygote.

EXAMPLE 1.5. There is a recessive autosomal gene in the Ayrshire breed of cattle in Britain which causes dropsy in the new-born calf. The frequency of this abnormality is about 1 in 300 births (Donald, Deas, and Wilson, 1952). A means of reducing the frequency of the defect would obviously be the avoidance of the use of bulls known or thought to be heterozygous. We might first want to know what proportion of bulls would be expected to be heterozygotes. In this case the conditions for Hardy-Weinberg equilibrium are certainly not all fulfilled: the breed is not a single random-breeding population, and the abnormal homozygotes are not fully viable up to the time of birth. So we can only get a rough idea of the frequency of heterozygotes by assuming the observations to refer to a population in Hardy-Weinberg equilibrium. On this assumption, $q^2 = 0.0033$, so $q = 0.057$; the frequency of heterozygotes is $2q(1-q) = 0.11$. So we should expect, very approximately, one bull in ten to be a heterozygote.

Mating frequencies and another proof of the Hardy-Weinberg law. Let us now look more closely into the breeding structure of a random-mating population, distinguishing the types of mating according to the genotypes of the pairs, and seeing what are the genotype frequencies among the progenies of the different types of mating. This provides a general method for relating genotype frequencies in successive generations, which we shall use in a later chapter. It also provides another proof of the Hardy-Weinberg law; a proof more cumbersome than that already given but showing more clearly how the Hardy-Weinberg frequencies arise from the Mendelian laws of segregation. The procedure is to obtain first the frequencies of all possible mating types according to the frequencies of the genotypes among the parents, and then to obtain the frequencies of genotypes among the progeny of each type of mating according to the Mendelian ratios.

Consider a locus with two alleles, and let the frequencies of genes and genotypes in the parents be, as before,

	Genes		Genotypes		
	A_1	A_2	A_1A_1	A_1A_2	A_2A_2
Frequencies	p	q	P	H	Q

There are altogether nine types of mating, and their frequencies when mating is random are found thus:

		Genotype and frequency of male parent		
		A_1A_1	A_1A_2	A_2A_2
		P	H	Q
Genotype and frequency of female parent	A_1A_1 $\quad P$	P^2	PH	PQ
	A_1A_2 $\quad H$	PH	H^2	HQ
	A_2A_2 $\quad Q$	PQ	HQ	Q^2

Since the sex of the parent is irrelevant in this context, some of the types of mating are equivalent, and the number of different types reduces to six. By summation of the frequencies of equivalent types, we obtain the frequencies of mating types in the first two columns of Table 1.1. Now we have to consider the genotypes of offspring produced by each type of mating, and find the frequency of each genotype in the total progeny, assuming, of course, that all types of mating are equally fertile and all genotypes equally viable. This is done in the right hand side of Table 1.1. Thus, for example, matings of the type $A_1A_1 \times A_1A_1$ produce only A_1A_1 offspring. So, of all the A_1A_1

genotypes in the total progeny, a proportion P^2 come from this type of mating. Similarly a quarter of the offspring of $A_1A_2 \times A_1A_2$ matings are A_1A_1. So this type of mating, which has a frequency of H^2, contributes a proportion $\frac{1}{4}H^2$ of the total A_1A_1 progeny. To find the frequency of each genotype in the total progeny we add the

TABLE I.I

Mating		Genotype and frequency of progeny		
Type	*Frequency*	A_1A_1	A_1A_2	A_2A_2
$A_1A_1 \times A_1A_1$	P^2	P^2	—	—
$A_1A_1 \times A_1A_2$	$2PH$	PH	PH	—
$A_1A_1 \times A_2A_2$	$2PQ$	—	$2PQ$	—
$A_1A_2 \times A_1A_1$	H^2	$\frac{1}{4}H^2$	$\frac{1}{2}H^2$	$\frac{1}{4}H^2$
$A_1A_2 \times A_2A_2$	$2HQ$	—	HQ	HQ
$A_2A_2 \times A_2A_2$	Q^2	—	—	Q^2
Sums		$(P+\frac{1}{2}H)^2$	$2(P+\frac{1}{2}H)(Q+\frac{1}{2}H)$	$(Q+\frac{1}{2}H)^2$
=		p^2	$2pq$	q^2

frequencies contributed by each type of mating. The sums, after simplification, are given at the foot of the table, and from the identity given in equation *1.1* they are seen to be equal to p^2, $2pq$, and q^2. These are the Hardy-Weinberg equilibrium frequencies, and we have shown that they are attained by one generation of random mating, irrespective of the genotype frequencies among the parents.

Multiple alleles. Restriction of the treatment to two alleles at a locus suffices for many purposes. If we are interested in one particular allele, as often happens, then all the other alleles at the locus can be treated as one. Formulation of the situation in terms of two alleles is therefore often possible even if there are in fact more than two. If we are interested in more than one allele we can still, if we like, treat the situation as a two-allele system by considering each allele in turn and lumping the others together. But the treatment can be easily extended to cover more than two alleles, and no new principle is introduced. In general, if q_1 and q_2 are the frequencies of any two alleles, A_1 and A_2, of a multiple series, then the genotype frequencies under Hardy-Weinberg equilibrium are as follows (Li, 1955a):

Genotype:	A_1A_1	A_1A_2	A_2A_2
Frequency:	q_1^2	$2q_1q_2$	q_2^2

These frequencies are also attained by one generation of random mating. This can readily be seen by reducing the situation to a two-allele system, and considering each allele in turn. Or it can be proved, though somewhat more laboriously, by the method explained above for the two-allele system.

EXAMPLE 1.6. The ABO blood groups in man are determined by a series of allelic genes. For the purpose of illustration we shall recognise three alleles, A, B, and O, and show how the gene frequencies can be estimated from the blood group frequencies. Let the frequencies of the A, B, and O genes be p, q, and r respectively, so that $p+q+r=1$. The following table shows (1) the genotypes, (2) the blood groups (i.e. phenotypes) corresponding to the different genotypes, (3) the expected frequencies of the blood groups in terms of p, q, and r, on the assumption of Hardy-Weinberg equilibrium, (4) observed frequencies of blood groups in a sample of 190,177 United Kingdom airmen, quoted by Race and Sanger (1954).

Genotype	AA AO	BB BO	OO	AB
Blood group	A	B	O	AB
Frequency (%)				
expected	p^2+2pr	q^2+2qr	r^2	$2pq$
observed	41·716	8·560	46·684	3·040

Calculation of the gene frequencies is rather more complicated than with two alleles. The following is the simplest method: a more refined method is described by Ceppellini *et al.* (1955). First, the frequency of the O gene is simply the square root of the frequency of the O group. Next it will be seen that the sum of the frequencies of the B and O groups is $q^2+2qr+r^2 = (q+r)^2 = (1-p)^2$. So $p = 1 - \sqrt{(\bar{B}+\bar{O})}$, where $\bar{B}$ and $\bar{O}$ are the frequencies of the blood groups B and O. In the same way $q = 1 - \sqrt{(\bar{A}+\bar{O})}$, and we have seen that $r = \sqrt{\bar{O}}$. This method gives the following gene frequencies in the sample:

$$\text{A gene:} \quad p = 0·2567$$
$$\text{B gene:} \quad q = 0·0598$$
$$\text{O gene:} \quad r = 0·6833$$

$$\text{Total} \qquad 0·9998$$

As a result of sampling errors these frequencies do not add up exactly to unity, but we shall not trouble to make an adjustment for so small a discrepancy. We may now calculate the expected frequency of the AB blood group, which has not been used in arriving at these gene frequencies, and see whether the observed frequency agrees satisfactorily. The expected frequency of AB from estimates of p and q is 3·070 per cent, which is in good agreement with the observed frequency of 3·040 per cent. ($\chi^2 = 0·7$, with 1 d.f., calculated by the method given by Race and Sanger.)

Sex-linked genes. With sex-linked genes the situation is rather more complex than with autosomal genes. The relationship between gene frequency and genotype frequency in the homogametic sex is the same as with an autosomal gene, but the heterogametic sex has only two genotypes and each individual carries only one gene instead of two. For this reason two-thirds of the sex-linked genes in the population are carried by the homogametic sex and one-third by the heterogametic. For the sake of brevity we shall now refer to the heterogametic sex as male. Consider two alleles, A_1 and A_2, with frequencies p and q, and let the genotypic frequencies be as follows:

Females			*Males*	
A_1A_1	A_1A_2	A_2A_2	A_1	A_2
P	H	Q	R	S

The frequency of A_1 among the females is then $p_f = P + \frac{1}{2}H$, and the frequency among the males is $p_m = R$. The frequency of A_1 in the whole population is

$$\bar{p} = \tfrac{2}{3}p_f + \tfrac{1}{3}p_m \qquad\qquad .....(1.3)$$
$$= \tfrac{1}{3}(2p_f + p_m)$$
$$= \tfrac{1}{3}(2P + H + R) \qquad\qquad ......(1.4)$$

Now, if the gene frequencies among males and among females are different, the population is not in equilibrium. The gene frequency in the population as a whole does not change, but its distribution between the two sexes oscillates as the population approaches equilibrium. The reason for this can be seen from the following considerations. Males get their sex-linked genes only from their mothers; therefore p_m is equal to p_f in the previous generation. Females get their sex-linked genes equally from both parents; therefore p_f is equal to the mean of p_m and p_f in the previous generation. Using primes to indicate the previous generation, we have

$$p_m = p'_f$$
$$p_f = \tfrac{1}{2}(p'_m + p'_f)$$

The difference between the frequencies in the two sexes is

$$p_f - p_m = \tfrac{1}{2}(p'_m + p'_f) - p'_f$$
$$= -\tfrac{1}{2}(p'_f - p'_m)$$

i.e. half the difference in the previous generation, but in the other direction. Therefore the distribution of the genes between the two sexes oscillates, but the difference is halved in successive generations and the population rapidly approaches an equilibrium in which the

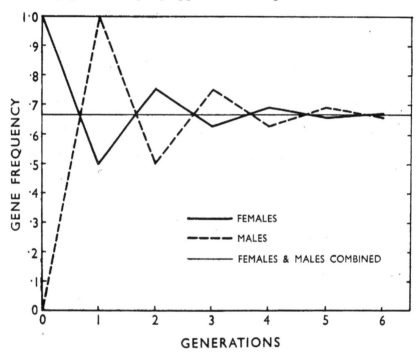

FIG. 1.2. Approach to equilibrium under random mating for a sex-linked gene, showing the gene frequency among females, among males, and in the two sexes combined. The population starts with females all of one sort ($q_f = 1$), and males all of the other sort ($q_m = 0$).

frequencies in the two sexes are equal. The situation is illustrated in Fig. 1.2, which shows the consequences of mixing females of one sort (all A_1A_1) with males of another sort (all A_2) and letting them breed at random.

EXAMPLE 1.7. Searle (1949) gives the frequencies of a number of genes in a sample of cats in London. The animals examined were sent to clinics for destruction; they were therefore not necessarily a random sample. Among the genes studied was "yellow" (y) which is sex-linked and for which all three genotypes in females are recognisable, the heterozygote being tortoise-shell. The data were used to test for agreement with Hardy-Weinberg equilibrium. The numbers observed in each phenotypic class are shown in table (i). We may first see whether the gene frequency

(i)

	Females			Males	
	+ +	+y	yy	+	y
Numbers observed	277	54	7	311	42
Numbers expected	269·6	64·5	3·9	315·2	37·8

is equal in the two sexes. The numbers of genes counted, and the frequency (q) of the gene y, in each sex are as given in table (ii). The

(ii)

	+	y	q_y
in females	608	68	0·101
in males	311	42	0·119
total	919	110	0·107

χ^2 testing difference in q between the sexes is 0·4 which is quite insignificant. There is therefore no reason to think the population is not in equilibrium, and we may take the estimate of gene frequency from both sexes combined: it is $q = 0·107$. From this estimate of q the expected numbers in the different phenotypic classes are calculated; they are shown in table (i). Only the females are relevant to the test of random mating. The χ^2 testing agreement between observed and expected numbers in females is 4·4, with 2 degrees of freedom. This has a probability of 0·1 and cannot be judged significant. The data are therefore compatible with the Hardy-Weinberg equilibrium, in spite of the deficiency of tortoise-shell females. If the deficiency of heterozygous females were real we might attribute it to the method of sampling and infer that the tortoise-shells were sent for destruction less often than the other colours, on account of human preference.

More than one locus. The attainment of the equilibrium in genotype frequencies after one generation of random mating is true of all autosomal loci considered separately. But it is not true of the genotypes with respect to two or more loci considered jointly. To illustrate the point, consider a population made up of equal numbers of $A_1A_1B_1B_1$ and $A_2A_2B_2B_2$ individuals, of both sexes. The gene frequency at both loci is then $\frac{1}{2}$, and if the individuals mated at random only three out of the nine genotypes would appear in the progeny; the genotype $A_1A_1B_2B_2$, for example, would be absent though its frequency in an equilibrium population would be $\frac{1}{16}$. The missing genotypes appear in subsequent generations, but not immediately at their equilibrium frequencies. The approach to equilibrium is described by Li (1955a) and here we shall only outline the conclusions.

Consider two loci each with two alleles, and let the frequencies of the four types of gamete formed by the initial population be as follows:

type of gamete	A_1B_1	A_1B_2	A_2B_1	A_2B_2
frequency	r	s	t	u

Then if the population is in equilibrium, $ru = st$, as may be seen by writing the gametic frequencies in terms of the gene frequencies. The difference, $ru - st$, gives a measure of the extent of the departure from equilibrium. This difference is halved in each successive generation of random mating, and the approach to equilibrium is thus fairly rapid (see Fig. 1.3). If, however, more than two loci are to be considered jointly the approach to equilibrium becomes progressively slower as the number of loci increases.

Linked loci. If two loci are linked the approach to equilibrium under random mating is slower in proportion to the closeness of the linkage. When equilibrium is reached the coupling and repulsion phases are equally frequent; the frequencies of the gametic types then depend only on the gene frequencies and not at all on the linkage. It is easy to suppose that association between two characters, as for example between hair colour and eye colour, is evidence of linkage between the genes concerned. Association between characters, however, is more often evidence of pleiotropy than of linkage. Linkage can give rise to association only after a mixture of populations, the length of time that the association persists depending on the closeness of the linkage.

The approach to equilibrium after the mixture of populations differing in respect of the genes at two linked loci can be described in the manner of the preceding section. The departure from equilibrium, d, is expressed as $d = ru - st$, where ru is the frequency of coupling heterozygotes and st that of repulsion heterozygotes. If c

is the frequency of recombination between the two loci then the difference, d, at generation t is

$$d_t = (1 - c)d_{t-1}$$

Thus if, for example, there is 25 per cent recombination the difference is reduced by one quarter in each generation; or if there is 10 per cent recombination the difference is reduced by 10 per cent in each

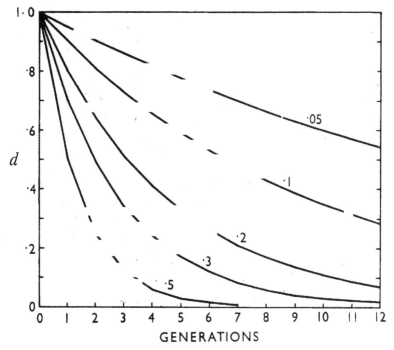

GENERATIONS

FIG. 1.3. Approach to equilibrium under random mating of two loci, considered jointly. The graphs show the difference of frequency (d) between coupling and repulsion heterozygotes in successive generations, starting with all individuals repulsion heterozygotes. The five graphs refer to different degrees of linkage between the two loci, as indicated by the recombination frequency shown alongside each graph. The graph marked .5 refers to unlinked loci.

generation. Closely linked loci will therefore continue for a considerable time to show the effects of a past mixture of populations. The approach to equality of coupling and repulsion phases with different degrees of linkage is illustrated in Fig. 1.3.

Assortative mating. Assortative mating is a form of non-random mating, but this is the most convenient place to mention it. If the mated pairs tend to be of the same genotype more often than would occur by chance this is called positive assortative mating, and if less often it is called negative assortative (or sometimes disassortative) mating. The consequences are described by Wright (1921) and summarised by Li (1955a) and will be only briefly outlined here. Positive assortative mating is of some importance in human populations, where it occurs with respect to intelligence and other mental characters. These however are not single gene differences such as can be discussed in the present context. The consequences of assortative mating with a single locus can be deduced from Table 1.1 by appropriate modification of the frequencies of the types of mating to allow for the increased frequency of matings between like genotypes. The effect on the genotype frequencies among the progeny is to increase the frequencies of homozygotes and reduce that of heterozygotes. In effect the population becomes partially subdivided into two groups, mating taking place more frequently within than between the groups.

CHAPTER 2

CHANGES OF GENE FREQUENCY

We have seen that a large random-mating population is stable with respect to gene frequencies and genotype frequencies, in the absence of agencies tending to change its genetic properties. We can now proceed to a study of the agencies through which changes of gene frequency, and consequently of genotype frequencies, are brought about. There are two sorts of process: *systematic processes*, which tend to change the gene frequency in a manner predictable both in amount and in direction; and the *dispersive process*, which arises in small populations from the effects of sampling, and is predictable in amount but not in direction. In this chapter we are concerned only with the systematic processes, and we shall consider only large random-mating populations in order to exclude the dispersive process from the picture. There are three systematic processes: *migration, mutation,* and *selection.* We shall study these separately at first, assuming that only one process is operating at a time, and then we shall see how the different processes interact.

MIGRATION

The effect of migration is very simply dealt with and need not concern us much here, though we shall have more to say about it later, in connexion with small populations. Let us suppose that a large population consists of a proportion, m, of new immigrants in each generation, the remainder, $1 - m$, being natives. Let the frequency of a certain gene be q_m among the immigrants and q_0 among the natives. Then the frequency of the gene in the mixed population, q_1, will be

$$q_1 = mq_m + (1 - m)q_0$$
$$= m(q_m - q_0) + q_0 \qquad \ldots\ldots(2.1)$$

The change of gene frequency, Δq, brought about by one generation of immigration is the difference between the frequency before immigration and the frequency after immigration. Therefore

$$\Delta q = q_1 - q_0$$
$$= m(q_m - q_0) \qquad \qquad \ldots\ldots(2.2)$$

Thus the rate of change of gene frequency in a population subject to immigration depends, as must be obvious, on the immigration rate and on the difference of gene frequency between immigrants and natives.

MUTATION

The effect of mutation on the genetic properties of the population differs according to whether we are concerned with a mutational event so rare as to be virtually unique, or with a mutational step that recurs repeatedly. The first produces no permanent change, whereas the second does.

Non-recurrent mutation. Consider first a mutational event that gives rise to just one representative of the mutated gene or chromosome in the whole population. This sort of mutation is of little importance as a cause of change of gene frequency, because the product of a unique mutation has an infinitely small chance of surviving in a large population, unless it has a selective advantage. This can be seen from the following consideration. As a result of the single mutation there will be one A_1A_2 individual in a population all the rest of which is A_1A_1. The frequency of the mutated gene, A_2, is therefore extremely low. Now according to the Hardy-Weinberg equilibrium the gene frequency should not change in subsequent generations. But with this situation we can no longer ignore the variation of gene frequency due to sampling. With a gene at very low frequency the sampling variation, even though very small, may take the frequency to zero, and the gene will then be lost from the population. Though at each generation a single gene has an equal chance of surviving or being lost, the loss is permanent and the probability of the gene still being present decreases with the passage of generations (see Li, 1955a). The conclusion, therefore, is that a unique mutation without selective advantage cannot produce a permanent change in the population.

Recurrent mutation. It is with the second type of mutation—

recurrent mutation—that we are concerned as an agent for causing change of gene frequency. Each mutational event recurs regularly with characteristic frequency, and in a large population the frequency of a mutant gene is never so low that complete loss can occur from sampling. We have, then, to find out what is the effect of this "pressure" of mutation on the gene frequency in the population.

Suppose gene A_1 mutates to A_2 with a frequency u per generation. (u is the proportion of all A_1 genes that mutate to A_2 between one generation and the next.) If the frequency of A_1 in one generation is p_0 the frequency of newly mutated A_2 genes in the next generation is up_0. So the new gene frequency of A_1 is $p_0 - up_0$, and the change of gene frequency is $-up_0$. Now consider what happens when the genes mutate in both directions. Suppose for simplicity that there are only two alleles, A_1 and A_2, with initial frequencies p_0 and q_0. A_1 mutates to A_2 at a rate u per generation, and A_2 mutates to A_1 at a rate v. Then after one generation there is a gain of A_2 genes equal to up_0 due to mutation in one direction, and a loss equal to vq_0 due to mutation in the other direction. Stated in symbols, we have the situation:

Mutation rate $\qquad\qquad\qquad A_1 \overset{u}{\underset{v}{\rightleftharpoons}} A_2$

Initial gene frequencies $\qquad p_0 \quad q_0$

Then the change of gene frequency in one generation is

$$\Delta q = up_0 - vq_0 \qquad\qquad \dots\dots(2.3)$$

It is easy to see that this situation leads to an equilibrium in gene frequency at which no further change takes place, because if the frequency of one allele increases fewer of the other are left to mutate in that direction and more are available to mutate in the other direction. The point of equilibrium can be found by equating the change of frequency, Δq, to zero. Thus at equilibrium

$$pu = qv$$

or $\qquad\qquad\qquad\qquad \dfrac{p}{q} = \dfrac{v}{u}$

and $\qquad\qquad\qquad\qquad q = \dfrac{u}{u+v}$

$\left.\begin{array}{r}\\\\\\\\\\\end{array}\right\} \dots\dots(2.4)$

Three conclusions can be drawn from the effect of mutation on gene frequency. Measurements of mutation rates indicate values ranging between about 10^{-4} and 10^{-8} per generation (one in ten thousand and one in a hundred million gametes). With normal mutation rates, therefore, mutation alone can produce only very slow changes of gene frequency; on an evolutionary time-scale they might be important, but they could scarcely be detected by experiment unless with micro-organisms. The second conclusion concerns the equilibrium between mutation in the two directions. Studies of reverse mutation (from mutant to wild type) indicate that it is usually less frequent than forward mutation (from wild type to mutant), on the whole about one tenth as frequent (Muller and Oster, 1957). The equilibrium gene frequencies for such loci, resulting from mutation alone, would therefore be about 0·1 of the wild-type allele and 0·9 of the mutant; in other words the "mutant" would be the common form and the "wild type" the rare form. Since this is not the situation we find in natural populations it is clear that the frequencies of such genes are not the product of mutation alone. We shall see in the next section that the rarity of mutant alleles is attributable to selection. The third conclusion concerns the effects of an increase of mutation rates such as might be caused by an increase of the level of ionising radiation to which the population is subjected. Any loci at which the gene frequencies are in equilibrium from the effects of mutation alone will not be affected by a change of mutation rate, provided the change affects forward and reverse mutation proportionately. This can be seen from consideration of the equilibrium gene frequencies given in equation 2.4.

SELECTION

Hitherto we have supposed that all individuals in the population contribute equally to the next generation. Now we must take account of the fact that individuals differ in viability and fertility, and that they therefore contribute different numbers of offspring to the next generation. The proportionate contribution of offspring to the next generation is called the *fitness* of the individual, or sometimes the *adaptive value*, or *selective value*. If the differences of fitness are in any way associated with the presence or absence of a particular gene in the individual's genotype, then *selection* operates on that gene. When a gene is subject to selection its frequency in the offspring is not the same as in the parents, since parents of different genotypes pass on their genes unequally to the next generation. In this way selection causes a change of gene frequency, and consequently also of genotype frequency. The change of gene frequency resulting from selection is more complicated to describe than that resulting from mutation, because the differences of fitness that give rise to the selection are an aspect of the phenotype. We therefore have to take account of the degree of dominance shown by the genes in question. Dominance, in this connexion, means dominance with respect to fitness, and this is not necessarily the same as the dominance with respect to the main visible effects of the gene. Most mutant genes, for example, are completely recessive to the wild type in their visible

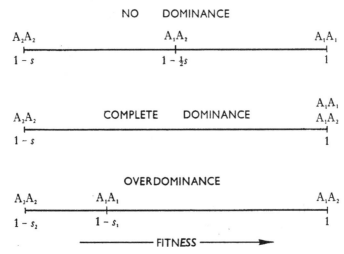

FIG. 2.1. Degrees of dominance with respect to fitness.

effects, but this does not necessarily mean that the heterozygote has a fitness equal to that of the wild-type homozygote. The meaning of the different degrees of dominance with which we shall deal is illustrated in Fig. 2.1.

It is most convenient to think of selection acting against the gene in question, in the form of selective elimination of one or other of the genotypes that carry it. This may operate either through reduced viability or through reduced fertility in its widest sense, including mating ability. In either case the outcome is the same: the genotype selected against makes a smaller contribution of gametes to form zygotes in the next generation. We may therefore treat the change of gene frequency as taking place between the counting of genotypes among the zygotes of the parent generation and the formation of zygotes in the offspring generation. The intensity of the selection is expressed as the *coefficient of selection*, s, which is the proportionate reduction in the gametic contribution of a particular genotype compared with a standard genotype, usually the most favoured. The contribution of the favoured genotype is taken to be 1, and the contribution of the genotype selected against is then $1 - s$. This expresses the fitness of one genotype compared with the other. Suppose, for example, that the coefficient of selection is $s = 0 \cdot 1$; this means that for every 100 zygotes produced by the favoured genotype, only 90 are produced by the genotype selected against.

The fitness of a genotype with respect to any particular locus is not necessarily the same in all individuals. It depends on the environmental circumstances in which the individual lives, and also on the genotype with respect to genes at other loci. When we assign a certain fitness to a genotype, this refers to the average fitness in the whole population. Though differences of fitness between individuals result in selection being applied to many, perhaps to all, loci simultaneously, we shall limit our attention here to the effects of selection on the genes at a single locus, supposing that the average fitness of the different genotypes remains constant despite the changes resulting from selection applied simultaneously to other loci. The conclusions we shall reach apply equally to natural selection occurring under natural conditions without the intervention of man, and to artificial selection imposed by the breeder or experimenter through his choice of individuals as parents and through the number of offspring he chooses to rear from each parent.

Change of gene frequency under selection. We have first to derive the basic formulae for the change of gene frequency brought about by one generation of selection. Then we can consider what they tell us about the effectiveness of selection. The different conditions of dominance have to be taken account of, but the method is the same for all, and we shall illustrate it by reference to the case of complete dominance with selection acting against the recessive homozygote. Let the genes A_1 and A_2 have initial frequencies p and q, A_1 being completely dominant to A_2, and let the coefficient of selection against A_2A_2 individuals be s. Multiplying the initial frequency by the fitness of each genotype we obtain the proportionate contribution of each genotype to the gametes that will form the next generation, thus:

Genotypes	A_1A_1	A_1A_2	A_2A_2	Total
Initial frequencies	p^2	$2pq$	p^2	1
Fitness	1	1	$1-s$	
Gametic contribution	p^2	$2pq$	$q^2(1-s)$	$1-sq^2$

Note that the total gametic contribution is no longer unity, because there has been a proportionate loss of sq^2 due to the selection. To find the frequency of A_2 gametes produced—and so the frequency of A_2 genes in the progeny—we take the gametic contribution of A_2A_2 individuals plus half that of A_1A_2 individuals and divide by the new total, i.e. we apply equation *1.1*. Thus the new gene frequency is

$$q_1 = \frac{q^2(1-s)+pq}{1-sq^2} \qquad \ldots \ldots (2.5)$$

The change of gene frequency, Δq, resulting from one generation of selection is

$$\Delta q = q_1 - q$$
$$= \frac{q^2(1-s)+pq}{1-sq^2} - q$$

which on simplification reduces to

$$\Delta q = -\frac{sq^2(1-q)}{1-sq^2} \qquad \ldots \ldots (2.6)$$

From this we see that the effect of selection on gene frequency depends not only on the intensity of selection, s, but also on the initial gene frequency. But both relationships are somewhat complex, and the examination of their significance will be postponed till after the other situations have been dealt with.

Selection may act against the dominant phenotype and favour the recessive: we then put $1-s$ for the fitness of A_1A_1 and of A_1A_2 genotypes. The expression for Δq is given in Table 2.1. The difference may best be appreciated by considering the effects of total elimination ($s=1$). The expression for selection against the dominant allele then reduces to $\Delta q = 1 - q$, which expresses the fact that if only the recessive genotype survives to breed the frequency of the recessive allele will become 1 after a single generation of selection. But, on the other hand, if there is complete elimination of the recessive genotype the frequency of the dominant allele does not reach 1 after a single generation. The difference between the effects of selection in opposite directions becomes less marked as the value of s decreases.

If there is incomplete dominance the expression for Δq is again different. The case of exact intermediate dominance is given in Table 2.1. Here we put $1 - \frac{1}{2}s$ for the fitness of A_1A_2, and $1 - s$ for the fitness of A_2A_2 genotype. For selection in the opposite direction in this case we need only interchange the initial frequencies of the two alleles, writing p in the place of q.

TABLE 2.1

Change of gene frequency, Δq, after one generation of selection under different conditions of dominance specified in Fig. 2.1.

Conditions of domin- ance and selection	Initial frequencies and fitness of the genotypes			Change of frequency, Δq, of gene A_2	
	A_1A_1 p^2	A_1A_2 $2pq$	A_2A_2 q^2		
No dominance selection against A_2	1	$1 - \frac{1}{2}s$	$1 - s$	$-\dfrac{\frac{1}{2}sq(1 - q)}{1 - sq}$	(1)
Complete dominance selection against A_2A_2	1	1	$1 - s$	$-\dfrac{sq^2(1 - q)}{1 - sq^2}$	(2)
Complete dominance selection against A_1 –	$1 - s$	$1 - s$	1	$+\dfrac{sq^2(1 - q)}{1 - s(1 - q^2)}$	(3)
Overdominance selection against A_1A_1 and A_2A_2	$1 - s_1$	1	$1 - s_2$	$+\dfrac{pq(s_1p - s_2q)}{1 - s_1p^2 - s_2q^2}$	(4)

When s is small the denominators differ little from 1, and the numerators alone can be taken to represent Δq sufficiently accurately for most purposes.

Finally, selection may favour the heterozygote, a condition known as overdominance. In this case we put $1 - s_1$ and $1 - s_2$ for the fitness of the two homozygotes. The expression for Δq is given in Table 2.1. This special case will be given more detailed attention later. The different conditions of dominance to which the expressions in Table 2.1 refer are illustrated diagrammatically in Fig. 2.1. Let us now see what these equations tell us about the effectiveness of selection.

Effectiveness of selection. We see from the formulae that the effectiveness of selection, i.e. the magnitude of Δq, depends on the initial gene frequency, q. The nature of this relationship is best appreciated from graphs showing Δq at different values of q. Fig. 2.2

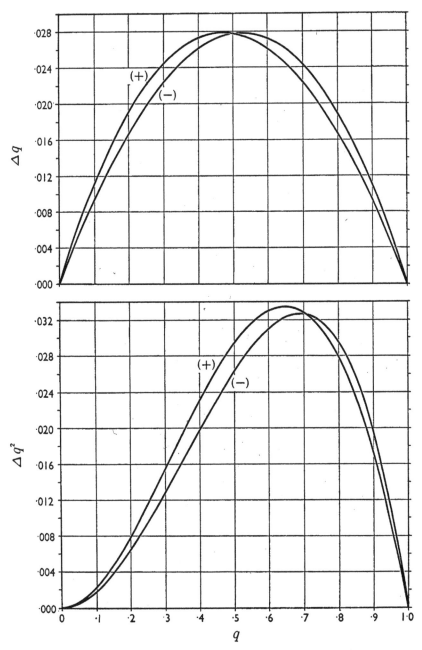

FIG. 2.2. Change of gene frequency, Δq, under selection of intensity $s = 0.2$, at different values of initial gene frequency, q. Upper figure: a gene with no dominance. Lower figure: a gene with complete dominance. The graphs marked $(-)$ refer to selection against the gene whose frequency is q, so that Δq is negative. The graphs marked $(+)$ refer to selection in favour of the gene, so that Δq is positive. (From Falconer, 1954a; reproduced by courtesy of the editor of the *International Union of Biological Sciences*.)

shows these graphs for the cases of no dominance and complete dominance. They also distinguish between selection in the two directions. A value of $s = 0.2$ was chosen for the coefficient of selection because, for reasons given in Chapter 12, this seems to be the right order of magnitude for the coefficient of selection operating on genes concerned with metric characters in laboratory selection experiments. First we may note that with this value of s there is never a great difference in Δq according to the direction of selection. The two important points about the effectiveness of selection that these graphs demonstrate are: (i) Selection is most effective at intermediate gene frequencies and becomes least effective when q is either large or small. (ii) Selection for or against a recessive gene is extremely ineffective when the recessive allele is rare. This is the consequence of the fact, noted earlier, that when a gene is rare it is represented almost entirely in heterozygotes.

Another way of looking at the effect of the initial gene frequency on the effectiveness of selection is to plot a graph showing the course of selection over a number of generations, starting from one or other extreme. Such graphs are shown in Fig. 2.3. They were constructed directly from those of Fig. 2.2, and refer again to a coefficient of selection, $s = 0.2$. They show that the change due to selection is at first very slow, whether one starts from a high or a low initial gene frequency; it becomes more rapid at intermediate frequencies and falls off again at the end. In the case of a fully dominant gene one is chiefly interested in the frequency of the homozygous recessive genotype, i.e. q^2. For this reason the graph shows the effect of selection on q^2 instead of on q.

It is often useful to express the change of gene frequency, Δq, under selection in a simplified form, which is a sufficiently good approximation for many purposes. If either the coefficient of selection, s, or the gene frequency, q, is small, then the denominators of the equations in Table 2.1 become very nearly unity, and we can use the numerators alone as expressions for Δq. Then for selection in either direction we have, with no dominance:

$$\Delta q = \pm \tfrac{1}{2}sq(1 - q) \quad \text{(approx.)} \qquad \ldots\ldots(2.7)$$

and with complete dominance:

$$\Delta q = \pm sq^2(1 - q) \quad \text{(approx.)} \qquad \ldots\ldots(2.8)$$

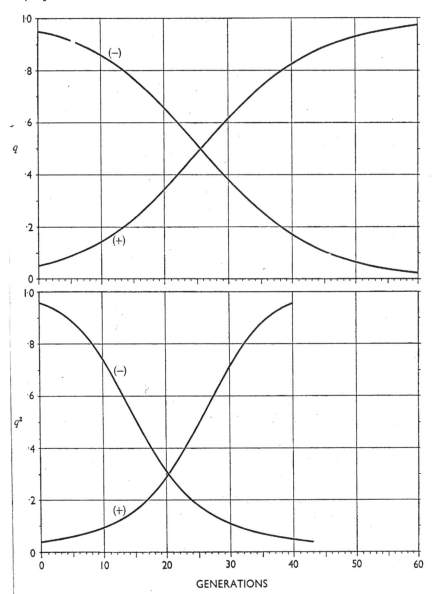

FIG. 2.3. Change of gene frequency during the course of selection from one extreme to the other. Intensity of selection, $s = 0.2$. Upper figure: a gene with no dominance. Lower figure: a gene with complete dominance, q being the frequency of the recessive allele and q^2 that of the recessive homozygote. The graphs marked (−) refer to selection against the gene whose frequency is q, so that q or q^2 decreases. The graphs marked (+) refer to selection in favour of the gene, so that q or q^2 increases. (From Falconer, 1954a; reproduced by courtesy of the editor of the *International Union of Biological Sciences*.)

EXAMPLE 2.1. As an example of the change of gene frequency under selection we shall take the case of a sex-linked gene, in spite of the added complication, because there is no well documented case of an autosomal gene. Fig. 2.4 shows the change of the frequency of the recessive sex-linked gene "raspberry" in *Drosophila melanogaster* over a period of about eighteen generations, described by Merrell (1953). The population was started with a gene frequency of 0·5 in both sexes, and was therefore in

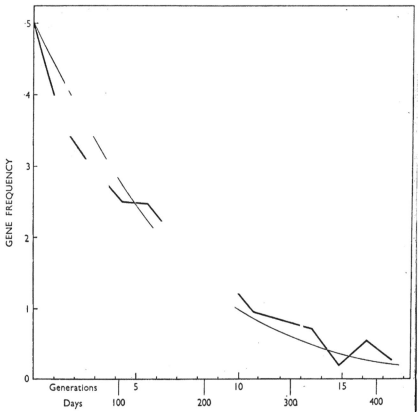

FIG. 2.4. Change of gene frequency under natural selection in the laboratory, as described in Example 2.1. (Data from Merrell, 1953.)

equilibrium at the beginning (see p. 17). Counts were made at about monthly intervals, and the gene frequency in both sexes combined (by equation *1.3*) is shown against the scale of days in the figure. Measurements of fitness were made by comparison of the relative viability of mutant and wild-type phenotypes, and of their relative success in mating. No differences of viability were detected, nor of the success of females in mating. But mutant males were only 50 per cent as successful as wild-type males in mating. The changes of gene frequency expected on the basis of this difference of fitness were then calculated generation by generation, and these calculated values are shown in the figure by the smooth curve, plotted against the scale of generations. From a similar experiment with a different mutant it was found that the calculated and observed curves coincided if a period of 24 days was taken as the interval between generations. For this reason 24 days to a generation was taken as the basis for superimposing the curves shown here. Since the calculated curve was to this extent made to fit the observed, the good agreement between the two cannot be taken as proof that selection operated only through the males' success in mating. But the similarity in their shapes illustrates well how the change of gene frequency is rapid at first, tails off as the gene frequency becomes lower, and becomes very slow when it approaches zero.

Number of generations required. How many generations of selection would be needed to effect a specified change of gene frequency? An answer to this question is sometimes required in connexion with breeding programmes or proposed eugenic measures. We shall here consider only the case of selection against a recessive when elimination of the unwanted homozygote is complete, i.e. $s = 1$. This would apply to natural selection against a recessive lethal, and artificial selection against an unwanted recessive in a breeding programme. We shall also, for the moment, suppose that there is no mutation. We had in equation 2.5 an expression for the new gene frequency after one generation of selection against a recessive. Substituting $s = 1$ in this equation and writing $q_0, q_1, q_2, \dots, q_t$ for the gene frequency after $0, 1, 2, \dots, t$ generations of selection we have

$$q_1 = \frac{q_0}{1 + q_0}$$

and

$$q_2 = \frac{q_1}{1 + q_1}$$

$$= \frac{q_0}{1 + 2q_0}$$

by substituting for q_1 and simplifying. So in general

$$q_t = \frac{q_0}{1 + tq_0} \qquad \qquad \dots\dots(2.9)$$

and the number of generations, t, required to change the gene frequency from q_0 to q_t is

$$t = \frac{q_0 - q_t}{q_0 q_t}$$

$$= \frac{1}{q_t} - \frac{1}{q_0} \qquad \ldots\ldots(2.10)$$

We may use this formula to illustrate the point already made, that when the frequency of a recessive gene is low selection is very slow to change it.

EXAMPLE 2.2. It is sometimes suggested, as a eugenic measure, that those suffering from serious inherited defects should be prevented from reproducing, since in this way the frequency of such defects would be reduced in future generations. Before deciding whether the proposal is a good one we ought to know what it would be expected to achieve. We cannot properly discuss this problem without taking mutation into account, as we shall do later; the answer we get ignoring mutation, as we do now, shows what is the best that could be hoped for. Let us take albinism as an example, though it cannot be regarded as a very serious defect, and ask the question: how long would it take to reduce its frequency to half the present value? The present frequency is about 1/20,000, and this makes $q_0 = 1/141$, as we saw in Example 1.4. The objective is $q^2 = 1/40,000$, which makes $q_t = 1/200$. So, from equation 2.10, $t = 200 - 141 = 59$ generations. With 25 years to a generation it would take nearly 1500 years to achieve this modest objective. More serious recessive defects are generally even less common than albinism and with them elimination would be still slower.

Balance between mutation and selection. Having described the effects of mutation and selection separately we must now compare them and consider them jointly. Which is the more effective process in causing change of gene frequency? Is it reasonable to attribute the low frequency of deleterious genes that we find in natural populations to the balance between mutation tending to increase the frequency and selection tending to decrease it? The expressions already obtained for the change of gene frequency under mutation or selection alone show that both depend on the initial gene frequency, but in different ways. Mutation to a particular gene is most effective in increasing its frequency when the mutant gene is rare (because there are more of the unmutated genes to mutate); but selection is least effective when the gene is rare. The relative effectiveness of the two processes depends therefore on the gene frequency, and if both processes operate for long enough a state of equilibrium will eventually be reached. So we must find what the gene frequency will be when equilibrium is reached. This is done by equating the two expressions for the change of gene frequency, because at equilibrium the change due to mutation will be equal and opposite to the change due to selection.

Let us consider first a fully recessive gene with frequency q, mutation rate to it u, and from it v, and selection coefficient against it s. Then from equations (*2.3*) and (*2.6*) we have at equilibrium

$$u(1-q) - vq = \frac{sq^2(1-q)}{1-sq^2} \qquad \ldots\ldots(2.11)$$

This equation is too complicated to give a clear answer to our question. But we can make two simplifications with only a trivial sacrifice of accuracy. We are specifically interested in genes at low equilibrium frequencies. If q is small the term vq representing back mutation is relatively unimportant and can be neglected; and we can use the approximate expression (equation *2.8*) for the selection effect. Making these simplifications we have the equilibrium condition for selection against a recessive gene

$$u(1-q) = sq^2(1-q) \quad \text{(approx.)}$$

$$u = sq^2 \qquad \text{(approx.)} \quad \ldots\ldots(2.12)$$

$$\hat{q} = \sqrt{\frac{u}{s}}. \qquad \text{(approx.)} \quad \ldots\ldots(2.13)$$

For a gene with no dominance similar reasoning from equation (1) in Table 2.1 gives the equilibrium condition

$$\hat{q} = \frac{u}{s} \quad \text{(approx.)} \qquad \ldots\ldots(2.14)$$

Finally, consider selection against a completely dominant gene, the frequency of the dominant gene being $1-q$, and the mutation rate to it being v. In this case $1-q$ is very small and the term $u(1-q)$ in equation *2.11* is negligible. We have therefore at equilibrium

$$vq = sq^2(1 - q) \quad \text{(approx.)}$$

$$q(1 - q) = \frac{v}{s} \quad \text{(approx.)}$$

or
$$H = \frac{2v}{s} \quad \text{(approx.)} \quad ......(2.15)$$

where H is the frequency of heterozygotes. If the mutant gene is rare H is very nearly the frequency of the mutant phenotype in the population.

EXAMPLE 2.3. If the equilibrium state is accepted as applicable, we can use it to get an estimate of the mutation rate of dominant abnormalities for which the coefficient of selection is known. Among some human examples described by Haldane (1949) is the case of dominant dwarfism (chondrodystrophy) studied in Denmark. The frequency of dwarfs was estimated at 10.7×10^{-5}, and their fitness $(1 - s)$ at 0.196. The estimate of fitness was made from the number of children produced by dwarfs compared with their normal sibs. The mutation rate, by equation (2.15), comes out at 4.3×10^{-5}. Though there is a possibility of serious error in the estimate of frequency owing to prenatal mortality of dwarfs, the mutation rate is almost certainly estimated within the right order of magnitude. For a discussion of the estimation of mutation rates in man see Crow (1956).

These expressions for the equilibrium gene frequency under the joint action of mutation and selection show that the gene frequency can have any value at equilibrium, depending on the relative magnitude of the mutation rate and the coefficient of selection. But if mutation rates are of the order of magnitude commonly accepted, i.e. 10^{-5}, or thereabouts, then only a mild selection against the mutant gene will be needed to hold it at a very low equilibrium frequency. For example, the following are the equilibrium frequencies of a recessive gene and of the recessive homozygote under various intensities of selection if the mutation rate is 10^{-5}:

$s =$	$\cdot 001$	$\cdot 01$	$\cdot 1$	$\cdot 5$
$q =$	$\cdot 1$	$\cdot 03$	$\cdot 01$	$\cdot 0045$
$q^2 =$	$\cdot 01$	$\cdot 001$	$\cdot 0001$	2×10^{-5}

Thus, if a gene mutates at the rate of 10^{-5}, a selective disadvantage of 10 per cent is enough to hold the frequency of the recessive homozygote at one in ten thousand; and a 50 per cent disadvantage will hold it at one in fifty thousand. It is quite clear therefore that the low frequency of deleterious mutants in natural populations is in accord with what would be expected from the joint action of mutation and selection. A further conclusion is that mutation alone is most unlikely to be a cause of evolutionary change. It is not mutation, but selection, that chiefly determines whether a gene spreads through the population or remains a rare abnormality, unless the mutation rate is very much higher than seems to be the rule.

Let us now briefly consider two questions of social importance concerning the balance between selection and mutation: the effect of an increase of mutation rate, and the effect of a change in the intensity of selection against deleterious mutants. These questions are more fully discussed by Crow (1957).

Increase of mutation rate. Since the products of mutation are predominantly deleterious, the process of mutation has a harmful effect on a proportion of the individuals in a population. When an individual dies or fails to reproduce in consequence of the reduced fitness of its genotype, we may refer to this as a "genetic death." An increase in the frequency of genetic deaths would reduce the potential reproductive rate and might thus reduce the speed with which a species could multiply in an unoccupied territory. But when the numbers of adults are held constant by density-dependent factors, even quite a high frequency of genetic deaths will not affect the ability of the population to perpetuate itself, especially if the reproductive rate is high, because the death of some individuals leaves room for others that would otherwise have died from lack of food or some other cause. There is a species of *Drosophila*, for example (*D. tropicalis*, from Central America), in which 50 per cent of individuals in a certain locality suffer genetic death, and yet the population flourishes (Dobzhansky and Pavlovsky, 1955). In species with low reproductive rates the frequency of genetic deaths is of greater consequence, particularly in ourselves, where the death of every individual is a matter of concern. Let us therefore consider what effect is to be expected from an increase of mutation rate such as might be caused by an increase in the amount of ionising radiation to which human populations are exposed.

Let us take the case of a recessive gene with a mutation rate (to it) of u, the gene being in equilibrium at a frequency of q. Then, if the coefficient of selection against the homozygote is s, the frequency of genetic deaths is sq^2. This is the proportionate loss due to selection, as shown on p. 29, and it is equal to u, by equation *2.12*. Thus the frequency of genetic deaths, when equilibrium has been attained, depends on the mutation rate alone, and is not influenced by the degree of harmfulness of the gene. The reason for this apparent paradox is that the more harmful genes come to equilibrium at lower frequencies.

Now, if the mutation rate is increased, and maintained at the new level, the gene will begin to increase toward a new point of equilibrium at which sq^2 will be equal to the new mutation rate. Thus if the mutation rate were doubled the frequency of genetic deaths would also be doubled, when the new equilibrium had been reached. But the approach to the new equilibrium would be very slow. The change of gene frequency in the first generation is approximately

$$\Delta q = u(1-q) - sq^2(1-q)$$

u being the new mutation rate (from equations *2.3* and *2.8*, but ctingback negle mutation). To see what this means let us take a mutation rate of 10^{-5} as being probably representative of many loci, and let us suppose that this was doubled. We may with sufficient accuracy take $1 - q$ as unity. Then

$$\Delta q = 2 \times 10^{-5} - 10^{-5}$$
$$= 10^{-5}$$

The immediate effect of the increase of mutation rate would therefore be very small indeed.

Change of selection intensity. Intensification of selection is sometimes advocated as a eugenic measure in human populations, on the grounds that if sufferers from genetic defects were prevented from breeding the frequency of the defects would be reduced. We saw from Example 2.2. that the effect of selection against a recessive defect is very slow indeed, even when mutation is i nored. The true situation is even worse. We cannot reduce the gfrequency of an abnormality, whether dominant or recessive, below the new equilibrium frequency. The serious defects have already a fairly strong natural selection working on them, and the addition of artificial selection can do no more than make the coefficient of selection, s, equal to 1. This would probably seldom do more than double the present coefficient of selection, and the incidence of defects would be reduced to not less than half their present values (equations *2.13*, *2.14*, *2.15*). With a dominant gene the effect would be immediate, but with a recessive the approach to the new equilibrium would be extremely slow.

The situation with respect to recessives is complicated by the fact that deleterious recessives are certainly not at their equilibrium frequencies in present-day human populations (Haldane, 1939). The reason is that modern civilisation has reduced the degree of subdivision (i.e. inbreeding) and so reduced the frequency of homozygotes, as will be explained in the next chapter. In consequence both the gene frequencies and the homozygote frequencies are below their equilibrium values, and must be presumed to be at present increasing slowly toward new equilibria at higher values.

Perhaps the converse of the question posed above is one that should give us more concern, namely the consequences of the reduced intensity of natural selection under modern conditions. Minor genetic defects, such as colour-blindness, must presumably have had some selective disadvantage in the past but now have very little, if any, effect on fitness. Moreover, the development and extension of medical treatment prolongs the lives of many people with diseases that have at least some degree of genetic causation through genes that increase susceptibility. This relaxation of the selection operating on minor genetic defects and against genes concerned in the causation of disease suggests that the frequencies of these genes will increase toward new equilibria at higher values. If this is true we must expect the incidence of minor genetic defects to increase in the future, and also the proportion of people who need medical treatment for a variety of diseases. By applying humanitarian principles for our own good now we are perhaps laying up a store of inconvenience for our descendants in the distant future.

Selection favouring heterozygotes. We have considered the effects of selection operating on genes that are partially or fully dominant with respect to fitness; but, though the appropriate formula was given in Table 2.1, we have not yet discussed the consequences of overdominance with respect to fitness; that is, when the heterozygote has a higher fitness than either homozygote. At first sight it may seem rather improbable that selection should favour the heterozygote of two alleles rather than one or other of the homozygotes, but there are reasons for thinking that this in fact is not at all an uncommon situation. Let us first examine the consequences of this form of selection, and then consider the evidence of its occurrence in nature.

Selection operating on a gene with partial or complete dominance tends toward the total elimination of one or other allele, the final gene frequency, in the absence of mutation, being o or 1. When selection favours the heterozygote, however, the gene frequency tends toward an equilibrium at an intermediate value, both alleles remaining in the population, even without mutation. The reason is as follows. The change of gene frequency after one generation was given in Table 2.1 as being

$$\Delta q = \frac{pq(s_1 p - s_2 q)}{1 - s_1 p^2 - s_2 q^2}$$

The condition for equilibrium is that $\Delta q = 0$, and this is fulfilled when $s_1 p = s_2 q$. The gene frequencies at this point of equilibrium are therefore

$$\frac{p}{q} = \frac{s_2}{s_1}$$

or

$$q = \frac{s_1}{s_1 + s_2} \qquad \qquad \ldots\ldots(2.16)$$

Now, if q is greater than its equilibrium value (but not 1), and p therefore less, $s_1 p$ will be less than $s_2 q$, and Δq will be negative; that is to say q will decrease. Similarly if q is less than its equilibrium value (but not o) it will increase. Therefore when the gene frequency has any value, except o or 1, selection changes it toward the intermediate point of equilibrium given in equation 2.16, and both alleles remain permanently in the population. Three or more alleles at a locus are maintained in the same way, provided the heterozygote of any pair is superior in fitness to both homozygotes of that pair (Kimura, 1956). A feature of the equilibrium worthy of note is that the gene frequency depends not on the degree of superiority of the heterozygote but on the relative disadvantage of one homozygote compared with that of the other. Therefore there is a point of equilibrium at some more or less intermediate gene frequency whenever a heterozygote is superior to both the homozygotes, no matter by how little.

Our previous consideration of genes with complete dominance showed that the balance between selection and mutation satisfactorily accounts for the presence of deleterious genes at low frequencies, causing the appearance of rare abnormal, or mutant, individuals. Genes at intermediate frequencies, however, are common in very many species, and the presence of these cannot satisfactorily be accounted for in this way. But the intermediate frequencies are just what would be expected if selection favoured the heterozygotes. The existence in a population of individuals with readily discernible differences caused by genes at intermediate frequencies is referred to as *polymorphism*. The blood group differences of man are perhaps the best known examples, but antigenic differences are found also in many other species and are probably universal in animals. More striking forms of polymorphism are the colour varieties found in many species, particularly among insects, snails, and fishes. The genes causing polymorphism have usually no obvious advantage of one allele over another, all the genotypes being essentially normal, or "wild-type," individuals. In these circumstances, as we noted above, only a very slight superiority of the heterozygote would be sufficient to establish an equilibrium at an intermediate gene frequency. The properties of the genes concerned with polymorphism seem, therefore, to accord well with the hypothesis that selection is operating on them in favour of the heterozygotes, and this is generally conceded to be the most probable reason for their intermediate frequencies. As a general cause of polymorphism, however, it cannot be taken as fully proved, because the superior fitness of heterozygotes has been demonstrated in relatively few cases, and there are other possible reasons for the existence of polymorphism. For example, the genes might be in a transitional stage of a change from one extreme to the other as a result of slow environmental change; or the intermediate frequencies might be the point of equilibrium between mutation in opposite directions, with virtually no selective advantage of one allele over the other. But these explanations seem improbable, particularly as some polymorphisms are known to be of very long standing. The polymorphism of shell colours in the land snail *Cepaea nemoralis*, for example, goes back to Neolithic times (Cain and Sheppard, 1954a). Another possible cause of polymorphism lies in the heterogeneity of the environment in which a population lives. If the differences of environment influence the selection coefficients in such a way that one allele is favoured in some conditions and another allele in other conditions, then polymorphism may result provided that mating is not entirely at random over the range of environments. (See Levene, 1953; Li, 1955b; Mather, 1955a; Waddington, 1957.)

If heterozygotes are indeed superior in fitness, one naturally wants to enquire into the nature of their superiority. Unfortunately, however, very little is known about this, though evidence is accumulating, in the case of the human blood groups, that certain blood groups are associated with an increased susceptibility to certain diseases (Roberts, 1957); group O, for example, with duodenal ulcer and group A with pernicious anaemia. If one states this the other way round and says that the other alleles confer increased resistance to these diseases, then it is not unreasonable to suppose that each allele increases resistance to different diseases, and that the presence of two alleles increases the resistance to two different diseases, thereby giving a selective advantage to the heterozygote.

Another question of interest concerns the evolutionary significance of polymorphism. Is it an "adaptive" feature of a species? Does it, in other words, confer some advantage over a population without it? Some think that it does. (See, particularly, Dobzhansky, 1951*b*). Others, however, point out that the average fitness of a population with polymorphism resulting from superior fitness of heterozygotes is less than that of a population in which a single allele performs the same function as the two different alleles in the heterozygote (Cain and Sheppard, 1954*b*). On this view, polymorphism is a situation that, once established, is perpetuated by selection between individuals within the population, but is a disadvantage to the population as a whole in competition with another population lacking the polymorphism.

The foregoing account of polymorphism leaves many problems unsolved, and does little more than sketch the outlines of a most interesting aspect of the genetics of populations. In particular, we have not mentioned the extensive and detailed investigations of polymorphism in respect of inverted segments of chromosomes found in species of *Drosophila* and, to a lesser extent, in some other animals and plants. For a description of these studies, and also for a fuller general account of polymorphism, the reader must be referred to Dobzhansky (1951*a*). We conclude by giving one example of polymorphism where the nature of the superiority of heterozygotes is clear. Other cases are described by Dobzhansky (1951*a*), Ford (1953), Lerner (1954), and Sheppard (1958).

EXAMPLE 2.4. *Sickle-cell anaemia* (Allison, 1955). There is a gene, found in American negroes and in the indigenous East Africans, which causes the formation of an abnormal type of haemoglobin. Homozygotes suffer from an anaemia, characterised by the "sickle" shape of the erythrocytes; it is a severe disease from which many die. All the haemoglobin of homozygotes is of the abnormal type, though there is a variable admixture of foetal haemoglobin. Heterozygotes do not suffer from anaemia, but they can be recognised by the presence of sickle cells if the haemoglobin is deoxygenated. About 35 per cent of their haemoglobin is of the abnormal type. With respect to haemoglobin synthesis, therefore, the sickle-cell gene is partially dominant, though with respect to the anaemia it is recessive, and with respect to fitness it has been proved to be over-dominant. In routine surveys the few surviving homozygotes are not readily distinguished from heterozygotes; we shall refer to the combined heterozygotes and surviving homozygotes as "abnormals." The frequency of abnormals varies very much with the locality: in American negroes it is about 9 per cent, and in different parts of Africa it varies from zero up to a maximum of about 40 per cent. In view of the severe disability of the homozygotes it is impossible to account for these high frequencies unless the heterozygotes have a quite substantial selective advantage over the normal homozygotes. The nature of this selective advantage has been shown to be connected with resistance to malaria. Heterozygotes are less susceptible to malaria than normal homozygotes, and the frequency of abnormals in different areas is correlated with the prevalence of malaria. Let us work out the gene frequency corresponding with the maximum frequency of 40 per cent abnormals, and then find the magnitude of the selective advantage of heterozygotes necessary to maintain this gene frequency in equilibrium.

If the gene frequency is in equilibrium it will be the same after selection has taken place as it was before. Therefore, if we assume that all the selection takes place before adulthood—an assumption that is not very far from the truth—we can estimate the gene frequency from the genotype frequencies in the adult population. But it is first necessary to know what proportion of abnormals are homozygotes. This has been estimated as being approximately 2·9 per cent (Allison, 1954). Thus, when the frequeney of abnormals is 0·4, the frequency of homozygotes is 0·012, and that of heterozygotes is 0·388. The gene frequency, then, by equation *1.1*, is the frequency of homozygotes plus half the frequency of heterozygotes, which comes to $q = 0·206$. If this gene frequency is the equilibrium value maintained by natural selection favouring the heterozygotes, and if we assume mating to be random, then the gene frequency is related to the selection coefficients by equation *2.16*. The fitness of sickle-cell homozygotes, relative to that of heterozygotes, has been estimated from a comparison of viability and fertility as being approximately 0·25. Therefore the coefficient of selection against homozygotes is $s_2 = 0·75$. Substituting this value of s_2, and the value of q found above, in equation *2.16* gives $s_1 = 0·197$. This is the coefficient of selection against normal homozygotes, relative to heterozygotes. If we want to express the selective advantage of heterozygotes as the superiority of heterozygotes, relative to normal homozygotes, we may do so, since the fitness of heterozygotes relative to normal homozygotes is $\dfrac{1}{1-s_1}$. This is 1·24. Thus the selective advantage to be attributed to the resistance of heterozygotes to malaria, if these are the forces holding the gene in equilibrium, is 24 per cent.

The presence of the sickle-cell gene in American negroes can be attributed to their African origin. The gene's present frequency of 0·046, deduced in the manner described above, can be accounted for partly by racial mixture and partly by the change of habitat which, removing the advantage of heterozygotes, has exposed the gene to the full power of the selection against homozygotes.

As an example of polymorphism the sickle-cell gene is not altogether typical, because the differences of fitness are rather large and one of the genotypes is clearly abnormal. But it illustrates in an exaggerated form the nature of the selective forces that are presumed to underlie the more usual forms of polymorphism.

CHAPTER 3

SMALL POPULATIONS:

I. Changes of Gene Frequency under Simplified Conditions

We have now to consider the last of the agencies through which gene frequencies can be changed. This is the dispersive process, which differs from the systematic processes in being random in direction, and predictable only in amount. In order to exclude this process from the previous discussions we have postulated always a "large" population, and we have seen that in a large population the gene frequencies are inherently stable. That is to say, in the absence of migration, mutation, or selection, the gene and genotype frequencies remain unaltered from generation to generation. This property of stability does not hold in a small population, and the gene frequencies are subject to random fluctuations arising from the sampling of gametes. The gametes that transmit genes to the next generation carry a sample of the genes in the parent generation, and if the sample is not large the gene frequencies are liable to change between one generation and the next. This random change of gene frequency is the dispersive process.

The dispersive process has, broadly speaking, three important consequences. The first is differentiation between sub-populations. The inhabitants of a large area seldom in nature constitute a single large population, because mating takes place more often between inhabitants of the same region. Natural populations are therefore more or less subdivided into local groups or sub-populations, and the sampling process tends to cause genetic differences between these, if the number of individuals in the groups is small. Domesticated or laboratory populations, in the same way, are often subdivided—for example, into herds or strains—and in them the subdivision and its resultant differentiation are often more marked. The second consequence is a reduction of genetic variation within a small population. The individuals of the population become more and more alike in genotype, and this genetic uniformity is the reason for the widespread use of inbred strains of laboratory animals in physiological and allied fields of research. (An inbred strain, it may be noted, is a small population.) The third consequence of the dispersive process is an increase in the frequency of homozygotes at the expense of heterozygotes. This, coupled with the general tendency for deleterious alleles to be recessive, is the genetic basis of the loss of fertility and viability that almost always results from inbreeding. To explain these three consequences of the dispersive process is the chief purpose of this chapter.

There are two different ways of looking at the dispersive process and of deducing its consequences. One is to regard it as a sampling process and to describe it in terms of sampling variance. The other is to regard it as an inbreeding process and describe it in terms of the genotypic changes resulting from matings between related individuals. Of these, the first is probably the simpler for a description of how the process works, but the second provides a more convenient means of stating the consequences. The plan to be followed here is first to describe the general nature of the dispersive process from the point of view of sampling. This will show how the three chief consequences come about. Then we shall approach the process afresh from the point of view of inbreeding, and show how the two viewpoints connect with each other. In all this we shall confine our attention to the simplest possible situation, excluding migration, mutation, and selection. Thus we shall see what happens in small populations in the absence of other factors influencing gene frequency. In the next chapter we shall extend the conclusions to more realistic situations, by removing the restrictive simplifications, and we shall in particular consider the joint effects of the dispersive process and the systematic processes. Finally, in Chapter 5, we shall consider the special cases of pedigreed populations, and very small populations maintained by regular systems of close inbreeding.

The Idealised Population

In order to reduce the dispersive process to its simplest form we imagine an idealised population as follows. We suppose there to be initially one large population in which mating is random, and this population becomes subdivided into a large number of sub-populations. The subdivision might arise from geographical or ecological causes under natural conditions, or from controlled breeding in domesticated or laboratory populations. The initial random-mating population will be referred to as the *base population*, and the sub-populations will be referred to as *lines*. All the lines together constitute the whole population, and each line is a "small population" in which gene frequencies are subject to the dispersive process. When a single locus is under discussion we cannot properly understand what goes on in one line except by considering it as one of a large number of lines. But what happens to the genes at one locus in a number of lines happens equally to those at a number of loci in one line, provided they all start at the same gene frequency. So the consequences of the process apply equally to a single line provided we consider many loci in it.

The simplifying conditions specified for the idealised population are the following:

1. Mating is restricted to members of the same line. The lines are thus isolated in the sense that no genes can pass from one line to another. In other words migration is excluded.

2. The generations are distinct and do not overlap.

3. The number of breeding individuals in each line is the same for all lines and in all generations. Breeding indviduals are those that transmit genes to the next generation.

4. Within each line mating is random, including self-fertilisation in random amount.

5. There is no selection at any stage.

6. Mutation is disregarded.

The situation implied by these conditions is represented diagrammatically in Fig. 3.1, and may be described thus: All breeding

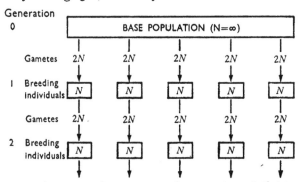

FIG. 3.1. Diagrammatic representation of the subdivision of a single large population—the base population—into a number of sub-populations, or lines.

individuals contribute equally to a pool of gametes from which zygotes will be formed. Union of gametes is strictly random. Out of a potentially large number of zygotes only a limited number survive to become breeding individuals in the next generation, and this is the stage at which the sampling of the genes transmitted by the gametes takes place. Survival of zygotes is random, and consequently the contribution of the parents to the next generation is not uniform, but varies according to the chances of survival of their progeny. Since the population size is constant from generation to generation, the average number of progeny that reach breeding age is one per individual parent or two per mated pair of parents. For any particular zygote the chance of survival is small, and therefore the number of progeny contributed by individual parents, or by pairs of parents, has a Poisson distribution.

The following symbols will be used in connexion with the idealised population.

N = the number of breeding individuals in each line and generation. This is the *population size*.

t = time, in generations, starting from the base population at t_0.

q = frequency of a particular allele at a locus.

$p = 1 - q$ = frequency of all other alleles at that locus. q and p refer to the frequencies in any one line; $\bar{q}$ and $\bar{p}$ refer to the frequencies in the whole population and are the means of q and p; q_0 and p_0 are the frequencies in the base population.

Sampling

Variance of gene frequency. The change of gene frequency resulting from sampling is random in the sense that its direction is unpredictable. But its magnitude can be predicted in terms of the variance of the change. Consider the formation of the lines from the base population. Each line is formed from a sample of N individuals drawn from the base population. Since each individual carries two genes at a locus, the sub-division of the population represents a series of samples each of $2N$ genes, drawn at random from the base population. The gene frequencies in these samples will have an average value equal to that in the base population, i.e. q_0, and will be distributed about this mean with a variance $p_0 q_0/2N$, which is simply the variance of a ratio, the sample size being in this case $2N$. Thus the change of gene frequency, Δq, resulting from sampling in one generation, can be stated in terms of its variance as

$$\sigma^2_{\Delta q} = \frac{p_0 q_0}{2N} \qquad \dots\dots(3.1)$$

This variance of Δq expresses the magnitude of the change of gene frequency resulting from the dispersive process. It expresses the expected change in any one line, or the variance of gene frequencies that would be found among many lines after one generation. Its effect is a dispersion of gene frequencies among the lines; in other words the lines come to differ in gene frequency, though the mean in the population as a whole remains unchanged.

In the next generation the sampling process is repeated, but each line now starts from a different gene frequency and so the second sampling leads to a further dispersion. The variance of the change now differs among the lines, since it depends on the gene frequency, q_1, in the first generation of each line separately. The effect of continued sampling through successive generations is that each line fluctuates irregularly in gene frequency, and the lines spread apart progressively, thus becoming differentiated. The erratic changes of gene frequency shown by the individual lines are exemplified in Fig. 3.2;

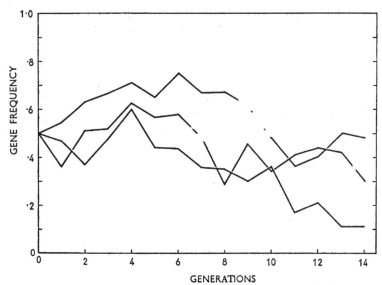

FIG. 3.2. Random drift of the colour gene "non-agouti" in three lines of mice, each maintained by 6 pairs of parents per generation. (Original data.)

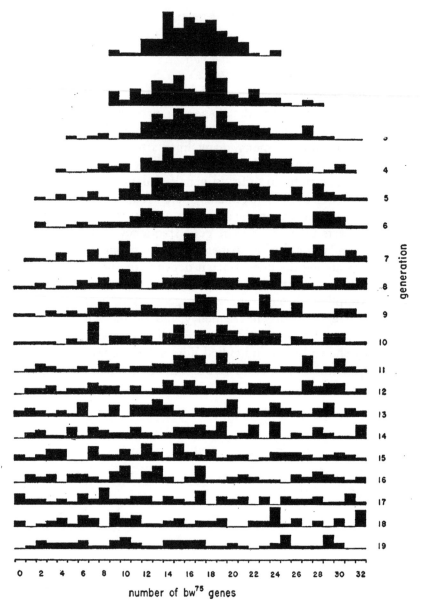

number of bw⁷⁵ genes

FIG. 3.3. Distributions of gene frequencies in 19 consecutive
generations among 105 lines of *Drosophila melanogaster*, each of 16
individuals. The gene frequencies refer to two alleles at the
"brown" locus (*bw⁷⁵* and *bw*), with initial frequencies of 0·5. The
height of each black column shows the number of lines having the
gene frequency shown on the scale below. (From Buri, 1956;
reproduced by courtesy of the author and the editor of *Evolution*.)

and the consequent differentiation, or spreading apart, of the lines in Fig. 3.3. These changes of gene frequency resulting from sampling in small populations are known as *random drift* (Wright, 1931).

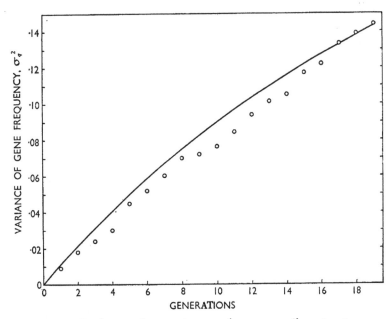

FIG. 3.4. Variance of gene frequencies among lines in the experiment illustrated in Fig. 3.3. The circles are the observed values, and the smooth curve shows the expected variance as given by equation *3.2*. The value taken for N is 11·5, which is the "effective number," N_e, as explained in the next chapter. (Data from Buri, 1956.)

As the dispersive process proceeds, the variance of gene frequency among the lines increases, as shown in Fig. 3.4. At any generation, t, the variance of gene frequencies, σ_p^2, among the lines is as follows (see Crow, 1954):

$$\sigma_q^2 = p_0 q_0 \left[1 - \left(1 - \frac{1}{2N} \right)^t \right] \qquad \dots\dots(3.2)$$

Since the mean gene frequency among all the lines remains unchanged, $\bar{q} = q_0$. We may note a fact that will be needed later, and is obvious from equation *3.2*, namely that $\sigma_p^2 = \sigma^2$. The dispersion of the gene frequencies, which we have described by reference to one locus in many lines, could equally well be described by reference to the frequencies at a number of different loci in one line, provided they all started from the same initial frequency, and were unlinked.

Fixation. There are limits to the spreading apart of the lines that can be brought about by the dispersive process. The gene frequency cannot change beyond the limits of 0 or 1, and sooner or later each line must reach one or other of these limits. Moreover, the limits are "traps" or points of no return, because once the gene frequency has reached 0 or 1 it cannot change any more in that line. When a particular allele has reached a frequency of 1 it is said to be *fixed* in that line, and when it reaches a frequency of 0 it is *lost*. When an allele reaches fixation no other allele can be present in that line, and the line may then be said to be fixed. When a line is fixed all individuals in it are of identical genotype with respect to that locus. Eventually all lines, and all loci in a line, become fixed. The individuals of a line are then genetically identical, and this is the basis of the genetic uniformity of highly inbred strains.

The proportion of the lines in which different alleles at a locus are fixed is equal to the initial frequencies of the alleles. If the base population contains two alleles A_1 and A_2 at frequencies p_0 and q_0 respectively, then A_1 will be fixed in the proportion p_0 of the lines, and A_2 in the remaining proportion, q_0. The variance of the gene frequency among the lines is then $p_0 q_0$, as may be seen from equation 3.2 by putting t equal to infinity. (In Fig. 3.3 the lines in which fixation or loss has just occurred are shown, but not those in which it occurred earlier.)

When concerned with the attainment of genetic uniformity one wants to know how soon fixation takes place; what is the probability of a particular locus being fixed, or what proportion of all loci in a line will be fixed, after a certain number of generations. Consideration of the progressive nature of the dispersion, as illustrated in Fig. 3.3, will show that fixation does not start immediately; the dispersion of gene frequencies must proceed some way before any line is likely to reach fixation. To deduce the probability of fixation is mathematically complicated (see particularly Wright, 1931; Kimura, 1955), and only an outline of the conclusions can be given here. There are two phases in the dispersive process: during the initial phase the gene frequencies are spreading out from the initial value; this leads to a steady phase, when the gene frequencies are evenly spread out over the range between the two limits, and all gene frequencies except the two limits are equally probable. The duration of the initial phase in generations is a small multiple of the population size, depending on the initial gene frequency. With $q_0 = 0·5$ it lasts about $2N$ generations, and with $q_0 = 0·1$ it lasts about $4N$ generations (Kimura, 1955). (In the experiment illustrated in Fig. 3.3 it lasted till about the seventeenth generation.) The theoretical distributions of gene frequency during the initial phase, with original frequencies of 0·5 and 0·1, are shown in Fig. 3.5.

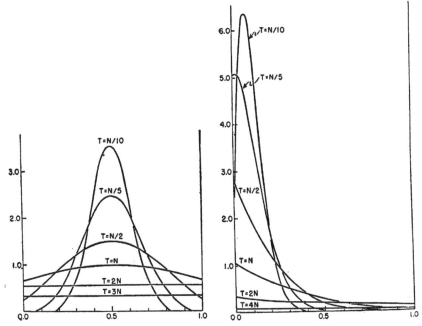

FIG. 3.5. Theoretical distributions of gene frequency among lines. The initial and mean gene frequency is 0·5 in the left hand figure, and 0·1 in the right hand figure. Previously fixed lines are excluded. N=population size; T=time in generations. Note the general agreement of the left hand figure with the observed distributions shown in Fig. 3.3. (From Kimura, 1955; reproduced by courtesy of the author and the editor of the *Proc. Nat. Acad. Sci. Wash.*)

To visualise the process one might think of a pile of dry sand in a narrow trough open at the two ends. Agitation of the trough will cause the pile to spread out along the trough, till eventually it is evenly spread along its length. Toward the end of the spreading out some of the sand will have fallen off the ends of the trough, and this represents fixation and loss. Continued agitation after the sand is

E   F.Q.G.

evenly spread will cause it to fall off the ends at a steady rate, and the depth of sand left in the trough will be continually reduced at a steady rate until in the end none is left. The initial gene frequency is represented by the position of the initial pile of sand. If it is near one end of the trough, much of the sand will have fallen off that end be-

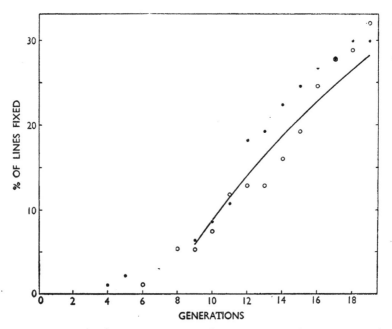

FIG. 3.6. Fixation and loss occurring among 107 lines of *Droso-phila melanogaster*, during 19 generations. This is not the same experiment as that illustrated in Figs. 3.3 and 3.4, but was similar in nature. There were 16 parents per generation in each line, and the effective number (see chapter 4) was 9. The closed circles show the percentage of lines in which the bw^{75} allele has become fixed; the open circles show the percentage in which it has been lost and the bw allele fixed. The smooth curve is the expected amount of fixation of one or other allele, computed from the effective number by equation 3.3. (Data from Buri, 1956.)

fore any reaches the other end, and the total amount falling off each end will be in proportion to the relative distance of the initial pile from the two ends. Relating this model to the diagram of the process in Fig. 3.5, the position along the trough represents the horizontal axis, or gene frequency, and the depth of the sand represents the vertical axis, or the probability of a line having a particular gene frequency. The graphs are thus analogous to longitudinal sections through the trough and its sand.

The probability of fixation at any time during the initial phase is too complicated for explanation here, and the reader is referred to the papers of Kimura (1954, 1955). After the steady phase has been reached fixation proceeds at a constant rate: a proportion $1/2N$ of the lines previously unfixed become fixed in each generation. The proportion of lines in which a gene with initial frequency q_0 is expected to be fixed, lost, or to be still segregating is as follows (Wright, 1952a):

$$\left.\begin{array}{ll} \text{fixed:} & q_0 - 3p_0q_0P \\ \text{lost:} & p_0 - 3p_0q_0P \\ \text{neither:} & 6p_0q_0P \end{array}\right\}\cdots\cdots(3.3)$$

where $P = \left(1 - \dfrac{1}{2N}\right)^t$.

Fig. 3.6 shows the progress of fixation and loss in an experiment with *Drosophila*.

Genotype frequencies. Change of gene frequency leads to change of genotype frequencies; so the genotype frequencies in small populations follow the changes of gene frequency resulting from the dispersive process. In the idealised population, which we are still considering, mating is random within each of the lines. Consequently the genotype frequencies in any one line are the Hardy-Weinberg frequencies appropriate to the gene frequency in the previous generation of that line. As the lines drift apart in gene frequency they become differentiated also in genotype frequencies. But differentiation is not the only aspect of the change: the general direction of the change is toward an increase of homozygous, and a decrease of heterozygous, genotypes. The reason for this is the dispersion of gene frequencies from intermediate values toward the extremes. Heterozygotes are most frequent at intermediate gene frequencies (see Fig. 1.1), so the drift of gene frequencies toward the extremes leads, on the average, to a decline in the frequency of heterozygotes.

The genotype frequencies in the population as a whole can be deduced from a knowledge of the variance of gene frequencies in the following way. If an allele has a frequency q in one particular line, homozygotes of that allele will have a frequency of q^2 in that line. The frequency of these homozygotes in the population as a whole will therefore be the mean value of q^2 over all lines. We shall write this mean frequency of homozygotes as $\overline{(q^2)}$. The value of $\overline{(q^2)}$ can be found from a knowledge of the variance of gene frequencies among the lines, by noting that the variance of a set of observations is found by deducting the square of the mean from the mean of the squared observations. Thus

$$\sigma_q^2 = \overline{(q^2)} - \bar{q}^2$$

and
$$\overline{(q^2)} = \bar{q}^2 + \sigma_q^2 \qquad \ldots\ldots(3.4)$$

where σ_q^2 is the variance of gene frequencies among the lines, as given in equation 3.2, and $\bar{q}^2$ is the square of the mean gene frequency. Since the mean gene frequency, $\bar{q}$, is equal to the original, q_0, it follows that $\bar{q}^2$ or q_0^2 is the original frequency of homozygotes in the base population. Thus in the population as a whole the frequency of homozygotes of a particular allele increases, and is always in excess of the original frequency by an amount equal to the variance of the gene frequency among the lines. In a two-allele system the same applies to the other allele, and the frequency of heterozygotes is reduced correspondingly. Noting from equation 3.2 that $\sigma_p^2 = \sigma_q^2$ we therefore find the genotypic frequencies for a locus with two alleles as follows:

Genotype	Frequency in whole population
A_1A_1	$p_0^2 + \sigma_q^2$
A_1A_2	$2p_0q_0 - 2\sigma^2$
A_2A_2	$q_0^2 + \sigma^2$

$$\left.\right\}\ldots\ldots(3.5)$$

These genotype frequencies are no longer the Hardy-Weinberg frequencies appropriate to the original or mean gene frequency. The Hardy-Weinberg relationships between gene frequency and genotype frequencies, though they hold good within each line separately, do not hold if the lines are taken together and regarded as a single population. This fact causes some difficulty in relating gene and genotype frequencies in natural populations, because they are often more or less subdivided and the degree of subdivision is seldom known. An example of the decrease of heterozygotes resulting from the dispersion of gene frequencies is shown in Fig. 3.7.

The foregoing account of genotype frequencies describes the situation in terms of one locus in many lines. It can be regarded equally as referring to many loci in one line; then the change in any one line or small population is an increase in the number of loci at which individuals are homozygous and a corresponding decrease in the number at which they are heterozygous—in short an increase of homozygotes at the expense of heterozygotes. This change of genotype frequencies resulting from the dispersive process is the genetic basis of the phenomenon of inbreeding depression, of which a full explanation will be found in Chapter 14.

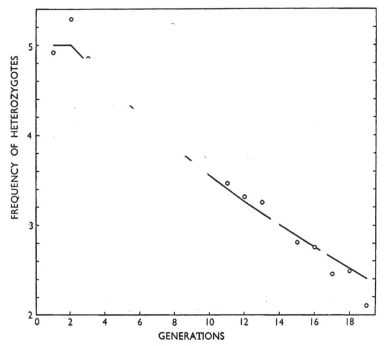

FIG. 3.7. Change of frequency of heterozygotes among 105 lines of *Drosophila melanogaster*, each with 16 parents. The same experiment as is illustrated in Figs. 3.3. and 3.4. The frequency of heterozygotes refers to the population as a whole, all lines taken together. The smooth curve is the expected frequency of heterozygotes. (Data from Buri, 1956.)

We have now surveyed the general nature of the dispersive process and its three major consequences—differentiation of sub-populations, genetic uniformity within sub-populations, and overall increase in the frequency of homozygous genotypes. Let us now look at the process from another viewpoint, as an inbreeding process. Instead of regarding the increase of homozygotes as a consequence of the dispersion of gene frequencies, we shall now look directly at the manner in which the additional homozygotes arise.

INBREEDING

Inbreeding means the mating together of individuals that are related to each other by ancestry. That the degree of relationship between the individuals in a population depends on the size of the population will be clear by consideration of the numbers of possible ancestors. In a population of bisexual organisms every individual has two parents, four grand-parents, eight great-grandparents, etc., and t generations back it has 2^t ancestors. Not very many generations back the number of individuals required to provide separate ancestors for all the present individuals becomes larger than any real population could contain. Any pair of individuals must therefore be related to each other through one or more common ancestors in the more or less remote past; and the smaller the size of the population in previous generations the less remote are the common ancestors, or the greater their number. Thus pairs mating at random are more closely related to each other in a small population than in a large one. This is why the properties of small populations can be treated as the consequences of inbreeding.

The essential consequence of two individuals having a common ancestor is that they may both carry replicates of one of the genes present in the ancestor; and if they mate they may pass on these replicates to their offspring. Thus inbred individuals—that is to say, offspring produced by inbreeding—may carry two genes at a locus that are replicates of one and the same gene in a previous generation. Consideration of this consequence of inbreeding shows that there are two sorts of identity among allelic genes, and two sorts of homozygote. The sort of identity we have hitherto considered is a functional identity. Two genes are regarded as being identical if they are not recognisably different in their phenotypic effects, or by any other functional criterion; in other words, if they have the same allelemorphic state. Following the terminology of Crow (1954) they may be called *alike in state*. An individual carrying a pair of such genes is a homozygote in the ordinary sense. The new sort of identity is one of replication. If two genes originated from the replication of one gene in a previous generation, they may be said to be *identical by descent*, or simply *identical*. An individual possessing two identical genes at a locus may be called an identical homozygote. Genes that are not identical by descent may be called *independent*, whether they are alike in state or different alleles; and homozygotes of independent genes may be called independent homozygotes.

Identity by descent provides the basis for a measure of the dispersive process, through the degree of relationship between the mating pairs. The measure is the *coefficient of inbreeding*, which is the probability that the two genes at any locus in an individual are identical by descent. It refers to an individual and expresses the degree of relationship between the individual's parents. If the parents mated at random then the coefficient of inbreeding of the progeny is the probability that two gametes taken at random from the parent generation carry identical genes at a locus. The coefficient of inbreeding, generally symbolised by F, was first defined by Wright (1922) as the correlation between uniting gametes; the definition given here, which follows that of Malécot (1948) and Crow (1954), is equivalent.

The degree of relationship expressed in the inbreeding coefficient is essentially a comparison between the population in question and some specified or implied base population. Without this point of reference it is meaningless, as the following consideration will show. On account of the limitation in the number of independent ancestors in any population not infinitely large, all genes now present at a locus in the population would be found to be identical by descent if traced far enough back into the remote past. Therefore the inbreeding coefficient only becomes meaningful if we specify some time in the past beyond which ancestries will not be pursued, and at which all genes present in the population are to be regarded as independent— that is, not identical by descent. This point is the base population and by its definition it has an inbreeding coefficient of zero. The inbreeding coefficient of a subsequent generation expresses the amount of the dispersive process that has taken place since the base population, and compares the degree of relationship between the individuals now, with that between individuals in the base population. Reference to the base population is not always explicitly stated, but is always implied. For example, we can speak of the inbreeding coefficient of a population subdivided into lines. The comparison of relationship is between the individuals of a line and individuals taken at random from the whole population. The base population implied is a hypothetical population from which all the lines were derived.

Inbreeding in the idealised population. Let us now return to the idealised population and deduce the coefficient of inbreeding in successive generations, starting with the base population and its progeny constituting generation 1. The situation may be visualised by thinking of a hermaphrodite marine organism, capable of self-fertilisation, shedding eggs and sperm into the sea. There are N individuals each shedding equal numbers of gametes which unite at random. All the genes at a locus in the base population have to be regarded as being non-identical; so, considering only one locus, among the gametes shed by the base population there are $2N$ different sorts, in equal numbers, bearing the genes A_1, A_2, A_3, etc. at the A locus. The gametes of any one sort carry identical genes; those of different sort carry genes of independent origin. What is the probability that a pair of gametes taken at random carry identical genes? This is the inbreeding coefficient of generation 1. Any gamete has a $1/2N$th chance of uniting with another of the same sort, so $1/2N$ is the probability that uniting gametes carry identical genes, and is thus the coefficient of inbreeding of the progeny. Now consider the second generation. There are now two ways in which identical homozygotes can arise, one from the new replication of genes and the other from the previous replication. The probability of newly replicated genes coming together in a zygote is again $1/2N$. The remaining proportion, $1 - 1/2N$, of zygotes carry genes that are independent in their origin from generation 1, but may have been identical in their origin from generation 0. The probability of their identical origin in generation 0 is what we have already deduced as the inbreeding coefficient of generation 1. Thus the total probability of identical homozygotes in generation 2 is

$$F_2 = \frac{1}{2N} + \left(1 - \frac{1}{2N}\right)F_1$$

where F_1 and F_2 stand for the inbreeding coefficients of generations 1 and 2 respectively. The same argument applies to subsequent generations, so that in general the inbreeding coefficient of individuals in generation t is

$$F_t = \frac{1}{2N} + \left(1 - \frac{1}{2N}\right)F_{t-1} \qquad \ldots\ldots(3.6)$$

Thus the inbreeding coefficient is made up of two parts: an "increment," $1/2N$, attributable to the new inbreeding, and a "remainder," attributable to the previous inbreeding and having the inbreeding coefficient of the previous generation. In the idealised population the "new inbreeding" arises from self-fertilisation, which brings together genes replicated in the immediately preceding generation. Exclusion of self-fertilisation simply shifts the replication one generation further back, so that the "new inbreeding" brings together genes replicated in the grand-parental generation; the coefficient of inbreeding is affected, but not very much, as we shall see later. The distinction between "new" and "old" inbreeding brings clearly to light a point which we note here in passing because it will be needed later and is often important in practice: if there is no "new inbreeding," as would happen if the population size were suddenly increased, the previous inbreeding is not undone, but remains where it was before the increase of population size.

Let us call the "increment" or "new inbreeding" ΔF, so that

$$\Delta F = \frac{1}{2N} \qquad \ldots\ldots(3.7)$$

Equation 3.6 may then be rewritten in the form

$$F_t = \Delta F + (1 - \Delta F)F_{t-1} \qquad \ldots\ldots(3.8)$$

Further rearrangement makes clearer the precise meaning of the "increment," ΔF.

$$\Delta F = \frac{F_t - F_{t-1}}{1 - F_{t-1}} \qquad \ldots\ldots(3.9)$$

From the equation written thus we see that the "increment," ΔF, measures the *rate of inbreeding* in the form of a proportionate increase. It is the increase of the inbreeding coefficient in one generation, relative to the distance that was still to go to reach complete inbreeding. This measure of the rate of inbreeding provides a convenient way of going beyond the restrictive simplifications of the idealised population, and it thus provides a means of comparing the inbreeding effects of different breeding systems. When the inbreeding coefficient is expressed in terms of ΔF, equation 3.8 is valid for any breeding system and is not restricted to the idealised population, though only in the idealised population is ΔF equal to $1/2N$.

So far we have done no more than relate the inbreeding coefficient in one generation to that of the previous generation. It remains to extend equation 3.8 back to the base population and so express the inbreeding coefficient in terms of the number of generations. This is made easier by the use of a symbol, P, for the complement of the inbreeding coefficient, $1 - F$, which is known as the *panmictic index*. Substitution of $P = 1 - F$ in equation 3.8 gives

$$\frac{P_t}{P_{t-1}} = 1 - \Delta F \qquad \qquad \ldots \ldots (3.10)$$

Thus the panmictic index is reduced by a constant proportion in each generation. Extension back to generation $t - 2$ gives

$$\frac{P_t}{P_{t-2}} = (1 - \Delta F)^2$$

and extension back to the base population gives

$$P_t = (1 - \Delta F)^t P_0 \qquad \qquad \ldots \ldots (3.11)$$

where P_0 is the panmictic index of the base population. The base population is defined as having an inbreeding coefficient of 0, and therefore a panmictic index of 1. The inbreeding coefficient in any generation, t, referred to the base population, is therefore

$$F_t = 1 - (1 - \Delta F)^t \qquad \qquad \ldots \ldots (3.12)$$

The consequences of the dispersive process were described earlier from the viewpoint of sampling variance. Let us now look again at them, applying the rate of inbreeding and the inbreeding coefficient as measures of the process. Strictly speaking we should refer still to the idealised population, but the equating of the two viewpoints can be regarded as generally valid except in some very special and unlikely circumstances (see Crow, 1954).

Variance of gene frequency. First, the variance of the change of gene frequency in one generation, taken from equation 3.1 and expressed in terms of the rate of inbreeding, becomes

$$\sigma_{\Delta q}^2 = \frac{p_0 q_0}{2N} = p_0 q_0 \Delta F \qquad \qquad \ldots (3.13)$$

Similarly, the variance of gene frequencies among the lines at generation t, taken from equation 3.2 and expressed in terms of the inbreeding coefficient from 3.12, becomes

$$\sigma_q^2 = p_0 q_0 \left[1 - \left(1 - \frac{1}{2N} \right)^t \right]$$
$$= p_0 q_0 F \qquad \qquad \ldots (3.14)$$

Thus ΔF expresses the rate of dispersion and F the cumulated effect of random drift.

Genotype frequencies. Leaving fixation aside for the moment, let us consider next the genotype frequencies in the population as a whole. The genotype frequencies expressed in terms of the variance of gene frequency in equations *3.5* can be rewritten in terms of the coefficient of inbreeding from equation 3.*14*. The frequency of A_2A_2, for example, is

$$(\overline{q^2}) = q_0^2 + \sigma_q^2 = q_0^2 + p_0q_0F$$

The genotype frequencies expressed in this way are entered in the left-hand side of Table 3.1. As was explained before, this way of writing the genotype frequencies shows how the homozygotes in-

TABLE 3.1

Genotype frequencies for a locus with two alleles, expressed
in terms of the inbreeding coefficient, F.

	Original fre-quencies		*Change due to inbreeding*		*Origin: Independent*		*Identical*
A_1A_1	p_0^2	+	p_0q_0F	or	$p_0^2(1-F)$	+	p_0F
A_1A_2	$2p_0q_0$	−	$2p_0q_0F$	or	$2p_0q_0(1-F)$		
A_2A_2	q_0^2	+	p_0q_0F	or	$q_0^2(1-F)$	+	q_0F

crease at the expense of the heterozygotes. Recognition of identity by descent to which the inbreeding viewpoint led us means that we can now distinguish the two sorts of homozygote, identical and independent, among both the A_1A_1 or A_2A_2 genotypes. The frequency of identical homozygotes among both genotypes together is by definition the inbreeding coefficient, F; and it is clear that the division between the two genotypes is in proportion to the initial gene frequencies. So p_0F is the frequency of A_1A_1 identical homozygotes, and q_0F that of A_2A_2 identical homozygotes. The remaining genotypes, both homozygotes and heterozygotes, carry genes that are independent in origin and are therefore the equivalent of pairs of gametes taken at random from the population as a whole. Their frequencies are therefore the Hardy-Weinberg frequencies. Thus, from the inbreeding viewpoint, we arrive at the genotype frequencies shown in the right-hand columns of Table 3.1. This way of writing the genotype frequencies shows how homozygotes are divided between those of independent and those of identical origin. The equivalence of the two ways of expressing the genotype frequencies can be verified from their algebraic identity. Both ways show equally clearly how the heterozygotes are reduced in frequency in proportion to $1 - F$. The term "heterozygosity" is often used to express the frequency of heterozygotes at any time, relative to their frequency in the base population. The heterozygosity is the same as the panmictic index, P. Thus if H_t and H_0 are the frequencies of heterozygotes for a pair of alleles at generation t and in the base population respectively, then the heterozygosity at generation t is

$$\frac{H_t}{H_0} = P_t \qquad \qquad \dots (3.15)$$

Fixation. There is little to add, from the inbreeding viewpoint, to the description of fixation given earlier. The rate of fixation—that is the proportion of unfixed loci that become fixed in any generation—is equal to ΔF, after the steady phase has been reached and the distribution of gene frequencies has become flat. The quantity P in equations *3.3* which give the probability of a gene having become fixed or lost, is equal to $1 - F$. We may note, however, that the probability of fixation is not very different from the inbreeding coefficient itself. The explanation comes more readily by considering the probability that a locus remains unfixed. This probability was given in equation *3.3* for a locus with two alleles after enough generations have passed to take the population into the steady phase. Expressed in terms of the inbreeding coefficient, from equation *3.12*, it is $6p_0q_0(1 - F)$. Now, the value of p_0q_0 does not change very much over quite a wide range of gene frequencies, and so the probability that a locus is still unfixed is not very sensitive to the initial gene frequency. The value of $6p_0q_0$ lies between 1·0 and 1·5 over a range of gene frequency from 0·2 to 0·8, a range that is likely to cover many situations. Consequently the probability that a line still segregates, or the proportion of loci expected to remain unfixed, is likely to lie between $(1 - F)$ and $1·5(1 - F)$. Thus the inbreeding coefficient gives a good idea of the approximate probability of fixation, even in the absence of a knowledge of the initial gene frequencies. That the approximation may be quite close enough for practical purposes may be seen by taking a specific example. In work involving immunological reactions it may be necessary to produce a strain in which all loci that determine the reactions have been fixed. One therefore wants to know the inbreeding coefficient necessary to raise the probability of fixation, or the proportion of loci expected to be fixed, to a certain level—say 90 per cent. The inbreeding coefficient needed to do this would, on the above considerations, lie between 0·90 and 0·93, and this would answer the question with quite enough accuracy for most purposes.

CHAPTER 4

SMALL POPULATIONS:

II. Less Simplified Conditions

In order to simplify the description of the dispersive process we confined our attention in the last chapter to an idealised population, and to do this we had to specify a number of restrictive conditions, which could seldom be fulfilled in real populations. The purpose of this chapter is to adapt the conclusions of the last chapter to situations in which the conditions imposed do not hold; in other words to remove the more serious restrictions and bring the conclusions closer to reality. The restrictive conditions were of two sorts, one sort being concerned with the breeding structure of the population and the other excluding mutation, migration, and selection from consideration. We shall first describe the effects of deviations from the idealised breeding structure, and then consider the outcome of the dispersive process when mutation, migration, or selection are operating at the same time.

Effective Population Size

If the breeding structure does not conform to that specified for the idealised population, it is still possible to evaluate the dispersive process in terms of either the variance of gene frequencies or the rate of inbreeding. This can be done by the same general methods and no new principles are involved. We shall therefore give the conclusions briefly and without detailed explanation. The most convenient way of dealing with any particular deviation from the idealised breeding structure is to express the situation in terms of the *effective number* of breeding individuals, or the *effective population size*. This is the number of individuals that would give rise to the sampling variance or the rate of inbreeding appropriate to the conditions under consideration, if they bred in the manner of the idealised population. Thus, by converting the actual number, N, to the effective number, N_e, we can apply the formulae deduced in the last chapter. The rate of inbreeding, for example, is

$$\Delta F = \frac{1}{2N_e} \qquad \qquad .....(4.1)$$

just as for the idealised population $\Delta F = 1/2N$ (equation 3.7).

The relationships between actual and effective numbers in the situations most commonly met with are given below. The exact expressions are often complicated, but in most circumstances an approximation can be used with sufficient accuracy. We should first note that the actual number, N, refers to breeding individuals—the breeding individuals of one generation—and it therefore cannot be obtained directly from a census, unless the different age-groups are distinguished.

Bisexual organisms: self-fertilisation excluded. The exclusion of self-fertilisation makes very little difference to the rate of inbreeding, unless N is very small, as with close inbreeding. The relationship of effective to actual numbers (Wright, 1931) is

$$N_e = N + \tfrac{1}{2} \quad \text{(approx.)} \qquad \qquad .....(4.2)$$

and the rate of inbreeding is

$$\Delta F = \frac{1}{2N+1} \quad \text{(approx.)} \qquad \qquad .....(4.3)$$

The exact expression for the inbreeding coefficient in a bisexual population, and its derivation, are given by Malécot (1948).

Different numbers of males and females. In domestic and laboratory animals the sexes are often unequally represented among the breeding individuals, since it is more economical, when possible, to use fewer males than females. The two sexes, however, whatever their relative numbers, contribute equally to the genes in the next generation. Therefore the sampling variance attributable to the two sexes must be reckoned separately. Since the sampling variance is proportional to the reciprocal of the number, the effective number is twice the harmonic mean of the numbers of the two sexes (Wright, 1931), so that

$$\frac{1}{N_e} = \frac{1}{4N_m} + \frac{1}{4N_f} \qquad \qquad .....(4.4)$$

where N_m and N_f are the actual numbers of males and females respectively. The rate of inbreeding is then

$$\Delta F = \frac{1}{8N_m} + \frac{1}{8N_f} \quad \text{(approx.)} \qquad \ldots\ldots(4.5)$$

This gives a close enough approximation unless both N_m and N_f are very small, as with close inbreeding. It should be noted that the rate of inbreeding depends chiefly on the numbers of the less numerous sex. For example, if a population were maintained with an indefinitely large number of females but only one male in each generation, the effective number would be only about 4.

Unequal numbers in successive generations. The rate of inbreeding in any one generation is given, as before, by $1/2N$. If the numbers are not constant from generation to generation, then the mean rate of inbreeding is the mean value of $1/2N$ in successive generations. The effective number is the harmonic mean of the numbers in each generation (Wright, 1939). Over a period of t generations, therefore,

$$\frac{1}{N_e} = \frac{1}{t}\left[\frac{1}{N_1} + \frac{1}{N_2} + \frac{1}{N_3} + \ldots + \frac{1}{N_t}\right] \quad \text{(approx.)} \qquad \ldots\ldots(4.6)$$

Thus the generations with the smallest numbers have the most effect. The reason for this can be seen by consideration of the "new" and "old" inbreeding referred to in connexion with equation 3.6. An expansion in numbers does not affect the previous inbreeding; it merely reduces the amount of new inbreeding. So, in a population with fluctuating numbers the inbreeding proceeds by steps of varying amount, and the present size of the population indicates only the present rate of inbreeding.

Non-random distribution of family size. This is probably the commonest and most important deviation from the breeding system of the idealised population. Its consequence is usually to render the effective number less than the actual, but in special circumstances it makes it greater. Family size means here the number of progeny of an individual parent or of a pair of parents, that survive to become breeding individuals. It will be remembered that each breeding individual in the idealised population contributes equally to the pool of gametes, and therefore equally also to the potential zygotes in the next generation. Survival of zygotes is random. The mean number of progeny surviving to breeding age is 1 for individual parents and 2 for pairs of parents. Since the chance of survival for any particular zygote is small, the variation of family size follows a Poisson distribution. The variance of family size is therefore equal to the mean family size, equality of mean and variance being a property of the Poisson distribution. Thus in a population of bisexual organisms, in which all other conditions of the idealised population are satisfied, family size will have a mean and a variance of 2. In natural populations the mean is not likely to differ much from 2, but the variance must be expected to be usually greater, for reasons of differing fertility between the parent individuals and differing viability between the families. If the variance of family size is increased, a greater proportion of the following generation will be the progeny of a smaller number of parents, and the effective number of parents will be less than the actual number. Conversely, if the variance of family size is reduced below that of the idealised population, the effective number will be greater than the actual number. It can be shown that, when the mean family size is 2, the effective number is as follows (Wright, 1940; Crow, 1954):

$$N_e = \frac{4N}{2 + \sigma_k^2} \qquad \dots(4.7)$$

where σ_k^2 is the variance of family size. (Strictly speaking this is the effective number as it affects variance of gene frequency and fixation: for its effects on the inbreeding coefficient, $N_e = \frac{4N-2}{2 + \sigma_k^2}$. The difference is small and we shall ignore it.) Thus, when there is equal fertility of the parents and random survival of the progeny $\sigma_k^2 = 2$, and $N_e = N$. When differences of fertility and viability make σ_k^2 greater than 2, as in most actual populations, then N_e is less than N. The effective number under consideration here refers to a population with equal numbers of males and females, and with monogamous mating. If males are not restricted to a single mate, then the families of males are likely to be more variable in size than those of females. In these circumstances the relationship of effective to actual numbers will differ for male and female parents.

It is possible by controlled breeding to make the variance of family size, σ_k^2, less than 2, and therefore to make the effective number greater than the actual. If two members of each family are deliberately chosen to be parents of the next generation, then the variance of family size is zero. Under these special circumstances, and if the sexes are equal in numbers, the effective number is twice the actual:

$$N_e = 2N \qquad \qquad \dots(4.8)$$

The rate of inbreeding is consequently half what it would be in an idealised population of equal size, and is usually less than half the rate of inbreeding under normal circumstances and random mating. Under this controlled breeding system the rate of inbreeding is the lowest possible with a given number of breeding individuals. The reduced variance of family size is the path through which the "deliberate avoidance of inbreeding" works. The problem often arises of keeping a stock with minimum inbreeding, but with a limitation of the actual population size imposed by the space or facilities available. A common practice under these circumstances is the deliberate avoidance of sib-matings and perhaps also of cousin-matings. One may go further and by the use of pedigrees (in the manner described in the next chapter) choose pairs for mating that have the least possible relationship with each other. Deliberate avoidance of inbreeding in this way has the effect of distributing the individuals chosen to be parents evenly over the available families, and thus reduces the variance of family size and the rate of inbreeding. The same result, however, can be achieved with less labour simply by ensuring that the available families are as far as possible equally represented among the individuals chosen to be the parents of the next generation. If, in addition, matings between close relatives are avoided, the inbreeding coefficient in any generation is slightly lower and is more uniform between the individuals in the generation than if matings between close relatives are allowed; but the rate of inbreeding is the same.

If the sexes are unequal in numbers, but the individuals chosen as parents are equally distributed, in numbers and sexes, between the families, so that the variance of family size is still zero, then the rate of inbreeding is given by the following formula (Gowe, Robertson, and Latter, 1959):

$$\Delta F = \frac{3}{32N_m} + \frac{1}{32N_f} \qquad \qquad \dots(4.9)$$

where N_m and N_f are the actual numbers of male and female parents respectively, and females are more numerous than males.

EXAMPLE 4.1. Several flocks of poultry in the United States and in Canada, which are used as controls for breeding experiments, are maintained by the following breeding system (Gowe, Robertson, and Latter, 1959).

There are 50 breeding males and 250 breeding females in each generation. Every male is the son of a different father, and every female the daughter of a different mother, so that the variance of family size is zero. One of the objectives of this breeding system is to minimise the rate of inbreeding. Let us therefore find what the rate of inbreeding is, and then see how much is achieved in this respect by the deliberate equalisation of family size. By equation 4.9 the rate of inbreeding in these flocks is $\Delta F = 0.002$. If there were no deliberate choice of breeding individuals, and family size conformed to a Poisson distribution, the rate of inbreeding by equation 4.5 would be $\Delta F = 0.003$. Thus, without the deliberate equalisation of family size the rate of inbreeding would be 50 per cent greater. If a low rate of inbreeding were the only objective, the number of females could be substantially reduced without much effect. For example, if there were no more females than males, with 50 of each sex ($N = 100$) and with equalisation of family size, the rate of inbreeding from equation 4.8 would be $\Delta F = 0.0025$, which is not very much greater than with five times as many females. This illustrates the point, mentioned earlier, that most of the inbreeding comes from the less numerous sex.

Ratio of effective to actual number. When matings are controlled and pedigree records kept, the rate of inbreeding can readily be computed, as will be explained in the next chapter. But pedigree records are not available for natural populations, nor for laboratory populations kept by mass culture, as for example *Drosophila* populations. How are we to estimate the rate of inbreeding in such populations? We know the effective number is likely to be less than the actual, but how much less? To estimate the effective number requires a special experiment, and only the actual number is likely to be known. Determinations of the ratio of effective to actual numbers, N_e/N, from data on man, *Drosophila*, and the snail *Lymnaea*, led to values ranging from 70 per cent to 95 per cent (Crow and Morton, 1955). In the absence of specific knowledge, therefore, it would seem reasonable to take the effective number as being, very roughly, about three-quarters of the actual number. There are two methods by which the ratio N_e/N may be determined: (1) by the estimation of the variance of family size, which yields N_e by equation 4.7 (though adjustment has to be made if the mean family size at the time of measurement is not 2); and (2) by the estimation of the variance of the changes of gene frequency during inbreeding, which yields N_e by equation *3.1*. Both methods have been applied to *Drosophila melanogaster* in laboratory cultures. The ratio N_e/N for female parents was 71 per cent by the first method and 76 per cent by the second; and for male parents, 48 per cent and 35 per cent (Crow and Morton, 1955). The ratio N_e/N for the sexes jointly, determined by the second method, ranged from 56 per cent to 83 per cent, with a mean of 70 per cent, in five experiments with equal actual numbers of males and females (Kerr and Wright, 1954*a*, *b*; Wright and Kerr, 1954; Buri, 1956). The low value of 56 per cent was found in rather poor culture conditions of crowding, where there was more competition (Buri, 1956).

EXAMPLE 4.2. As an illustration of the use of the ratio N_e/N let us find the expected rate of inbreeding in a population of *Drosophila* maintained by 20 pairs of parents in each generation. The actual number is $N=40$. If the effective number were equal to the actual, the rate of inbreeding, by equation *4.1*, would be $\Delta F = 1/80 = 1 \cdot 25$ per cent. If we take $N_e = 0 \cdot 7N$, from the experimental results cited above, then $N_e = 28$, and the rate of inbreeding is $\Delta F = 1/56 = 1 \cdot 786$ per cent. The coefficient of inbreeding after 10, 50, and 100 generations would then be (by equation *3.12*) 17 per cent, 59 per cent, and 84 per cent.

MIGRATION, MUTATION, AND SELECTION

The description of the dispersive process given so far in this chapter and the previous one is conditional on the systematic process of mutation, migration, and selection being absent, and its relevance to real populations is therefore limited. So let us now consider the effects of the dispersive and systematic processes when acting jointly. The systematic processes, as we have seen in Chapter 2, tend to bring the gene frequencies to stable equilibria at particular values which would be the same for all populations under the same conditions. The dispersive process, in contrast, tends to scatter the gene frequencies away from these equilibrium values, and if not held in check by the systematic processes it would in the end lead to all genes being either fixed or lost in all populations not infinite in size. The tendency of the systematic processes to change the gene frequency toward its equilibrium value becomes stronger as the frequency deviates further from this value. For this reason the opposing tendencies of the dispersive and systematic processes reach a point of balance: a point at which the dispersion of the gene frequencies is held in check by the systematic processes. When this point of balance is reached there will be a certain degree of differentiation between sub-populations, but it will neither increase nor decrease so long as the conditions remain unchanged. The problem is therefore to find the distribution of gene frequencies among the lines of a subdivided population when this steady state has been reached. The solution is complicated mathematically, and we shall give only the main conclusions, explaining their meaning but not their derivation. For details of the joint action of the dispersive and systematic processes, see Wright (1931, 1942, 1948, 1951).

Mutation and migration. Mutation and migration can be dealt with together because they change the gene frequency in the same manner. Consider again a population subdivided into many lines, all with an effective size N_e; and let a proportion, m, of the breeding individuals of every generation in each line be immigrants coming at random from all other lines. Consider two alleles at a locus, with mean frequencies $\bar{p}$ and $\bar{q}$ in the population as a whole, and with mutation rates u and v in the two directions. Then, when the balance between dispersion on the one hand and mutation and migration on the other is reached, the variance of the gene frequency among the lines is given by the following expression (Wright, 1931; Malécot, 1948):

$$\sigma^2 = \frac{\bar{p}\bar{q}}{1 + 4N_e(u+v+m)} \quad \text{(approx.)} \qquad \ldots\ldots(4.10)$$

The degree of dispersion represented here by the variance of the gene frequency can also be expressed as a coefficient of inbreeding, by putting $\sigma_q^2 = F\bar{p}\bar{q}$, from equation 3.14. Then

$$F = \frac{1}{1 + 4N_e(u+v+m)} \quad \text{(approx.)} \qquad \ldots\ldots(4.11)$$

The theoretical distributions of the gene frequency appropriate to four different values of F, when the mean gene frequency is 0·5, are shown in Fig. 4.1. These distributions show how high F must be for there to be a substantial amount of fixation or of differentiation between sub-populations. What the distributions depict can be stated in three ways: (a) If we had a large number of sub-populations and we determined the frequency of a particular gene in all of them, the distribution curve is what we should obtain by plotting the percentage of sub-populations showing each gene frequency. Or, in other words, the height of the curve at a particular gene frequency shows the probability of finding that gene frequency in any one sub-population. (b) If we had one sub-population and measured the gene frequencies at a large number of loci, all of which started with the same initial frequency, the curve is the distribution of frequencies that we should

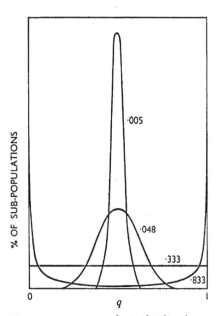

find. (c) If we had one sub-population and measured the frequency of one particular gene repeatedly over a long period of time, the curve is the distribution of frequencies that we should find. The distributions describe the state of affairs when equilibrium between the systematic and dispersive processes has been reached, and the population as a whole is in a steady state. From the distributions shown in Fig. 4.1 it will be seen that when F is 0·005 there is very little differentiation, and when F is 0·048 there is a fair amount of differentiation but still no fixation. When F is 0·333 the distribution is flat, which means that all gene frequencies are equally probable (including 0 and 1); thus there is much differentiation, and in addition a substantial amount of fixation and loss occurs.

FIG. 4.1. Theoretical distributions of gene frequency among sub-populations, when dispersion is balanced by mutation or migration. The states of dispersion to which the curves refer are indicated by the values of F in the figure. (Redrawn from Wright, 1951.)

When F exceeds this critical value intermediate gene frequencies become rarer, and a greater proportion of sub-populations have the gene either fixed or lost. When mutation or migration occurs, fixation or loss is not a permanent state in any one sub-population; the amount of fixation or loss is what would be found at any one time.

Let us return now to the expression, 4.11, relating the coefficient of inbreeding to the rates of mutation and migration when the population has reached the steady state; and let us consider the rates of mutation or migration, in relation to the effective population size, that would just allow the dispersive process to go to the critical point corresponding to the value of $F=0.333$. Putting this value of F in equation *4.11* yields

$$u+v+m=\frac{1}{2N_e} \quad \text{(approx.)} \quad \quad \dots(4.12)$$

First let us consider mutation alone. If the sum of the mutation rates in the two directions $(u+v)$ were 10^{-5}, which is a realistic value to take according to what is known of mutation rates, then the critical state of dispersion will be reached in sub-populations of effective size $N_e=50,000$. In other words, mutation rates of this order of magnitude will arrest the dispersive process before the critical state only in populations with effective numbers greater than 50,000. Populations smaller than this will show a substantial amount of fixation of genes having this mutation rate. In practice, therefore, mutation may be discounted as a force opposing dispersion in populations that would commonly be regarded as "small"; populations, that is, with effective numbers of the order of 100, or even 1,000.

With migration the picture is different, because what would be considered a high rate of mutation would be judged a low rate of migration. The critical value of $F=0.333$ will occur when $m=1/2N_e$. With this rate of migration there would be only one immigrant individual in every second generation, irrespective of the population size. Thus we see that only a small amount of interchange between sub-populations will suffice to prevent them from differentiating appreciably in gene frequency.

The situation to which this consideration of migration refers is known as the "island model." It pictures a discontinuous population such as might be found inhabiting widely separated islands, interchange taking place by occasional migrants from one sub-population to another. But differentiation of sub-populations by random drift can take place also in a continuous population if the motility of the organism is small in relation to the population density. This is known as "isolation by distance" or the "neighbourhood model" (Wright, 1940; 1943; 1946; 1951). Clearly, if there is little dispersal over the territory between one generation and the next the choice of mates is restricted and mating cannot be at random. The population is then subdivided into "neighbourhoods" (Wright, 1946) within which individuals find mates. A neighbourhood is an area within which mating is effectively random. The size of a neighbourhood depends on the distance covered by dispersal between one generation and the next. If the distances between localities inhabited by offspring and parents at corresponding stages of the life cycle are distributed with a variance σ_d^2, then the area of a neighbourhood is the area enclosed by a circle of radius $2\sigma_d$, which is $\pi(2\sigma_d)^2$. The effective population size of a neighbourhood is the number of breeding individuals in the area of a neighbourhood. The subdivision of a population into neighbourhoods leads to random drift, but the amount of local differentiation depends on the size of the whole population as well as on the effective number in the neighbourhood. If the whole population is not very much larger than the neighbourhood then the whole population will drift, and there will be little local differentiation within it. The conclusion to which the neighbourhood model leads is that a great amount of local differentiation will take place if the effective number in a neighbourhood is of the order of 20, and a moderate amount if it is of the order of 200; but with larger neighbourhoods it will be negligible. There will be much more local differentiation in a population inhabiting a linear territory, such as a river or shore line, because a neighbourhood is then open to immigration only from two directions instead of from all round. The extent of a neighbourhood in a population distributed in one dimension is the square root of the area of a neighbourhood in a population distributed in two dimensions. The effective population size is therefore the number of breeding individuals in a distance $2\sigma_d\sqrt{\pi}$ of territory.

EXAMPLE 4.3. As an illustration of the computation of the effective population size of a neighbourhood we may take some observations from the detailed studies by Lamotte (1951) of the snail *Cepaea nemoralis* in France. Marked individuals were released in spring and the distance travelled from the point of release by those recaptured in the autumn was noted. Since the snails are inactive in winter this represented the displacement occurring in one year. The mean displacement was 8·1 metres, and its standard deviation 9·4 m. The standard deviation of the displacement between birth and mating, which usually takes place in the second year of life, was estimated as $\sigma_d = 15$ m. The area occupied by a neighbourhood is therefore $\pi(2\sigma_d)^2 = 12\cdot5\sigma_d^2 = 2{,}813$ sq. m. The density of individuals in two large colonies was found to be 2 per sq. m., and in another 3 per sq. m. The effective population size of the neighbourhoods in these colonies was therefore about 5,600 and 8,400. These figures are a good deal larger than the size of neighbourhoods from which we would expect differentiation within the colonies. Five colonies inhabiting linear territories had densities ranging from 4·5 to 20 individuals per metre. The effective population size of the neighbourhoods in these colonies ranged from 236 to 1,050. These are approaching the size from which differentiation within a colony would be expected.

Selection. Selection operating on a locus in a large population brings the gene frequency to an equilibrium; when selection against a recessive or semidominant gene is balanced by mutation the equilibrium is at a low gene frequency, and when selection favours the heterozygote the equilibrium is more likely to be at an intermediate frequency. The question we have now to consider is: How much can the dispersive process disturb these equilibria and cause small populations to deviate from the point of equilibrium? The importance of this question lies in the fact that an increase of the frequency of a deleterious gene will reduce the fitness—that is, will increase the frequency of "genetic deaths"—and the dispersive process may therefore lead to non-adaptive changes in small populations. We shall not attempt to cover the joint effects of selection and dispersion in detail, but shall merely illustrate their general nature by reference to a particular case of selection

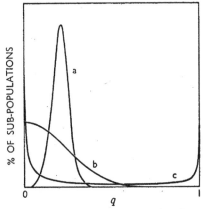

FIG. 4.2. Theoretical distributions of gene frequency among sub-populations when the dispersion is balanced by mutation and selection. The graphs refer to a recessive gene with $u = v = \frac{1}{2}vs$, in populations of size: (a) $N_e = 50/s$, (b) $N_e = 5/s$, and (c) $N_e = 0.5/s$. (Redrawn from Wright, 1942.)

against a recessive gene balanced by mutation. The effects of selection in favour of heterozygotes will be discussed in the next chapter, because they have more importance in connexion with close inbreeding.

Fig. 4.2 shows the state of dispersion of a gene among subpopulations of three sizes under the following conditions. Mutation is supposed to be the same in both directions, and the coefficient of selection against the homozygote is supposed to be twenty times the mutation rate. In a large population the balance between the mutation and the selection would bring the gene frequency to equilibrium at about 0·2. The population sizes to which the graphs refer are (a) $N_e = 50/s$, (b) $N_e = 5/s$, and (c) $N_e = 0·5/s$. If we assumed a mutation rate of 10^{-5} in both directions then the intensity of selection would be $s = 20 \times 10^{-5}$, and the effective population sizes to which the graphs refer would be (a) 250,000 (b) 25,000 and (c) 2,500. These graphs show that with the largest value of N_e there is little differentiation between sub-populations; with the intermediate value of N_e random drift is strong enough to cause a good deal of differentiation; with the smallest value of N_e the effects of random drift predominate over those of mutation and selection, intermediate gene frequencies are almost absent, and in the majority of sub-populations the allele is either fixed or lost. In this case, moreover, a fair proportion of the sub-populations have the deleterious allele fixed in them. This illustrates how random drift can overcome relatively weak selection and lead to fixation of a deleterious gene.

This particular case illustrates in principle what will happen when the processes of random drift, selection, and mutation are all operating. But we need to have some idea of how intense the selection must be before it overcomes the effects of random drift. If we are content not to be very precise we can say that selection begins to be more important than random drift when the coefficient of selection, s, is of the order of magnitude of $1/4N_e$. For example, in a population of effective size 100, the critical value of s would be about 0·0025. This is a very low intensity of selection, quite beyond the reach of experimental detection. The conclusion to be drawn, therefore, is that in all but very small populations, even a very slight selective advantage of one allele over another will suffice to check the dispersive process before it causes an appreciable amount of fixation or of differentiation between sub-populations.

EXAMPLE 4.4. The opposing forces of dispersion and selection are illustrated in Fig. 4.3, from an experiment with *Drosophila melanogaster* (Wright and Kerr, 1954). The frequency of the sex-linked gene "Bar" was followed for 10 generations in 108 lines each maintained by 4 pairs of parents. (On account of the complication of sex-linkage, which increases the rate of dispersion, the theoretical effective number was 6·765: the effective number as judged from the actual rate of dispersion was $N_e = 4·87$.) The initial gene frequency was 0·5. The circles in the figure show the distribution of the gene frequency among the lines in the fourth to tenth generations, when the distribution had reached its steady form. The smooth curve shows the theoretical distribution based on $N_e = 5$ and a coefficient of selection against Bar of $s = 0.17$. Previously fixed lines are not included in the distributions. Altogether, at the tenth generation, 95

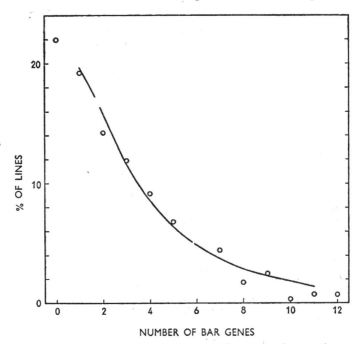

Fig. 4.3. Distribution of gene frequencies under inbreeding and selection, as explained in Example 4.4. (Data from Wright and Kerr, 1954.)

of the 108 lines had become fixed for the wild-type allele and 3 for Bar while 10 remained unfixed. Thus, despite a 17 per cent selective dis-advantage, the deleterious allele was fixed in about 3 per cent of the lines.

RANDOM DRIFT IN NATURAL POPULATIONS

Having described the dispersive process and its theoretical consequences, we may now turn to the more practical question of how far these consequences are actually seen in natural populations. The answering of this question is beset with difficulties, and the following comments are intended more to indicate the nature of these difficulties than to answer the question.

The theory of small populations, outlined in this and the preceding chapter, is essentially mathematical in nature and is un-
questionably valid: given only the Mendelian mechanism of inheritance, the conclusions arrived at are a necessary consequence under
the conditions specified. The question at issue, then, is whether the
conditions in natural populations are often such as would allow the
dispersion of gene frequencies to become detectable. The phenomena which would be expected to result from the dispersive process,
if the conditions were appropriate, are differentiation between the
inhabitants of different localities, and differences between successive
generations. Both these phenomena are well known in subdivided or
small isolated populations, and it is tempting to conclude that because
they are the expected consequences of random drift, random drift
must be their cause. But there are other possible causes: the environmental conditions probably differ from one locality to another
and from one season to another; so the intensity, or even the direction
of selection may well vary from place to place and from year to year,
and the differences observed could equally well be attributed to
variation of the selection pressure. Before we can justifiably attribute
these phenomena to random drift, therefore, we have to know (a)
that the effective population size is small enough, (b) that the sub-
populations are well enough isolated (or the size of the "neighbourhoods" sufficiently small), and (c) that the genes concerned are subject
to very little selection.

The estimation of the present size of a population, though not technically easy, presents no difficulties of principle. But the present
state of differentiation depends on the population size in the past,
and this can generally only be guessed at. It is difficult to know how
often the population may have been drastically reduced in size in
unfavourable seasons, and the dispersion taking place in these
generations of lowest numbers is permanent and cumulative. There
is less difficulty in deciding whether the sub-populations are sufficiently well isolated. With a discontinuous population inhabiting
widely separated islands, it is often possible to be reasonably sure
that there is not too much immigration; and with a continuous
population the size of the "neighbourhoods" is, at least in principle,
measurable. The greatest difficulty lies in estimating the intensity of
natural selection acting on the genes concerned. Selection of an
intensity far lower than could be detected experimentally is sufficient
to check dispersion in all but the smallest populations. It seems

rather unlikely—though this is no more than an opinion—that any gene that modifies the phenotype enough to be recognised would have so little effect on fitness. The genes concerned with quantitative differences, which are not individually recognisable, may however be nearly enough neutral for random drift to take place. There is no doubt at all that genes of this sort do show random drift, at least in laboratory populations, as will be shown in Chapter 15. Of the individually recognisable genes, those concerned with polymorphism seem the most likely to show the effects of random drift. At intermediate frequencies a small displacement from the equilibrium would be detectable, and therefore a relatively small amount of dispersion of the gene frequency might well lead to recognisable differentiation. The following example will serve to illustrate the observed differentiation of a natural population, as well as the difficulties of its interpretation.

EXAMPLE 4.5. The polymorphism in respect of the banding of the shell in the snail *Cepaea nemoralis* has been extensively studied by Lamotte (1951) in France. The population is subdivided into colonies with a high degree of isolation between them. The absence of dark-coloured bands on the shell is caused by a single recessive gene. The mean frequency of bandless snails is 29 per cent, but individual colonies range between the two extremes, some being entirely bandless and a few entirely banded. The colonies vary in the number of individuals that they contain, and 291 colonies were divided into three groups according to their population size. The variation in the frequency of bandless snails was then compared in the three groups, as shown in Fig. 4.4. The variation between the colonies, which measures the degree of differentiation, was found to be greater among the small colonies than among the large. The variance of the frequency of bandless between colonies was 0·067 among colonies of 500–1,000 individuals, 0·048 among colonies of 1,000–3,000, and 0·037 among colonies of 3,000–10,000 individuals. This dependence of the degree of differentiation on the population size is interpreted by Lamotte as evidence that the differentiation is caused by random drift.

Cain and Sheppard (1954a), on the other hand, offer a different interpretation, sustained by an equally thorough study of colonies in England. They show that predation by birds—chiefly thrushes—exerts a strong selection in favour of shell colours matching the background of the habitat. Though the polymorphism is maintained by selection, of an unknown nature, in favour of heterozygotes, the frequency of the different types in any colony is determined by selection in relation to the nature of the habitat. In the areas occupied by small colonies, they argue, there is less variation of habitat than in the areas occupied by large colonies. Therefore the variation of habitat between small colonies is greater than between large. This they regard as the cause of the greater differentiation among small colonies than among large, selection bringing the frequency of bandless forms to a value appropriate to the mean habitat of the colony. It is not for us here to attempt an assessment of these two conflicting interpretations.

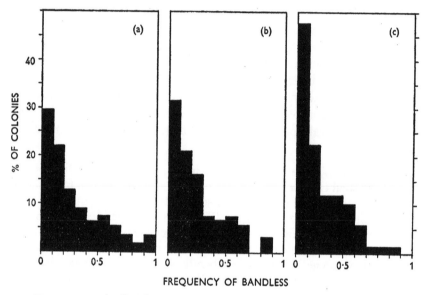

FIG. 4.4.　Distribution of the frequency of bandless snails among colonies of three sizes. (Data from Lamotte, 1951.)

	(a)	(b)	(c)
Population size	500–1,000	1,000–3,000	3,000–10,000
Mean frequency of bandless	0·292	0·256	0·211
Variance between colonies	0·067	0·048	0·037

CHAPTER 5

SMALL POPULATIONS:

III. Pedigreed Populations and Close Inbreeding

In the two preceding chapters the genetic properties of small populations were described by reference to the effective number of breeding individuals; and expressions were derived, in terms of the effective number, by means of which the state of dispersion of the gene frequencies could be expressed as the coefficient of inbreeding. The coefficient of inbreeding, which is the probability of any individual being an identical homozygote, was deduced from the population size and the specified breeding structure. It expressed, therefore, the average inbreeding coefficient of all individuals of a generation. When pedigrees of the individuals are known, however, the coefficient of inbreeding can be more conveniently deduced directly from the pedigrees, instead of indirectly from the population size. This method has several advantages in practice. Knowledge is often required of the inbreeding coefficient of individuals, rather than of the generation as a whole, and this is what the calculation from pedigrees yields. In domestic animals some individuals often appear as parents in two or more generations, and this overlapping of generations causes no trouble when the pedigrees are known. (Non-overlapping of generations was one of the conditions of the idealised population which we have not yet removed.) The first topic for consideration in this chapter is therefore the computation of inbreeding coefficients from pedigrees. The second topic concerns regular systems of close inbreeding. When self-fertilisation is excluded the rate of inbreeding expressed in terms of the population size is only an approximation, and the approximation is not close enough if the population size is very small. Under systems of close inbreeding, therefore, the rate of inbreeding must be deduced differently, and this is best done also by consideration of the pedigrees.

When the coefficient of inbreeding is deduced from the pedigrees of real populations it does not necessarily describe the state of dispersion of the gene frequencies. It is essentially a statement about the pedigree relationships, and its correspondence with the state of dispersion is dependent on the absence of the processes that counteract dispersion, in particular on selection being negligible. We were able to use the coefficient of inbreeding as a measure of dispersion in the preceding chapters because the necessary conditions for its relationship with the variance of gene frequencies were specified.

PEDIGREED POPULATIONS

The inbreeding coefficient of an individual is the probability that the pair of alleles carried by the gametes that produced it were identical by descent. Computation of the inbreeding coefficient therefore requires no more than the tracing of the pedigree back to common ancestors of the parents and computing the probabilities at each segregation. Consider the pedigree in Fig. 5.1. X is the individual we are interested in, whose parents are P and Q. We want to know what is the probability that X receives identical alleles transmitted through P and Q from A. Consider first B and C. The probability that they receive replicates of the same gene from A is $\frac{1}{2}$, and the probability that they receive different genes is $\frac{1}{2}$. But if they receive different genes from A, then the probability of these being identical as a result of previous inbreeding is the inbreeding coefficient of A. Therefore the total probability of B and C receiving identical genes from A is $\frac{1}{2}(1 + F_A)$. Put in other words, this is the probability that two gametes taken at random from A will contain identical alleles. Now consider the rest of the path through B. The probability that B passes the gene it got from A on to D is $\frac{1}{2}$; from D to P is $\frac{1}{2}$, and from P to X is $\frac{1}{2}$. Similarly for the other side of the ancestry through C and Q. Putting all this together we find the probability that X receives identical alleles descended from A is $\frac{1}{2}(1 + F_A)(\frac{1}{2})^{3+2}$, or $\frac{1}{2}(1 + F_A)(\frac{1}{2})^{n_1 + n_2}$, where n_1 is the number of generations from one parent back to the common ancestor and n_2 from the other. If the two parents have more than one ancestor in common the separate probabilities for each of the common ancestors have to be summed to give the inbreeding coefficient of the progeny of these parents. Thus the general expression for the inbreeding coefficient of an

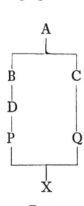

FIG. 5.1

individual is

$$F_X = \Sigma[(\tfrac{1}{2})^{n_1+n_2+1}(\mathrm{I} + F_A)] \qquad \ldots\ldots(5.1)$$

(Wright, 1922). When inbreeding coefficients are computed in this way it is necessary, of course, to define the base population to which the present inbreeding is referred. The base population might be the individuals from which an experiment was started or a herd founded; or it might be those born before a certain date. The designation of an individual as belonging to the base population means that it will be assumed to have zero inbreeding coefficient. When pedigrees are long and complicated there may be very many common ancestors, but it is not necessary to trace back all lines of descent. A sufficiently accurate estimate can be got by sampling a limited number of lines of descent (Wright and McPhee, 1925).

EXAMPLE 5.1. As an illustration of the use of formula *5.1* let us consider the hypothetical pedigree in Fig. 5.2. The relevant individuals in the pedigree are indicated by letters. Individual Z is the one whose inbreeding coefficient is to be computed. Its parents are X and Y, so we have to trace the paths of common ancestry connecting X with Y. There are four common ancestors, A, B, C, and H, and five paths connecting X with Y through them. We assume A, B, and C to have zero inbreeding coefficients, since the pedigree tells us nothing about their ancestry. Individual H, however, has parents that are half sibs, and the inbreeding coefficient of H is therefore

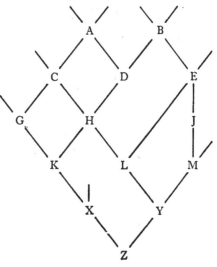

FIG. 5.2

Common ancestor	Path from X to Y	Generations to common ancestor: from X	from Y	Inbreeding coeff. of common ancestor	Contribution to inbreeding of Z	
A	KGCADHL	4	4	o	$(\tfrac{1}{2})^9$	= ·00195
B	KHDBEJM	4	4	o	$(\tfrac{1}{2})^9$	= ·00195
B	KHDBEL	4	3	ᴗ	$(\tfrac{1}{2})^8$	= ·00391
C	KGCHL	3	3	o	$(\tfrac{1}{2})^7$	= ·00781
H	KHL	2	2	$\tfrac{1}{8}$	$(\tfrac{1}{2})^5 \cdot \tfrac{9}{8}$	= ·03516
	Total by summation					0·05078

$(\frac{1}{2})^{(1+1+1)} = \frac{1}{8}$. The computation of the separate paths may now be made as shown in the table. By addition of the contributions from the five paths we get the inbreeding coefficient of Z as $F_Z = 0.05078$, or 5.1 per cent.

"Coancestry." There is another method of computing inbreeding coefficients (Cruden, 1949; Emik and Terrill, 1949) which is more convenient for many purposes, and is also more readily adapted to a variety of problems. We shall use it later to work out the inbreeding coefficients under regular systems of close inbreeding. The method does not differ in principle from the formula *5.1* given above, but instead of working from the present back to the common ancestors we work forward, keeping a running tally generation by generation, and compute the inbreeding that will result from the matings now being made. The inbreeding coefficient of an individual depends on the amount of common ancestry in its two parents. Therefore, instead of thinking about the inbreeding of the progeny, we can think of the degree or relationship by descent between the two parents. This we shall call the *coancestry* of the two parents, and symbolise it by *f*. It is identical with the inbreeding coefficient of the progeny, and is the probability that two gametes taken one from one parent and one from the other will contain alleles that are identical by descent. (Malécot, 1948, calls this the "coefficient de parenté," but the translation "coefficient of relationship" cannot be used because Wright (1922) has used this term with a different meaning.)

Consider the generalised pedigree in Fig. 5.3. X is an individual with parents P and Q and grandparents A, B, C, and D. Now, the coancestry of P with Q is fully determined by the coancestries relating A and B with C and D, and if these are known we need go no further back in the pedigree. It can be shown that the coancestry of P with Q is simply the mean of the four coancestries AC, AD, BC, and BD.

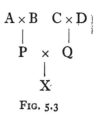

FIG. 5.3

This will be clearer if stated in the form of probabilities, though the explanation is cumbersome when put into words. Take one gamete at random from P and one from Q, and repeat this many times. In half the cases P's gamete will carry a gene from A and in half from B: similarly for Q's gamete. So the two gametes, one from P and one from Q, will carry genes from A and C in a quarter of the cases, from A and D in a quarter, from B and C in a quarter, and from B and D in a quarter of the cases. Now the probability that two gametes taken at random, one from A and the other from C, are identical by descent is the coancestry of A with C, i.e. f_{AC} etc. So, reverting now to symbols,

$$f_{PQ} = \tfrac{1}{4}f_{AC} + \tfrac{1}{4}f_{AD} + \tfrac{1}{4}f_{BC} + \tfrac{1}{4}f_{BD}$$

This gives the basic rule relating coancestries in one generation with those in the next:

$$F_X = f_{PQ} = \tfrac{1}{4}(f_{AC} + f_{AD} + f_{BC} + f_{BD}) \qquad \dots\dots(5.2)$$

With this rule the experimenter can tabulate the coancestries generation by generation, and this gives a basis for planning matings and computing inbreeding coefficients. More detailed accounts of the operation are given by Cruden (1949), Emik and Terrill (1949), and Plum (1954).

If there is overlapping of generations it may happen that we must find the coancestry between individuals belonging to different generations. This situation is covered by the following supplementary rules, which can readily be deduced by a consideration of probabilities in the manner explained above. Referring to the same pedigree (Fig. 5.3),

$$\left.\begin{aligned} f_{PC} &= \tfrac{1}{2}(f_{AC} + f_{BC}) \\ f_{PD} &= \tfrac{1}{2}(f_{AD} + f_{BD}) \\ \text{and} \qquad f_{PQ} &= \tfrac{1}{2}(f_{PC} + f_{PD}) \end{aligned}\right\} \quad \dots\dots(5.3)$$

which by substitution reduces to the basic rule.

Before we can apply this method to systems of close inbreeding we have to see how the basic rule is to be applied when there are fewer than four grandparents. As an example we shall consider the coancestry between a pair of full sibs. The pedigree can be written as in Fig. 5.4: A and B are parents of both P_1 and P_2, which are full sibs and have an offspring X. Applying the basic rule (equation 5.2), and noting that $f_{BA} = f_{AB}$, we have

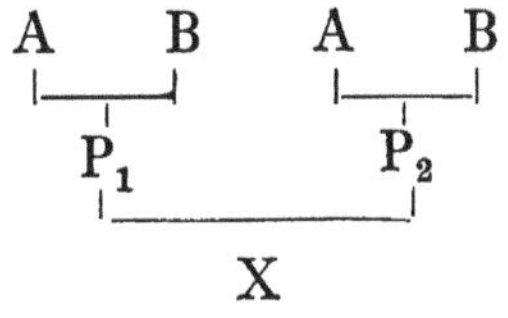

FIG. 5.4

$$F_X = f_{P_1P_2} = \tfrac{1}{4}(f_{AA} + f_{BB} + 2f_{AB}) \qquad \dots\dots(5.4)$$

The meaning of f_{AA}, the coancestry of an individual with itself, is the probability that two gametes taken at random from A will contain identical alleles, and we have already seen that this probability is equal to $\frac{1}{2}(1+F_A)$. The value of F_A will be known from the coancestry of A's parents. The coancestry between offspring and parent can be found in a similar way, by application of the supplementary rules in 5.3. Substituting the individuals in Fig. 5.4 for those in Fig. 5.3 and applying the first two equations of 5.3 gives

$$\left.\begin{aligned} f_{PA} &= \tfrac{1}{2}(f_{AA}+f_{AB}) \\ f_{PB} &= \tfrac{1}{2}(f_{BB}+f_{AB}) \end{aligned}\right\} \dots\dots(5.5)$$

where P is equivalent to either P_1 or P_2; and applying the third equation of 5.3 gives the coancestry between full sibs

$$\begin{aligned} f_{P_1P_2} &= \tfrac{1}{2}(f_{PA}+f_{PB}) \\ &= \tfrac{1}{4}(f_{AA}+f_{BB}+2f_{AB}) \end{aligned}$$

as above. We now have all the rules needed for computing the inbreeding coefficients in successive generations under regular systems of inbreeding.

REGULAR SYSTEMS OF INBREEDING

The consequences of regular systems of inbreeding have been the subject of much study. They were first described in detail by Wright (1921) in a series of papers which form the foundation of the whole theory of small populations. Wright's studies were based on the method of path coefficients (Wright, 1934, 1954). Haldane (1937, 1955) and Fisher (1949) derived the consequences by the method of matrix algebra. The inbreeding coefficients in successive generations can, however, be more simply derived by application of the rules of coancestry explained in the previous section, and this is the method we shall follow here. We shall illustrate the application of the method for consecutive full-sib mating, which is one of the most commonly used systems, and give the results for some other systems. The inbreeding coefficients refer to autosomal genes; the results for sex-linked genes are described by Wright (1933) in a paper which also contains a useful summary of the results for autosomal genes in a great variety of mating systems.

Full-sib mating. The equation 5.4 given above for the coancestry between full sibs can be applied to successive generations to

TABLE 5.1

Inbreeding coefficients under various systems of close inbreeding

Generation (t)	A	B (1)	B (2)	C	D
0	0	0	0	0	0
1	·500	·250	0	·125	·250
2	·750	·375	0	·219	·375
3	·875	·500	·063	·305	·438
4	·938	·594	·172	·381	·469
5	·969	·672	·293	·449	·484
6	·984	·734	·409	·509	·492
7	·992	·785	·512	·563	·496
8	·996	·826	·601	·611	·498
9	·998	·859	·675	·654	·499
10	·999	·886	·736	·691	
11		·908	·785	·725	
12		·926	·826	·755	
13		·940	·859	·782	
14		·951	·886	·806	
15		·961	·908	·827	
16		·968	·925	·846	
17		·974	·940	·863	
18		·979	·951	·878	
19		·983	·960	·891	
20		·986	·968	·903	

Column	System of mating	Recurrence equation
A	Self-fertilisation, or repeated backcrosses to highly inbred line.	$\frac{1}{2}(1 + F_{t-1})$
B	Full brother × sister, or offspring × younger parent:	
(1)	Inbreeding coefficient.	$\frac{1}{4}(1 + 2F_{t-1} + F_{t-2})$
(2)	Probability of fixation (from Schäfer, 1937).	
C	Half sib (females half sisters).	$\frac{1}{8}(1 + 6F_{t-1} + F_{t-2})$
D	Repeated backcrosses to random-bred individual.	$\frac{1}{4}(1 + 2F_{t-1})$

give the inbreeding coefficients under continued full-sib mating. But it is more convenient to rearrange the equation so that the inbreeding coefficient is given in terms of the inbreeding coefficients of the previous generations. Note first that, because the mating system is regular, contemporaneous individuals have the same inbreeding coefficients and coancestries: so, referring again to the pedigree in Fig. 5.4, $f_{AA} = f_{BB}$, and $F_A = F_B$. Now, if we let t be the generation to which individual X belongs, then $f_{AB} = F_{t-1}$, and $f_{AA} = f_{BB} = \frac{1}{2}(1 + F_{t-2})$. The coancestry equation can therefore be rewritten to give the inbreeding coefficient in any generation, t, in terms of the inbreeding coefficients of the previous two generations, thus:

$$F_t = \tfrac{1}{4}(1 + 2F_{t-1} + F_{t-2}) \qquad \ldots\ldots(5.6)$$

This recurrence equation enables us to write down the inbreeding coefficients in successive generations. In the first generation F_{t-1} and F_{t-2} are both zero and so $F_{(t=1)} = 0.25$. The inbreeding coefficients in the first four generations are 0.25, 0.375, 0.50, and 0.59. The rate of inbreeding is not constant in the first few generations, as may be seen by computing $\varDelta F$ from equation 3.9. For the first four generations $\varDelta F$ is 0.25, 0.17, 0.20, and 0.19. It later settles down to a constant value of 0.191 (Wright, 1931). The inbreeding coefficients over the first 20 generations of full-sib mating are given in Table 5.1.

Some other systems of mating may now be mentioned briefly. **Self-fertilisation** gives the most rapid inbreeding. If X is the offspring of P, we have from the coancestry identities

$$F_X = f_{PP} = \tfrac{1}{2}(1 + F_P)$$

and the recurrence equation is therefore

$$F_t = \tfrac{1}{2}(1 + F_{t-1}) \qquad \ldots\ldots(5.7)$$

The inbreeding coefficients over the first ten generations of self-fertilisation are given in Table 5.1. The rate of inbreeding is constant from the beginning; $\varDelta F = 0.5$ exactly.

Parent-offspring mating, in which offspring are mated to the younger parent, gives the same series of inbreeding coefficients as full-sib mating for autosomal genes, but for sex-linked genes it gives a slightly higher rate of inbreeding. For sex-linked genes $\varDelta F$ is 0.293 after the first few generations (Wright, 1933).

Half-sib mating is usually between paternal half sibs, one male being mated to two or more of his half sisters. If these females are half sisters of each other the recurrence equation is

$$F_t = \tfrac{1}{8}(1 + 6F_{t-1} + F_{t-2}) \qquad \dots(5.8)$$

The first 20 generations are given in Table 5.1. There are, however, practical difficulties in the way of maintaining this system regularly, and sometimes females that are full sisters of each other have to be used. The inbreeding will then go a little faster. If full-sister females are always used the recurrence equation is

$$F_t = \tfrac{1}{16}(3 + 8F_{t-1} + 4F_{t-2} + F_{t-3}) \qquad \dots(5.9)$$

Repeated backcrosses to an individual or to a highly inbred line are often made, for a variety of purposes. The resulting inbreeding is as follows. The pedigree (Fig. 5.5) shows an individual, A, which will probably be a male, mated to his daughter, C, his granddaughter, D, etc. From the supplementary rule (5.5)

$$F_X = f_{AD} = \tfrac{1}{2}(f_{AA} + f_{AC})$$
$$= \tfrac{1}{2}\{\tfrac{1}{2}(1 + F_A) + F_D\}$$

A × B

A × C

A × D

X

Fɪɢ. 5.5

The recurrence equation is therefore

$$F_t = \tfrac{1}{4}(1 + F_A + 2F_{t-1}) \qquad \dots\dots(5.10)$$

where F_A is the inbreeding coefficient of the individual to which the repeated backcrosses are made. If A is an individual from the base population and $F_A = 0$, the equation becomes

$$F_t = \tfrac{1}{4}(1 + 2F_{t-1}) \qquad \dots(5.11)$$

The inbreeding coefficients over the first 9 generations are given in Table 5.1. If A is an individual from a highly inbred line and $F_A = 1$, the equation becomes

$$F_t = \tfrac{1}{2}(1 + F_{t-1}) \qquad \dots(5.12)$$

which is identical with self-fertilisation. In this case A need not be the same individual in successive generations: it can be any member of the inbred line.

EXAMPLE 5.2. As an example of the use of coancestry for computing inbreeding coefficients let us consider populations derived from "2-way" and from "4-way" crosses between highly inbred lines. In a 2-way cross two inbred lines are crossed and the population is maintained by random mating among the cross-bred individuals and subsequently among their progeny. In a 4-way cross four inbred lines are crossed in two pairs, and the two cross-bred groups are again crossed, subsequent generations being maintained by random mating. If the base population is taken to be a real, or hypothetical, random-bred population from which the inbred lines were derived, we may compute the inbreeding coefficients of the population derived from the cross, referring it to this base. The crosses and subsequent generations are shown schematically in the diagram below.

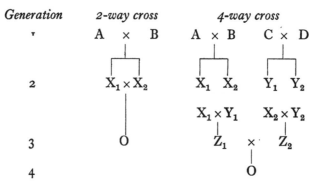

The inbred lines are represented by A, B, C, and D. If they are fully inbred, as we shall take them to be, the coefficient of inbreeding of the individuals from the lines is 1, and the coancestry of an individual with another of the same line is also 1. Therefore only one individual of each line need be represented in the scheme, even though any number may actually be used. The progeny of the crosses between the inbred lines are represented by X and Y, the suffices 1 and 2 indicating different individuals. In the 2-way cross the progeny of these cross-bred individuals are the foundation generation whose inbreeding coefficient we are to compute. They are represented by O. In the 4-way cross the two sorts of cross-bred individuals, X and Y, are crossed, one sort with the other. Two such matings are represented in the scheme. They produce the "double-cross" individuals, Z, whose progeny constitute the foundation generation represented by O, whose inbreeding coefficient we are to compute.

In the computation of the coancestries we shall omit the symbol f, writing for example AB for f_{AB}, the coancestry of individual A with individual B. The coancestries of the parents in generation 1 are

$$AA = BB = CC = DD = 1$$

and $\qquad$ $AB = AC = AD = BC = BD = CD = 0$

The coancestries in the second generation of the 2-way cross are

$$X_1X_2 = \tfrac{1}{4}(AA + BB + AB + BA) \quad \text{(by equation 5.2)}$$
$$= \tfrac{1}{4}(\quad 1 \quad + \quad 1 \quad + \quad 0 \quad + \quad 0 \quad)$$
$$= \tfrac{1}{2}$$

Therefore $F_0 = 0.5$, which is the required inbreeding coefficient of the foundation generation of the population derived from the 2-way cross. The subsequent matings between the O individuals need produce no further inbreeding provided enough 2nd generation matings are made.

The coancestries in the second generation of the 4-way cross are

$$X_1X_2 = Y_1Y_2 = \tfrac{1}{2} \quad \text{(as shown for the 2-way cross)}$$

and $\qquad$ $X_1Y_2 = X_2Y_1 = \tfrac{1}{4}(AC + AD + BC + BD) = 0$

The coancestries in the third generation are

$$Z_1Z_2 = \tfrac{1}{4}(X_1X_2 + Y_1Y_2 + X_1Y_2 + X_2Y_1)$$
$$= \tfrac{1}{4}(\quad \tfrac{1}{2} \quad + \quad \tfrac{1}{2} \quad + \quad 0 \quad + \quad 0$$
$$= \tfrac{1}{4}$$

Therefore the inbreeding coefficient of the foundation generation is $F_0 = 0.25$. Again, the inbreeding need not increase further, provided enough third generation matings are made.

The meaning of these coefficients of inbreeding, with the base population as stated, may be clarified thus. If we made a large number of 2-way, or of 4-way, crosses each with a different set of inbred lines, the populations derived from the crosses would constitute a set of lines or sub-populations. The inbreeding coefficients would then indicate the expected amount of dispersion of gene frequencies among these lines. Populations derived from 2-way crosses are equivalent to progenies of one generation of self-fertilisation. The gene frequencies can therefore have only three values, 0, $\tfrac{1}{2}$, and 1. Populations derived from 4-way crosses are equivalent to progenies of one generation of full-sib mating, and the gene frequencies can have only five values, 0, $\tfrac{1}{4}$, $\tfrac{1}{2}$, $\tfrac{3}{4}$, and 1.

Reference to a different base population. Having computed a coefficient of inbreeding with reference to a certain group of individuals as the base population, one may then want to change the base and refer the inbreeding coefficient to another group of individuals. One might, for example, compute the inbreeding coefficient of a herd of cattle referred to the foundation animals of the herd as the base, and then want to recompute the inbreeding coefficient so as to refer to the breed as a whole with a base popula-. tion in the more remote past. Let X represent the group of individuals whose inbreeding coefficient is required, and let A and B represent ancestral groups, A being more remote than B, as shown in Fig. 5.6. Then it follows from equation $3.II$ that

A

$\downarrow$

B

$\downarrow$

X

Fig. 5.6

$$P_{X.A} = P_{X.B} P_{B.A} \qquad \ldots \ldots (5.I3)$$

where $P_{X.A} = I - F_{X.A}$; $F_{X.A}$ being the inbreeding coefficient of, X referred to A as base, and similarly for the other subscripts.

EXAMPLE 5.3. A selection experiment with mice was started from a foundation population made by a 4-way cross of highly inbred lines (Falconer, 1953). According to the computation given above in Example 5.2, the inbreeding coefficient of this foundation population was reckoned to be 25 per cent. On this basis the inbreeding coefficients of subsequent generations were computed from the pedigrees by the coancestry method. The inbreeding coefficient at generation 24, computed thus, was 58·8 per cent. What would the inbreeding coefficient be if referred to the foundation population as base, instead of to the more remote hypothetical population from which the inbred lines were derived? The figures to be substituted in equation $5.I3$ are $P_{X.A} = 0·412$ and $P_{B.A} = 0·75$. Therefore $P_{X.B} = \dfrac{0·412}{0·75}$ = 0·549. The inbreeding coefficient at generation 24, referred to the foundation population as base, is therefore 45·1 per cent.

We may use this population of mice also to compare the rate of inbreeding when computed by the two methods, from the pedigrees and from the effective population size. Computed from the pedigrees, the average rate of inbreeding over the 24 generations is found from equation $3.I2$ thus: $0·451 = I - (I - \Delta F)^{24}$, whence $\Delta F = 2·47$ per cent. The population was maintained by six pairs of parents in each generation. Matings were made between individuals with the lowest coancestries and this has the effect of equalising family size, as explained in the previous chapter. Therefore, by equation 4.8, the effective number was twice the actual, i.e. $N_e = 24$. The rate of inbreeding, by equation $4.I$, is therefore $\Delta F = \dfrac{I}{48} = 2·08$ per cent. The slightly higher rate of inbreeding as computed directly from the pedigrees can be attributed to some irregularities in the mating system, resulting from the sterility of some parents and the death of some whole litters. The random drift of a colour gene in this line, and two others maintained in the same manner, was shown in Fig. 3.2.

Fixation. One is often more interested in the probability of fixation as a consequence of inbreeding than in the inbreeding coefficient. The inbreeding coefficient gives the probability of an individual being a homozygote, which is $1 - 2p_0q_0(1 - F)$ from Table 3.1. But one wants to know also how soon all individuals in a line can be expected to be homozygous for the same allele. This is the "purity" implied by the term "pure line" which is often used to mean highly inbred line. The degree of "purity" is the probability of fixation. The probability of fixation has been worked out by Haldane (1937, 1955), Schäfer (1937) and Fisher (1949). It depends on the number of alleles and their arrangement in the initial mating of the line. The probabilities of fixation over the first 20 generations of full-sib mating are given in Table 5.1, when 4 alleles were present in the initial mating. There cannot, of course, be more than 4 alleles in a sib-mated line, and when there are fewer the probability of fixation is greater (see Haldane, 1955).

Linkage. Linkage introduces a problem in connexion with the consequences of inbreeding of which a solution is sometimes needed. Individuals heterozygous at a particular locus will also be heterozygous for a segment of chromosome in which the locus lies, and it may be of interest to know the average length of heterozygous segments. The form in which this problem most commonly arises is connected with the transference of a marker gene to an inbred line by repeated backcrosses, when one wants to know how much of the foreign chromosome is transferred along with the marker. This problem has been worked out by Bartlett and Haldane (1935). A dominant gene can be transferred by successive crosses of the heterozygote to the strain into which it is to be introduced. In this case the mean length of chromosome introduced with the gene after t crosses is $1/t$ cross-over units on each side of the gene. A recessive gene is commonly transferred by alternating backcrosses and intercrosses from which the homozygote is extracted. The mean length of foreign chromosome in this case is $2/t$ cross-over units on each side, after t cycles. Other cases are described in the paper cited. From this and a knowledge of the total map length of the organism we can arrive at the expected proportion of the total chromatin that is still heterogeneous.

EXAMPLE 5.4. What percentage of the total chromatin is expected to be still heterogeneous after a dominant gene has been transferred to an inbred strain of mice by five, and by ten successive backcrosses? The

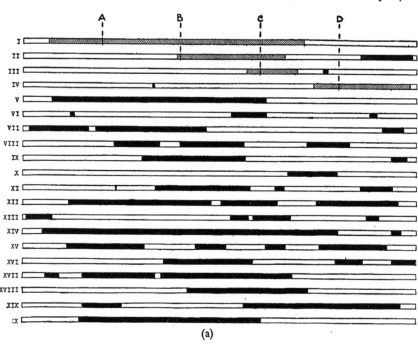

(a)

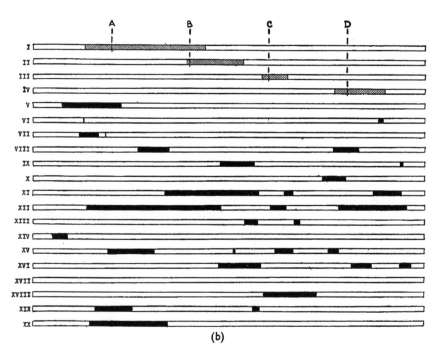

(b)

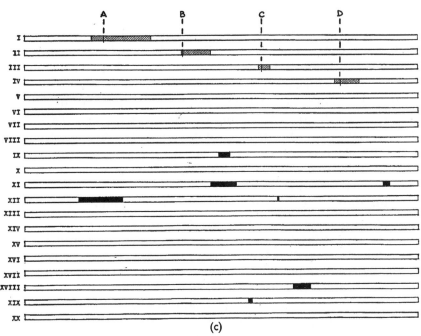

(c)

Fig. 5.7. Theoretical models illustrating the distribution of heterozygous segments of chromosome (shown black) after (*a*) 5 generations, (*b*) 10 generations, and (*c*) 20 generations of full-sib mating, in an organism with twenty chromosomes, such as the mouse. The total map-length is taken to be 2500 centimorgans, and the chromosomes are assumed to be of equal genetic length. The points marked *A*, *B*, *C*, *D*, in chromosomes I to IV are loci held heterozygous by forced segregation, and the associated hetero-zygous segments are cross-hatched. (From Fisher, *The Theory of Inbreeding*, Oliver and Boyd, 1949; reproduced by courtesy of the author and publishers.)

expected length of heterogeneous chromosome associated with the gene is 0·2. centimorgans after five crosses, and 0·1 cM after ten. The average map length of the 20 chromosomes in male mice is 977 cM (Slizynski, 1955). Therefore 0·2 per cent of the chromosome will be heterogeneous after five crosses, and 0·1 per cent after ten, assuming that the gene is transferred through males, and taking the average as being the length of the chromosome carrying the gene. The percentage of chromatin not associated with the gene that is expected still to be heterogeneous can be taken as approximately $1 - F_t$ from column A of Table 5.1: that is, 3·1 per cent after five crosses and 0·1 per cent after ten. The total percentage of heterogeneous chromatin is therefore 3·4 per cent after five crosses, and 0·2 per cent after ten.

The more general problem of the mean length of heterozygous segments during inbreeding has been treated by Haldane (1936) and by Fisher (1949). It need not be discussed in detail here. The conclusions are well illustrated in Fig. 5.7, which is Fisher's diagrammatic representation of the situation in an organism with 20 chromosomes, such as the mouse, after five, ten, and twenty generations of full-sib mating. The diagrams show the expected number and lengths of unfixed segments. The first four chromosomes are supposed to carry loci at which segregation is maintained by mating always heterozygotes with homozygotes. The slower reduction of the lengths of these unfixed segments can be seen.

Mutation. After a long period of inbreeding mutation may become an important factor in determining the frequency of heterozygotes. If u is the mutation rate of a gene that has reached near-fixation in the line, then the frequency of heterozygotes at this locus due to mutation is $4u$ under self-fertilisation, and $12u$ under full-sib mating, for autosomal loci (Haldane, 1936). These are very small frequencies if we are concerned with only one locus, but if the effects of all loci are taken together mutation is not entirely negligible as a source of heterozygosis in long inbred strains such as the widely used strains of mice. The practical consequences of the origin of heterogeneity by mutation are that the characteristics of a line will slowly change through the fixation of mutant alleles, and that sub-lines will become differentiated. Examples are given in Chapter 15.

Selection favouring heterozygotes. When close inbreeding is practised the object is generally to produce fixation, or homozygosis within the lines, and the experimenter is not usually interested in the differentiation between lines. It is therefore a matter of little concern which allele is fixed, so long as fixation occurs. Selection against a deleterious recessive may prevent the deleterious allele becoming fixed, but it will not prevent or delay the fixation of the more favourable allele. Therefore the conclusions about selection reached in the previous chapter are of little relevance to close inbreeding. Selection that favours heterozygotes, however, is another matter. A consequence of inbreeding almost universally observed is a reduction of fitness, the reasons for which will be given in Chapter 14. Thus selection resists the inbreeding, since the more homozygous individuals are the less fit, and this can only mean that selection favours heterozygotes—not necessarily heterozygotes of the loci taken singly, but heterozygotes of segments of chromosome. It is only necessary to have two deleterious genes, recessive or partially recessive, linked in repulsion, to confer a selective advantage on the heterozygote of the segment of chromosome within which the genes are located. It is therefore important to find out how the opposing tendencies of inbreeding and selection in favour of heterozygotes balance each other, in order to assess the reliability of the computed inbreeding coefficient as a measure of the probability of fixation.

The outcome of the joint action of inbreeding and selection in favour of heterozygotes depends on whether there is replacement of the less fit lines by the more fit; in other words, on whether selection operates between lines or only within lines. Within any one line, selection against homozygotes only delays the progress toward fixation and cannot arrest it, the delay being roughly in proportion to the intensity of the selection (Reeve, 1955*a*). Table 5.2 shows the

TABLE 5.2

Rate of inbreeding, ΔF, with selection favouring the heterozygote. (Except with self-fertilisation, the rates are only approximate over the first few generations of inbreeding.)

Coefficient of selection against the homozygotes	ΔF (%)		
	Self-fertilisation	*Full sib*	*Half sib**
(*s*)			
0	50·00	19·10	13·01
0·2	44·44	14·88	9·32
0·4	37·50	10·32	5·67
0·6	28·57	5·71	2·48
0·75	20·00	2·62	0·82
0·8	16·67	1·76	0·46

* Females full sisters to each other.

rates of inbreeding with various intensities of selection, when there are two alleles and selection acts equally against both homozygotes. (The rate of inbreeding, ΔF, is used here to mean the rate of dispersion of gene frequencies and, after the first few generations when the distribution of gene frequencies has become flat, it measures the rate of fixation—i.e. the proportion of unfixed loci that become fixed in each generation—as explained in Chapter 3.) The delay of fixation caused by selection is least under the closest systems of inbreeding.

Thus the rate is halved under self-fertilisation when the coefficient of selection is 0·67; under full-sib mating when it is 0·44; and under half-sib mating when it is 0·35. It will be seen from the table that the rate of inbreeding, though much reduced by intense selection, does not become zero until the coefficient of selection rises to 1. If there is only one line, therefore, fixation eventually goes to completion, unless both homozygotes are entirely inviable or sterile.

If there are many lines, however, selection may arrest the progress of fixation and lead to a state of equilibrium, for the following reason. The amount by which the inbreeding has changed the frequency of a particular gene from its original value differs at any one time from line to line. In other words, the state of dispersion of the locus has gone further in some lines than in others. Now, if those lines in which the dispersion has gone furthest, and which are consequently most reduced in fitness, die out or are discarded, and if they are replaced by sub-lines taken from the lines in which it has gone least far, then the progress of the dispersive process will have been set back. When there is replacement of lines in this way, and the selection is sufficiently intense, a state of balance between the opposing tendencies of inbreeding and selection is reached. The intensity of selection needed to arrest the dispersive process has been worked out for regular systems of close inbreeding (Hayman and Mather, 1953). Some of the conclusions, for the case of two alleles with equal selection against the two homozygotes, are given in Table 5.3, which shows the intensity of selection against the homozygotes which will (a) just allow fixation to go eventually to completion, and (b) arrest

TABLE 5.3

Balance between inbreeding and selection in favour of heterozygotes, when selection operates between lines. The figures are the selective disadvantages of homozygotes, s, expressed as percentages. Column (a) shows the highest value of s compatible with complete fixation. Column (b) shows the value of s that leads to a steady state at $P = 1 - F = 0.5$.

Mating system	(a) (P=0)	(b) (P=0·5)
Self-fertilisation	50·0	66·7
Full-sib	23·7	44·6
Half-sib	18·8	47·2
(females half sisters)		

the dispersive process at a point of balance where the frequency of heterozygotes is half its original value, i.e. where $P = 1 - F = 0·5$. These figures show that only a moderate advantage of heterozygotes will suffice to prevent complete fixation. Under full-sib mating, for example, loci, or segments of chromosomes that do not recombine, with a 25 per cent disadvantage in homozygotes will not all go to fixation. And, of those with a 50 per cent disadvantage, only about half will become fixed, no matter for how long the inbreeding is continued.

It must be stressed, however, that prevention of fixation in this way can only take place when there is replacement of lines and sub-lines. The following breeding methods, for example, would allow replacement of lines: if seed, set by self-fertilisation, were collected in bulk and a random sample taken for planting, and this were repeated in successive generations; or, if sib pairs of mice were taken at random from all the surviving progeny, so that the same amount of breeding space was occupied in successive generations.

The conclusions outlined above refer to a single locus. If there were more than a few loci on different chromosomes all subject to selection against homozygotes of an intensity sufficient to arrest or seriously delay the progress of inbreeding, the total loss of fitness from all the loci would be very severe. Inbred lines of organisms with a high reproductive rate, such as plants and *Drosophila*, might well stand up to a total loss of fitness sufficient to keep several loci or segments of chromosome permanently unfixed. But the loss of fitness involved in preventing the fixation of more than two or three loci in an organism such as the mouse would be crippling. Under laboratory conditions the highly inbred strains of mice, after 100 or more generations of sib-mating, have a fitness not much less than half that of non-inbred strains. It is conceivable that they might have one locus permanently unfixed, but it is difficult to believe that they can have more. Complete lethality or sterility of both homozygotes at one locus means a 50 per cent loss of progeny; at two unlinked loci, a 75 per cent loss. A mouse strain with a mortality or sterility of 50 per cent can be kept going, but hardly one with 75 per cent.

H

CHAPTER 6

CONTINUOUS VARIATION

It will be obvious, to biologist and layman alike, that the sort of variation discussed in the foregoing chapters embraces but a small part of the naturally occurring variation. One has only to consider one's fellow men and women to realise that they all differ in countless ways, but that these differences are nearly all matters of degree and seldom present clear-cut distinctions attributable to the segregation of single genes. If, for example, we were to classify individuals according to their height, we could not put them into groups labelled "tall" and "short," because there are all degrees of height, and a division into classes would be purely arbitrary. Variation of this sort, without natural discontinuities, is called *continuous variation*, and characters that exhibit it are called *quantitative characters* or *metric characters*, because their study depends on measurement instead of on counting. The genetic principles underlying the inheritance of metric characters are basically those outlined in the previous chapters, but since the segregation of the genes concerned cannot be followed individually, new methods of study have had to be developed and new concepts introduced. A branch of genetics has consequently grown up, concerned with metric characters, which is called variously *population genetics*, *biometrical genetics* or *quantitative genetics*. The importance of this branch of genetics need hardly be stressed; most of the characters of economic value to plant and animal breeders are metric characters, and most of the changes concerned in microevolution are changes of metric characters. It is therefore in this branch that genetics has its most important application to practical problems and also its most direct bearing on evolutionary theory.

How does it come about that the intrinsically discontinuous variation caused by genetic segregation is translated into the continuous variation of metric characters? There are two reasons: one is the simultaneous segregation of many genes affecting the character, and the other is the superimposition of truly continuous variation arising from non-genetic causes. Consider, for example, a simplified situation. Suppose there is segregation at six unlinked loci, each with two alleles at frequencies of o·5. Suppose that there is complete dominance of one allele at each locus and that the dominant alleles each add one unit to the measurement of a certain character. Then if the segregation of these genes were the only cause of variation there would

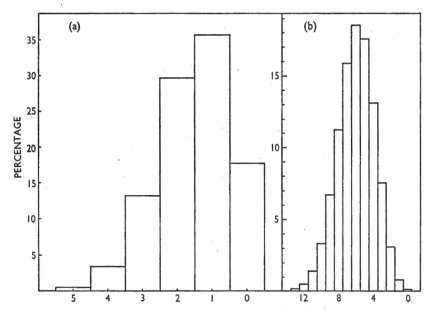

FIG. 6.1. Distributions expected from the simultaneous segregation of two alleles at each of several or many loci: (*a*) 6 loci, (*b*) 24 loci. There is complete dominance of one allele over the other at each locus, and the gene frequencies are all o·5. Each locus, when homozygous for the recessive allele, is supposed to reduce the measurement by 1 unit in (*a*), and by $\frac{1}{4}$ unit in (*b*). The horizontal scale, representing the measurement, shows the number of loci homozygous for the recessive allele, and the vertical axis shows the probability, or the percentage of individuals expected in each class. The probabilities are derived from the binomial expansion of $(\frac{1}{4} + \frac{3}{4})^n$, where *n* is the number of loci, and they are taken from the tables of Warwick (1932).

be 7 discrete classes in the measurements of the character, according to whether the individual had the dominant allele present at o, 1, 2, . . . or 6 of the loci. The frequencies of the classes would be according to the binomial expansion of $(\frac{1}{4} + \frac{3}{4})^6$, as shown in Fig. 6.1 (*a*). If our measurements were sufficiently accurate we should recognise these classes as being distinct and we should be able to place any individual unambiguously in its class. If there were more genes segregating but each had a smaller effect, there would be more classes with smaller differences between them, as in Fig. 6.1 (*b*). It would then be more difficult to distinguish the classes, and if the difference between the classes became about as small as the error of measurement we should no longer be able to recognise the discontinuities. In addition, metric characters are subject to variation from non-genetic causes, and this variation is truly continuous. Its effect is, as it were, to blur the edges of the genetic discontinuity so that the variation as we see it becomes continuous, no matter how accurate our measurements may be.

Thus the distinction between genes concerned with Mendelian characters and those concerned with metric characters lies in the magnitude of their effects relative to other sources of variation. A gene with an effect large enough to cause a recognisable discontinuity even in the presence of segregation at other loci and of non-genetic variation can be studied by Mendelian methods, whereas a gene whose effect is not large enough to cause a discontinuity cannot be studied individually. This distinction is reflected in the terms *major gene* and *minor gene*. There are, however, all intermediate grades, genes that cannot properly be classed as major or as minor, such as the "bad genes" of Mendelian genetics. And, furthermore, as a result of pleiotropy the same gene may be classed as major with respect to one character and minor with respect to another character. The distinction, though convenient, is therefore not a fundamental one, and there is no good evidence that there are two sorts of genes with different properties. Variation caused by the simultaneous segregation of many genes may be called *polygenic* variation, and the minor genes concerned are sometimes referred to as *polygenes* (see Mather, 1949).

METRIC CHARACTERS

The metric characters that might be studied in any higher organism are almost infinitely numerous. Any attribute that varies continuously and can be measured might in principle be studied as a metric character—anatomical dimensions and proportions, physiological functions of all sorts, and mental or psychological qualities. The essential condition is that they should be measureable. The technique of measurement, however, sets a practical limitation on what can be studied. Usually rather large numbers of individuals

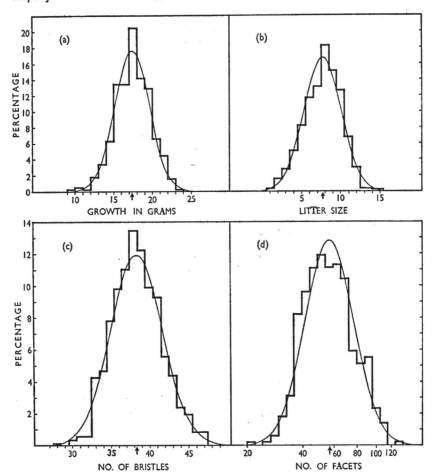

FIG. 6.2.　Frequency distributions of four metric characters, with normal curves superimposed. The means are indicated by arrows. The characters are as follows, the number of observations on which each histogram is based being given in brackets:

(a) Mouse (♂♂): growth from 3 to 6 weeks of age. (380)

(b) Mouse: litter size (number of live young in 1st litters). (689)

(c) *Drosophila melanogaster* (♀♀): number of bristles on ventral surface of 4th and 5th abdominal segments, together. (900)

(d) *Drosophila melanogaster* (♀♀): number of facets in the eye of the mutant "Bar". (488)

(a), (b), and (c) are from original data: (d) is from data of Zeleny (1922).

have to be measured and the study of any character whose measurement requires an elaborate technique therefore becomes impracticable. Consequently the characters that have been used in studies of quantitative genetics are predominantly anatomical dimensions, or physiological functions measured in terms of an end-product, such as lactation, fertility, or growth rate.

Some examples of metric characters are illustrated in Fig. 6.2. The variation is represented graphically by the frequency distribution of measurements. The measurements are grouped into equally spaced classes and the proportion of individuals falling in each class is plotted on the vertical scale. The resulting histogram is discontinuous only for the sake of convenience in plotting. If the class ranges were made smaller and the number of individuals measured were increased indefinitely the histogram would become a smooth curve. The variation of some metric characters, such as bristle number or litter size, is not strictly speaking continuous because, being measured by counting, their values can only be whole numbers. Nevertheless, one can regard the measurements in such cases as referring to an underlying character whose variation is truly continuous though expressible only in whole numbers, in a manner analogous to the grouping of measurements into classes. For example, litter size may be regarded as a measure of the underlying, continuously varying character, fertility. For practical purposes such characters can be treated as continuously varying, provided the number of classes is not too small. When there are too few classes, as for example when susceptibility to disease is expressed as death or survival, different methods have to be employed, as will be explained in Chapter 18.

The frequency distributions of most metric characters approximate more or less closely to normal curves. This can be seen in Fig. 6.2, where the smooth curves drawn through the histograms are normal curves having means and variances calculated from the data. In the study of metric characters it is therefore possible to make use of the properties of the normal distribution and to apply the appropriate statistical techniques. Sometimes, however, the scale of measurement must be modified if a distribution approximating to the normal is to be obtained. The distribution in Fig. 6.2 (d), for example, would be skewed if measured and plotted simply as the number of facets. But it becomes symmetrical, and approximates to a normal distribution, if measured and plotted in logarithmic units. The criteria on which the choice of a scale of measurement rests cannot be fully appreciated at this stage, and will be explained in Chapter 17. Meantime it will be assumed that any metric character under discussion is measured on an appropriate scale and has a distribution that is approximately normal.

GENERAL SURVEY OF THE SUBJECT-MATTER

There are two basic genetic phenomena concerned with metric characters, both more or less familiar to all biologists, and each forms the basis of a breeding method. The first is the resemblance between relatives. Everyone is familiar with the fact that relatives tend to resemble each other, and the closer the relationship, in general the closer the resemblance. Though it is only in our own species that resemblances are readily discernible without measurement, the phenomenon is equally present in other species. The degree of resemblance varies with the character, some showing more, some less. The resemblance between offspring and parents provides the basis for selective breeding. Use of the more desirable individuals as parents brings about an improvement of the mean level of the next generation, and just as some characters show more resemblance than others, so some are more responsive to selection than others. The degree of resemblance between relatives is one of the properties of a population that can be readily observed, and it is one of the aims of quantitative genetics to show how the degree of resemblance between different sorts of relatives can be used to predict the outcome of selective breeding and to point to the best method of carrying out the selection. This problem will form the central theme of the next seven chapters, the resemblance between relatives being dealt with in Chapters 9 and 10, and the effects of selection in Chapters 11–13.

The second basic genetic phenomenon is inbreeding depression, with its converse hybrid vigour, or heterosis. This phenomenon is less familiar to the layman than the first, since the laws against incest prevent its more obvious manifestations in our own species; but it is well known to animal and plant breeders. Inbreeding tends to reduce the mean level of all characters closely connected with fitness in animals and in naturally outbreeding plants, and to lead in consequence to loss of general vigour and fertility. Since most characters of economic value in domestic animals and plants are aspects of vigour or fertility, inbreeding is generally deleterious. The reduced vigour and fertility of inbred lines is restored on crossing, and in certain circumstances this hybrid vigour can be made use of as a means of improvement. The enormous improvement of the yield of commercially grown maize has been achieved by this means and represents probably the greatest practical achievement of genetics (see Mangelsdorf, 1951). The effects of inbreeding and crossing will be described in Chapters 14–16.

The properties of a population that we can observe in connexion with a metric character are means, variances, and covariances. The natural subdivision of the population into families allows us to analyse the variance into components which form the basis for the measurement of the degree of resemblance between relatives. We can in addition observe the consequences of experimentally applied breeding methods, such as selection, inbreeding or cross-breeding. The practical objective of quantitative genetics is to find out how we can use the observations made on the population as it stands to predict the outcome of any particular breeding method. The more general aim is to find out how the observable properties of the population are influenced by the properties of the genes concerned and by the various non-genetic circumstances that may influence a metric character. The chief properties of genes that have to be taken account of are the degree of dominance, the manner in which genes at different loci combine their effects, pleiotropy, linkage, and fitness under natural selection. To take account of all these properties simultaneously, in addition to a variety of non-genetic circumstances, would make the problems unmanageably complex. We therefore have to simplify matters by dealing with one thing at a time, starting with the simpler situations.

The plan to be followed in the succeeding chapters is this: we shall first show what determines the population mean, and then introduce two new concepts—average effect and breeding value— which are necessary to an understanding of the variance. Then we shall discuss the variance, its analysis into components, and the covariance of relatives, which will lead us to the degree of resemblance between relatives. In all this we shall take full account of dominance from the beginning: the other complicating factors will be more briefly discussed when they become relevant. The most important simplification that we shall make concerns the effect of genes on fitness: we shall assume that Mendelian segregation is undisturbed by differential fitness of the genotypes. The description of means, variances, and covariances will refer to a random breeding population, with Hardy-Weinberg equilibrium genotype frequencies, with no selection and no inbreeding. That is to say, we shall describe the population before any special breeding method is applied to it. Then in Chapters 11–13 we shall describe the effects of selection, and in Chapters 14–16 the effects of inbreeding. This will cover the fundamentals of quantitative genetics, and in the final chapters we shall discuss some special topics.

CHAPTER 7

VALUES AND MEANS

We have seen in the early chapters that the genetic properties of a population are expressible in terms of the gene frequencies and genotype frequencies. In order to deduce the connexion between these on the one hand and the quantitative differences exhibited in a metric character on the other, we must introduce a new concept, the concept of *value*, expressible in the metric units by which the character is measured. The value observed when the character is measured on an individual is the *phenotypic value* of that individual. All observations, whether of means, variances, or covariances, must clearly be based on measurements of phenotypic values. In order to analyse the genetic properties of the population we have to divide the phenotypic value into component parts attributable to different causes. Explanation of the meanings of these components is our chief concern in this chapter, though we shall also be able to find out how the population mean is influenced by the array of gene frequencies.

The first division of phenotypic value is into components attributable to the influence of genotype and environment. The *genotype* is the particular assemblage of genes possessed by the individual, and the *environment* is all the non-genetic circumstances that influence the phenotypic value. Inclusion of all non-genetic circumstances under the term environment means that the genotype and the environment are by definition the only determinants of phenotypic value. The two components of value associated with genotype and environment are the *genotypic value* and the *environmental deviation*. We may think of the genotype conferring a certain value on the individual and the environment causing a deviation from this, in one direction or the other. Or, symbolically,

$$P = G + E \qquad \dots\dots(7.1)$$

where P is the phenotypic value, G is the genotypic value, and E is the environmental deviation. The mean environmental deviation in the population as a whole is taken to be zero, so that the mean phenotypic value is equal to the mean genotypic value. The term *population mean* then refers equally to phenotypic or to genotypic values. When dealing with successive generations we shall assume for simplicity that the environment remains constant from generation to generation, so that the population mean is constant in the absence of genetic change. If we could replicate a particular genotype in a number of individuals and measure them under environmental conditions normal for the population, their mean environmental deviations would be zero, and their mean phenotypic value would consequently be equal to the genotypic value of that particular genotype. This is the meaning of the genotypic value of an individual. In principle it is measurable, but in practice it is not, except when we are concerned with a single locus where the genotypes are phenotypically distinguishable, or with the genotypes represented in highly inbred lines.

For the purposes of deduction we must assign arbitrary values to the genotypes under discussion. This is done in the following way. Considering a single locus with two alleles, A_1 and A_2, we call the genotypic value of one homozygote $+a$, that of the other homozygote $-a$, and that of the heterozygote d. (We shall adopt the convention that A_1 is the allele that increases the value.) We thus have a scale of genotypic values as in Fig. 7.1. The origin, or point of zero value, on this scale is mid-way between the values of the two homozygotes.

Genotype	A_2A_2		A_1A_2	A_1A_1
Genotypic value	$-a$	o	d	$+a$

FIG. 7.1. Arbitrarily assigned genotypic values.

The value, d, of the heterozygote depends on the degree of dominance. If there is no dominance, $d = o$; if A_1 is dominant over A_2, d is positive, and if A_2 is dominant over A_1, d is negative. If dominance is complete, d is equal to $+a$ or $-a$, and if there is overdominance, d is greater than $+a$ or less than $-a$. The degree of dominance may be expressed as d/a.

EXAMPLE 7.1. For the purposes of illustration in this chapter, and also later on, we shall refer to a dwarfing gene in the mouse, known as "pygmy" (symbol pg), described by King (1950, 1955), and by Warwick and Lewis (1954). This gene reduces body-size and is nearly, but not quite, recessive in its effect on size. It was present in a strain of small mice (MacArthur's) at the time the studies cited above were made. The weights of mice of the three genotypes at 6 weeks of age were approximately as follows (sexes averaged):

	+ +	+pg	pg pg
Weight in grams	14	12	6

(The weight of heterozygotes given here is to some extent conjectural, but it is unlikely to be more than 1 gm. in error.) These are average weights obtained under normal environmental conditions, and they are therefore the genotypic values. The mid-point in genotypic value between the two homozygotes is 10 gm., and this is the origin, or zero-point, on the scale of values assigned as in Fig. 7.1. The value of a on this scale is therefore 4 gm., and that of d is 2 gm.

POPULATION MEAN

We can now see how the gene frequencies influence the mean of the character in the population as a whole. Let the gene frequencies of A_1 and A_2 be p and q respectively. Then the first two columns of Table 7.1 show the three genotypes and their frequencies in a random breeding population, from formula *1.2*. The third column shows the genotypic values as specified above. The mean value in the whole

TABLE 7.1

Genotype	Frequency	Value	freq. × val.
A_1A_1	p^2	$+a$	p^2a
A_1A_2	$2pq$	d	$2pqd$
A_2A_2	q^2	$-a$	$-q^2a$
		Sum =	$a(p-q)+2dpq$

population is obtained by multiplying the value of each genotype by its frequency and summing over the three genotypes. The reason why this yields the mean value may be understood by converting frequencies to numbers of individuals. Multiplying the value by the number of individuals in each genotype and summing over genotypes gives the sum of values of all individuals. The mean value would then be this sum of values divided by the total number of individuals. The procedure in working with frequencies is the same, but since the sum of the frequencies is 1, the sum of values × frequencies is the mean value. In other words, the division by the total number has already been made in obtaining the frequencies. Multiplication of values by frequencies to obtain the mean value is a procedure that will be often used in this chapter and subsequent ones. Returning to the population mean, multiplication of the value by the frequency of each genotype is shown in the last column of Table 7.1. Summation of this column is simplified by noting that $p^2 - q^2 = (p+q)(p-q) = p - q$. The population mean, which is the sum of this column, is thus

$$M = a(p-q) + 2dpq \qquad \ldots(7.2)$$

This is both the mean genotypic value and the mean phenotypic value of the population with respect to the character.

The contribution of any locus to the population mean thus has two terms: $a(p-q)$ attributable to homozygotes, and $2dpq$ attributable to heterozygotes. If there is no dominance ($d=0$) the second term is zero, and the mean is proportional to the gene frequency: $M=a(1-2q)$. If there is complete dominance ($d=a$) the mean is proportional to the square of the gene frequency: $M=a(1-2q^2)$. The *total range* of values attributable to the locus is $2a$, in the absence of overdominance. That is to say, if A_1 were fixed in the population ($p=1$) the population mean would be a, and if A_2 were fixed ($q=1$) it would be $-a$. If the locus shows overdominance, however, the mean of an unfixed population is outside this range.

EXAMPLE 7.2. Let us take again the pygmy gene in mice, as described in Example 7.1, and see what effect this gene would have on the population mean when present at two particular frequencies. First, the total range is from 6 gm. to 14 gm.: a population consisting entirely of pygmy homozygotes would have a mean of 6 gm., and one from which the gene was entirely absent would have a mean of 14 gm. (These values refer specifically to MacArthur's Small Strain at the time the observations were made.) Now suppose the gene were present at a frequency of 0·1, so that under random mating homozygotes would appear with a frequency of 1 per cent. The values to be substituted in equation 7.2 are $p=0.9$, $q=0.1$, and $a=4$ gm., $d=2$ gm., as shown in Example 7.1. The population mean, by equation 7.2, is therefore: $M=4 \times 0.8 + 2 \times 0.18 = 3.56$. This value of the mean, however, is measured from the mid-homozygote point, which is 10 gm., as origin. Therefore the actual value of the population mean is 13·56 gm. Next suppose the gene were present at a frequency of 0·4. Substituting in the same way, we find $M=1.76$, to which must be added 10 gm. for the origin, giving a value of 11·76 gm. Rough corroboration of these figures is given by the records of the strain carrying the gene. When the gene was present at a frequency of about 0·4 the mean weight was about 12 gm. Two generations later, when the pygmy gene had been deliberately eliminated, the mean weight rose to about 14 gm.

. Now we have to put together the contributions of genes at several loci and find their joint effect on the mean. This introduces the question of how genes at different loci combine to produce a joint effect on the character. For the moment we shall suppose that combination is by addition, which means that the value of a genotype with respect to several loci is the sum of the values attributable to the separate loci. For example, if the genotypic value of A_1A_1 is a_A and that of B_1B_1 is a_B, then the genotypic value of $A_1A_1B_1B_1$ is $a_A + a_B$. The consequences of non-additive combination will be explained at the end of this chapter. With additive combination, then, the population mean resulting from the joint effects of several loci is the sum of the contributions of each of the separate loci, thus:

$$M = \Sigma a(p-q) + 2\Sigma dpq \qquad \ldots\ldots(7.3)$$

This is again both the genotypic and the phenotypic mean value. The total range in the absence of overdominance is now $2\Sigma a$. If all alleles that increase the value were fixed the mean would be $+\Sigma a$, and if all alleles that decrease the value were fixed it would be $-\Sigma a$. These are the theoretical limits to the range of potential variation in the population. The origin from which the mean value in equation 7.3 is measured is the mid-point of the total range. This is equivalent to the average mid-homozygote point of all the loci separately.

EXAMPLE 7.3. As an example of two loci that combine additively, and also of their joint effects on the population mean, we shall refer to two colour genes in mice, whose effects on the number of pigment granules have been described by Russell (1949). This is a metric character which reflects the intensity of pigmentation in the coat. The two genes are "brown" (b) and "extreme dilution" (c^e), an allele of the albino series. Measurements were made of the number of melanin granules per unit volume of hair, in wild-type homozygotes, in the two single mutant homozygotes, and in the double mutant homozygote. We shall assume both wild-type alleles to be completely dominant, so that only these four genotypes need be considered. The mean numbers of granules in the four genotypes were as follows:

	B –	bb	$2a_B$
C –	95	90	5
$c^e c^e$	38	34	4
$2a_C$	57	56	

The difference between the two figures in each row and in each column measures the homozygote difference, or $2a$ on the scale of values assigned as in Fig. 7.1. Apart from the trivial discrepancy of 1 unit, these differences are independent of the genotype at the other locus. In other words, the difference of value between B – and bb is the same among C – genotypes as it is among $c^e c^e$ genotypes; and similarly the difference between C – and $c^e c^e$ is the same in B – as it is in bb. Thus the two loci combine additively, and the value of a composite genotype can be rightly predicted from knowledge of the values of the single genotypes. For example: the bb genotype is 5 units less than the wild-type, and the $c^e c^e$ is 57 units less; therefore bb $c^e c^e$ should be 62 units less, namely 33, which is almost identical with the observed value of 34.

We may use this example further to illustrate the effect of the two loci jointly on the population mean. Let us work out, from the effects of the loci taken separately, what would be the mean granule number in a population in which the frequency of bb was $q_B^2 = 0.4$, and that of $c^e c^e$ was $q_C^2 = 0.2$. For the effects of the loci separately we shall take $a_B = 2$ and $a_C = 28$. The population mean, considering one locus, is $M = a(1 - 2q^2)$, when there is complete dominance. For the B locus this is $M_B = 2 \times 0.2 = 0.4$; and for the C locus $M_C = 28 \times 0.6 = 16.8$. The mean, considering both loci together, is $M_B + M_C = 17.2$ (by equation 7.3). The point of origin from which this is measured is the mid-point between the two double homozygotes, which is $\frac{1}{2}(95 + 34) = 64.5$. Thus the mean granule number in this population would be $64.5 + 17.2 = 81.7$. We may check this from the observations of the values of the joint genotypes. The four genotypes would have the following frequencies and observed values:

Genotype	B – C –	B – $c^e c^e$	bb C –	bb $c^e c^e$
Frequency	0.48	0.12	0.32	0.08
Observed value	95	38	90	34

The mean value is obtained by multiplying the values by the frequencies and summing over the four genotypes. This yields a mean granule number of 81.68.

AVERAGE EFFECT

In order to deduce the properties of a population connected with its family structure we have to deal with the transmission of value from parent to offspring, and this cannot be done by means of genotypic values alone, because parents pass on their genes and not their genotypes to the next generation, genotypes being created afresh in each generation. A new measure of value is therefore needed which will refer to genes and not to genotypes. This will enable us to assign a "breeding value" to individuals, a value associated with the genes carried by the individual and transmitted to its offspring. The new measure is the "average effect." We can assign an average effect to a gene in the population, or to the difference between one gene and another of an allelic pair. The *average effect of a gene* is the mean deviation from the population mean of individuals which received that gene from one parent, the gene received from the other parent having come at random from the population. This may be stated in another way. Let a number of gametes all carrying A_1 unite at random with gametes from the population; then the mean deviation from the population mean of the genotypes so produced is equal to the average effect of the gene A_1. The concept of average effect is perhaps easier to grasp in the form of the *average effect of a gene-substitution*, which can more conveniently be used when only two alleles at a locus are under consideration. If we could change, say, A_2 genes into A_1 at random in the population, and could then note the resulting change of value, this would be the average effect of the gene-substitution. It is equal to the difference between the average effects of the two genes involved in the substitution. A graphical representation of the average effect of a gene-substitution is given later in Fig. 7.2.

It is important to realise that the average effect of a gene or a gene-substitution depends on the gene frequency, and that the average effect is therefore a property of the population as well as of the gene. The reason for this can be seen in the words "taken at random" in the definitions, because the content of the random sample depends on the gene frequency in the population. The point may perhaps be more easily understood from a specific example. Consider the substitution of a recessive gene, a, for its dominant allele, A. The substitution will change the value only when the individual already carries one recessive allele, in other words in heterozygotes. Changing AA into Aa will not affect the value, but changing Aa into aa will. Now, when the frequency of the recessive allele, a, is low there will be many AA individuals, which the substitution will not affect; but when the recessive is at high frequency there will be very few AA individuals, and most of the individuals in which a substitution can be made will be affected by it. Therefore the average effect of the substitution will be small when the frequency of the recessive allele is low and large when it is high.

Let us see how the average effect is related to the genotypic values, a and d, in terms of which the population mean was expressed. This will help to make the concept clearer. The reasoning is set out in Table 7.2. Consider a locus with two alleles, A_1 and A_2, at frequencies p and q respectively, and take first the average effect of the

TABLE 7.2

Type of gamete	Values and frequencies of genotypes produced			Mean value of genotypes produced	Population mean to be deducted	Average effect of gene
	A_1A_1	A_1A_2	A_2A_2			
	a	d	$-a$			
A_1	p	q		$pa+qd$	$-[a(p-q)+2dpq]$	$q[a+d(q-p)]$
A_2		p	q	$-qa+pd$	$-[a(p-q)+2dpq]$	$-p[a+d(q-p)]$

gene A_1, for which we shall use the symbol α_1. If gametes carrying A_1 unite at random with gametes from the population, the frequencies of the genotypes produced will be p of A_1A_1 and q of A_1A_2. The genotypic value of A_1A_1 is $+a$ and that of A_1A_2 is d, and the mean of these, taking account of the proportions in which they occur, is $pa+qd$. The difference between this mean value and the population mean is the average effect of the gene A_1. Taking the value of the population mean from equation 7.2 we get

$$\alpha_1 = pa+qd - [a(p-q)+2dpq]$$
$$= q[a+d(q-p)] \qquad\qquad .....(7.4a)$$

Similarly the average effect of the gene A_2 is

$$\alpha_2 = -p[a+d(q-p)] \qquad\qquad ......(7.4b)$$

Now consider the average effect of the gene-substitution, letting A_1 be substituted for A_2. Of the A_2 genes taken at random from the population for substitution, a proportion p will be found in A_1A_2 genotypes and a proportion q in A_2A_2 genotypes. In the former the substitution will change the value from d to $+a$, and in the latter from $-a$ to d. The average change is therefore $p(a-d)+q(d+a)$, which on rearrangement becomes $a+d(q-p)$. Thus the average effect of the gene-substitution (written as α, without subscript) is

$$\alpha = a+d(q-p) \qquad\qquad .....(7.5)$$

The relation of α to α_1 and α_2 can be seen by comparing equations 7.5 and 7.4, whence

I

$$\alpha = \alpha_1 - \alpha_2 \qquad \dots (7.6)$$

and

$$\left.\begin{array}{l} \alpha_1 = q\alpha \\ \alpha_2 = -p\alpha \end{array}\right\} \dots (7.7)$$

EXAMPLE 7.4. Consider again the pygmy gene and its effect on body weight, for which $a = 4$ gm. and $d = 2$ gm. If the frequency of the pg gene were $q = 0.1$, the average effect of substituting + for pg would be, by equation 7.5, $\alpha = 4 + 2 \times -0.8 = 2.4$ gm. And if the frequency were $q = 0.4$, the average effect of the gene-substitution would be: $\alpha = 4 + 2 \times -0.2 = 3.6$ gm. Thus, the average effect is greater when the gene frequency is greater. The average effects of the genes separately are, by equation 7.7:

	$q = 0.1$	$q = 0.4$
Average effect of + : $\alpha_1 =$	+0.24	+1.44
Average effect of pg : $\alpha_2 =$	−2.16	−2.16

(The identity of the average effects of pg at the two gene frequencies is only a coincidence.)

BREEDING VALUE

The usefulness of the concept of average effect arises from the fact, already noted, that parents pass on their genes and not their genotypes to their progeny. It is therefore the average effects of the parent's genes that determine the mean genotypic value of its progeny. The value of an individual, judged by the mean value of its progeny, is called the *breeding value* of the individual. Breeding value, unlike average effect, can therefore be measured. If an individual is mated to a number of individuals taken at random from the population then its breeding value is twice the mean deviation of the progeny from the population mean. The deviation has to be doubled because the parent in question provides only half the genes in the progeny, the other half coming at random from the population. Breeding values can be expressed in absolute units, but are usually more conveniently expressed in the form of deviations from the population mean, as defined above. Just as the average effect is a property of the gene and the population so is the breeding value a property of the individual and the population from which its mates are drawn. One cannot speak of an individual's breeding value without specifying the population in which it is to be mated.

Defined in terms of average effects, the breeding value of an individual is equal to the sum of the average effects of the genes it carries, the summation being made over the pair of alleles at each locus and over all loci. Thus, for a single locus with two alleles the breeding values of the genotypes are as follows:

$$
\begin{array}{cc}
\textit{Genotype} & \textit{Breeding value} \\
A_1A_1 & 2\alpha_1 = 2q\alpha \\
A_1A_2 & \alpha_1 + \alpha_2 = (q - p)\alpha \\
A_2A_2 & 2\alpha_2 = -2p\alpha
\end{array}
$$

EXAMPLE 7.5. Let us illustrate breeding values by reference to the pygmy gene in mice. The average effects of the + and pg genes were given in the last example. From these we may find the breeding values of the three genotypes as explained above. These breeding values, which are given below, are deviations from the population mean. The population means with gene frequencies of 0·1 and 0·4 were found in Example 7·2 and are shown again below in the column headed M.

	M	Breeding values		
		+ +	+ pg	pg pg
$q = 0 \cdot 1$	13·56	+0·48	−1·92	−4·32
$q = 0 \cdot 4$	11·76	+2·88	−0·72	−4·32

(The breeding values of pygmy homozygotes are only hypothetical because in fact pygmy homozygotes are nearly all sterile: but this complication may be overlooked in the present context.)

Extension to a locus with more than two alleles is straightforward, the breeding value of any genotype being the sum of the average effects of the two alleles present. If all loci are to be taken into account, the breeding value of a particular genotype is the sum of the breeding values attributable to each of the separate loci. If there is non-additive combination of genotypic values a slight complication arises. We have given two definitions of breeding value, a practical one in terms of the measured value of the progeny and a theoretical one in terms of average effects. Non-additive combination renders these two definitions not quite equivalent. This point will be more fully explained in Chapter 9.

Consideration of the definition of breeding value will show that in a population in Hardy-Weinberg equilibrium the mean breeding value must be zero; or if breeding values are expressed in absolute units the mean breeding value must be equal to the mean genotypic value and to the mean phenotypic value. This can be verified from the breeding values listed above. Multiplying the breeding value by the frequency of each genotype and summing gives the mean breeding value (expressed as a deviation from the population mean) as

$$2p^2q\alpha + 2pq(q-p)\alpha - 2q^2p\alpha = 2pq\alpha(p+q-p-q) = 0$$

The breeding value is sometimes referred to as the "additive genotype," and variation in breeding value ascribed to the "additive effects" of genes. Though we shall not use these terms we shall follow custom in using the term "additive" in connexion with the variation of breeding values to be discussed in the next chapter, and we shall use the symbol A to designate the breeding value of an individual.

Dominance Deviation

We have separated off the breeding value as a component part of the genotypic value of an individual. Let us consider now what makes up the remainder. When a single locus only is under consideration, the difference between the genotypic value, G, and the breeding value, A, of a particular genotype is known as the *dominance deviation D*, so that

$$G = A + D \qquad \qquad \ldots\ldots(7.8)$$

The dominance deviation arises from the property of dominance among the alleles at a locus, since in the absence of dominance breeding values and genotypic values coincide. From the statistical point of view the dominance deviations are interactions between alleles, or within-locus interactions. They represent the effect of putting genes together in pairs to make genotypes; the effect not accounted for by the effects of the two genes taken singly. Since the average effects of genes and the breeding values of genotypes depend on the gene frequency in the population, the dominance deviations are also dependent on gene frequency. They are therefore partly properties of the population and are not simply measures of the degree of dominance.

EXAMPLE 7.6. Continuing with the example of the pygmy gene, we may now list the genotypic values and the breeding values, and so obtain the dominance deviations of the three genotypes, by equation *7.8*. These values, all now expressed as deviations from the population mean, M, are as follows:

	$q=0\cdot1: M=13\cdot56$			$q=0\cdot4: M=11\cdot76$		
	+ +	+pg	pg pg	+ +	+pg	pg pg
Frequency	0·81	0·18	0·01	0·36	0·48	0·16
Genotypic value, G	+0·44	−1·56	−7·56	+2·24	+0·24	−5·76
Breeding value, A	+0·48	−1·92	−4·32	+2·88	−0·72	−4·32
Dominance dev., D	−0·04	+0·36	−3·24	−0·64	+0·96	−1·44

The relations between genotypic values, breeding values and dominance deviations can be illustrated graphically, as in Fig. 7.2,

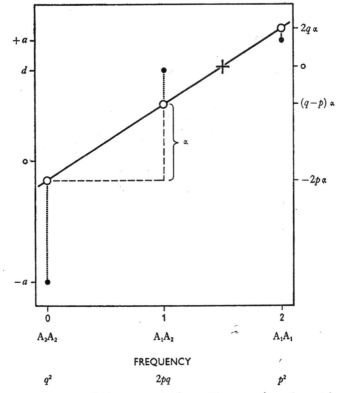

FIG. 7.2. Graphical representation of genotypic values (closed circles), and breeding values (open circles), of the genotypes for a locus with two alleles, A_1 and A_2, at frequencies p and q, as explained in the text. Horizontal scale: number of A_1 genes in the genotype. Vertical scales of value: on left—arbitrary values assigned as in Fig. 7.1; on right—deviations from the population mean. The figure is drawn to scale for the values: $d=\frac{3}{4}a$, and $q=\frac{1}{4}$.

and the meaning of the dominance deviation is perhaps more easily understood in this way. In the figure the genotypic value (black dots) is plotted against the number of A_1 genes in the genotype. A straight regression line is fitted by least squares to these points, each point being weighted by the frequency of the genotype it represents. The position of this line gives the breeding values of each genotype, as shown by the open circles. The differences between the breeding values and the genotypic values are the dominance deviations, indicated by vertical dotted lines. The cross marks the population mean. The average effect, α, of the gene-substitution is given by the difference in breeding value between A_2A_2 and A_1A_2, or between A_1A_2 and A_1A_1, as indicated. The original definition of the average effect of a gene-substitution was given by Fisher (1918, 1941) in terms of this linear regression of genotypic value on number of genes.

The dominance deviation can be expressed in terms of the arbitrarily assigned genotypic values a and d, by subtraction of the breeding value from the genotypic value, as shown in Table 7.3. The

TABLE 7.3

Values of genotypes in a two-allele system, measured as deviations from the population mean.
Population mean: $M = a(p - q) + 2dpq$
Average effect of gene-substitution: $\alpha = a + d(q - p)$

Genotypes	A_1A_1	A_1A_2	A_2A_2
Frequencies	p^2	$2pq$	q^2
Assigned values	a	d	$-a$
Deviations from population-mean:			
Genotypic value $\left\{ \vphantom{\begin{matrix}1\\1\end{matrix}}\right.$	$2q(a - pd)$	$a(q - p) + d(1 - 2pq)$	$-2p(a - qd)$
	$2q(\alpha - qd)$	$(q - p)\alpha + 2pqd$	$-2p(\alpha + pd)$
Breeding value	$2q\alpha$	$(q - p)\alpha$	$-2p\alpha$
Dominance deviation	$-2q^2d$	$2pqd$	$-2p^2d$

genotypic values must first be converted to deviations from the population mean, because the breeding values have been expressed in this way. The genotypic values, so converted, are given in two forms: in terms of a and in terms of α. Let us take the genotype A_1A_1 to show how these are obtained and how the dominance deviation is obtained by subtraction of the breeding value. The arbitrarily assigned genotypic value of A_1A_1 is $+a$, and the population mean is

$a(p-q)+2dpq$. Expressed as a deviation from the population mean, the genotypic value is therefore

$$a - [a(p-q)+2dpq] = a(1 - p + q) - 2dpq = 2qa - 2dpq = 2q(a - dp).$$

This may be expressed in terms of the average effect, α, by substituting $a = \alpha - d(q - p)$ (from equation 7.5), and the genotypic value then becomes $2q(\alpha - qd)$. Subtraction of the breeding value, $2q\alpha$, gives the dominance deviation as $-2q^2d$. By similar reasoning the dominance deviation of A_1A_2 is $2pqd$, and that of A_2A_2 is $-2p^2d$. Thus all the dominance deviations are functions of d. If there is no dominance d is zero and the dominance deviations are also all zero. Therefore in the absence of dominance, breeding values and genotypic values are the same. Genes that show no dominance ($d=0$) are sometimes called "additive genes," or are said to "act additively."

Since the mean breeding value and the mean genotypic value are equal, it follows that the mean dominance deviation is zero. This can be verified by multiplying the dominance deviation by the frequency of each genotype and summing. The mean dominance deviation is thus

$$- 2p^2q^2d + 4p^2q^2d - 2p^2q^2d = 0$$

Another fact, which will be needed later when we deal with variances, may be noted here: there is no correlation between the dominance deviation and the breeding value of the different genotypes. This can be shown by multiplying together the dominance deviation, the breeding value and the frequency of each genotype. Summation gives the sum of cross-products, and it works out to be zero, thus:

$$- 4p^2q^3\alpha d + 4p^2q^2(q - p)\alpha d + 4p^3q^2\alpha d = 4p^2q^2\alpha d(- q + q - p + p) = 0$$

Since the sum of cross-products is zero, breeding values and dominance deviations are uncorrelated.

INTERACTION DEVIATION

When only a single locus is under consideration the genotypic value is made up of the breeding value and the dominance deviation only. But when the genotype refers to more than one locus the genotypic value may contain an additional deviation due to non-additive combination. Let G_A be the genotypic value of an individual attributable to one locus, G_B that attributable to a second locus, and G the aggregate genotypic value attributable to both loci together. Then

$$G = G_A + G_B + I_{AB} \qquad \ldots\ldots(7.9)$$

where I_{AB} is the deviation from additive combination of these genotypic values. In dealing with the population mean, earlier in this chapter, we assumed that I was zero for all combinations of genotypes. If I is not zero for any combination of genes at different loci, those genes are said to "interact" or to exhibit "epistasis," the term epistasis being given a wider meaning in quantitative genetics than in Mendelian genetics. The deviation I is called the *interaction deviation* or *epistatic deviation*. Loci may interact in pairs or in threes or higher numbers, and the interactions may be of many different sorts, as the behaviour of major genes shows. The complex nature of the interactions, however, need not concern us, because in the aggregate genotypic value interactions of all sorts are treated together as a single interaction deviation. So for all loci together we can write

$$G = A + D + I \qquad \ldots\ldots(7.10)$$

where A is the sum of the breeding values attributable to the separate loci, and D is the sum of the dominance deviations.

If the interaction deviation is zero the genes concerned are said to "act additively" between loci. Thus "additive action" may mean two different things. Referred to genes at one locus it means the absence of dominance, and referred to genes at different loci it means the absence of epistasis.

EXAMPLE 7.7. As an example of non-additive combination of two loci we shall take the same two colour genes in mice that were used in Example 7.3 to illustrate additive combination; but this time we refer to their effects on the size of the pigment granules, instead of their number (Russell, 1949). The mean size (diameter in μ) of the granules in the four genotypes was as follows:

	B –	bb	Diff.
C –	1·44	0·77	0·67
$c^e c^e$	0·94	0·77	0·17
Diff.	0·50	0·00	

This time the differences are not independent of the other genotype: the c^e gene for example has quite a large effect on the B – genotype, but none at all on the bb genotype. Thus the two loci show epistatic interaction and do not combine additively. Let us therefore work out the interaction deviations. This is not altogether a straightforward matter because the deviations depend on the gene frequencies in the population under discussion; it does, however, help to clarify the meaning of the interaction deviations.

If we were to measure the homozygote differences of these two loci with the object of estimating the value of a for each, the results would depend on the gene frequency at the other locus. For example, the difference between B – and bb would be 0·67 if measured in C – genotypes, but 0·17 if measured in $c^e c^e$ genotypes. The value of a therefore depends on the population in which it is measured. Let us take, for the sake of illustration, a population in which the frequency of bb genotypes is $q_B^2 = 0\cdot 4$ and the frequency of $c^e c^e$ genotypes is $q_C^2 = 0\cdot 2$. Then the mean homozygote difference for the B locus will be $2a_B = (0\cdot 67 \times 0\cdot 8) + (0\cdot 17 \times 0\cdot 2) = 0\cdot 57$. Similarly, for the C locus, $2a_C = 0\cdot 30$. The object now is to find for each genotype the aggregate genotypic value, G, for the two loci combined (i.e. the observed values given above); then the genotypic values, G_B and G_C, derived from consideration of the two loci separately; and, finally, the interaction deviation, I_{BC}, according to equation 7.9. The procedure is simplified if all these values are expressed as deviations from the population mean. The table gives, in line (1), the four genotypes (assuming again complete dominance at both loci); in line (2), the frequency of each genotype in the population; and in line (3), the observed value of granule size in each genotype. The population mean is found by multiplying the value by the frequency of each genotype and summing over the four genotypes. This yields $M = 1\cdot 112$. Subtracting the population mean from the observed value gives the aggregate genotypic value, G, as a deviation from the population mean, shown in line (4). Now consider each locus separ-

		B – C –	B – $c^e c^e$	bb C –	bb $c^e c^e$	Mean
(1)	Genotypes	B – C –	B – $c^e c^e$	bb C –	bb $c^e c^e$	Mean
(2)	Frequencies	0·48	0·12	0·32	0·08	
(3)	Observed values	1·44	0·94	0·77	0·77	1·112
(4)	G	+0·328	–0·172	–0·342	–0·342	0
(5)	$G_B + G_C$	+0·288	–0·012	–0·282	–0·582	0
(6)	I_{BC}	+0·040	–0·160	–0·060	+0·240	0

ately, paying no regard to the other locus. The genotypic values for a single locus, expressed as deviations from the population mean, were given in Table 7.3. With complete dominance these reduce to $2aq^2$ for the two dominant genotypes combined, and $-2a(1-q^2)$ for the recessive homozygote. Take the B – genotype for example: the value of $2a_B$ in the population under consideration was shown above to be 0·57, and the value of q^2 assumed is 0·4; therefore the genotypic value is $0\cdot 57 \times 0\cdot 4 = +0\cdot 228$.

This is the average value of the B − genotype irrespective of the other locus.
The other single-locus values, found in a similar way, are as follows:

	B −	bb		C −	$c^e c^e$
G_B:	+0·228	−0·342	G_C:	+0·060	−0·240

The values given in line (5) of the table as $G_B + G_C$ are found by summation of the two appropriate single-locus values. For example, the B − C − genotype is +0·228 + 0·060 = +0·288. These are the genotypic values expected if there were additive combination. It may be noted that their mean, obtained by summation of (value × frequency) is zero, as is the mean of the aggregate genotypic values in line (4). Finally, the interaction deviations, I_{BC}, given in line (6) are obtained by subtracting the "expected" values in line (5) from the "actual" values in line (4). The mean interaction deviation is also zero.

CHAPTER 8

VARIANCE

The genetics of a metric character centres round the study of its variation, for it is in terms of variation that the primary genetic questions are formulated. The basic idea in the study of variation is its partitioning into components attributable to different causes. The relative magnitude of these components determines the genetic properties of the population, in particular the degree of resemblance between relatives. In this chapter we shall consider the nature of these components and how the genetic components depend on the gene frequency. Then, in the next chapter, we shall show how the degree of resemblance between relatives is determined by the magnitudes of the components.

The amount of variation is measured and expressed as the variance: when values are expressed as deviations from the population mean the variance is simply the mean of the squared values. The components into which the variance is partitioned are the same as the components of value described in the last chapter; so that, for example, the genotypic variance is the variance of genotypic values, and the environmental variance is the variance of environmental deviations. The total variance is the phenotypic variance, or the variance of phenotypic values, and is the sum of the separate components. The components of variance and the values whose variance they measure are listed in Table 8.1.

TABLE 8.1

COMPONENTS OF VARIANCE

Variance component	Symbol	Value whose variance is measured
Phenotypic	V_P	Phenotypic value
Genotypic	V_G	Genotypic value
Additive	V_A	Breeding value
Dominance	V_D	Dominance deviation
Interaction	V_I	Interaction deviation
Environmental	V_E	Environmental deviation

The total variance is then, with certain qualifications to be mentioned below, the sum of the components, thus:

$$V_P = V_G + V_E$$
$$= V_A + V_D + V_I + V_E \qquad \dots\dots(8.1)$$

Let us now consider these components of variance in detail.

GENOTYPIC AND ENVIRONMENTAL VARIANCE

The first division of phenotypic value that we made in the last chapter was into genotypic value and environmental deviation, $P = G + E$. The corresponding partition of the variance into genotypic and environmental components formulates the problem of "heredity versus environment" or "nature and nurture"; or, to put the question more precisely, the relative importance of genotype and environment in determining the phenotypic value. The "relative importance" of a cause of variation means the amount of variation that it gives rise to, as a proportion of the total. So the relative importance of genotype as a determinant of phenotypic value is given by the ratio of genotypic to phenotypic variance, V_G/V_P. The genotypic and environmental components cannot be estimated directly from observations on the population, but in certain circumstances they can be estimated in experimental populations. If one or other component could be completely eliminated, the remaining phenotypic variance would provide an estimate of the remaining component. Environmental variance cannot be removed because it includes by definition all non-genetic variance, and much of this is beyond experimental control. Elimination of genotypic variance can, however, be achieved experimentally. Highly inbred lines, or the F_1 of a cross between two such lines, provide individuals all of identical genotype and therefore with no genotypic variance. If a group of such individuals is raised under the normal range of environmental circumstances, their phenotypic variance provides an estimate of the environmental variance V . Subtraction of this from the phenotypic variance of a genetically mixed population then gives an estimate of the genotypic variance of this population.

EXAMPLE 8.1. Partitioning of the phenotypic variance into its genotypic and environmental components has been done for several characters in *Drosophila melanogaster*. The results are given later, in Table 8.2, but here we may describe the results for one character in more detail in order to show how the partitioning is made. The character is the length of the thorax (in units of 1/100 mm.), which may be regarded as a measure of body-size. The phenotypic variance was measured first in a genetically mixed— i.e. a random-bred—population, and then in a genetically uniform population, consisting of the F_1 generation of three crosses between highly inbred lines. The first estimates the genotypic and environmental variance together, and the second estimates the environmental variance alone. So, by subtraction, an estimate of the genotypic variance is obtained. The results, obtained by F. W. Robertson (1957*b*), were as follows:

Population	*Components*	*Observed variance*
Mixed	$V_G + V_E$	0·366
Uniform	V_E	0·186
Difference	V_G	0·180
	$V_G/V_P = 49\%$	

Thus 49 per cent of the variation of thorax length in this genetically mixed population is attributable to genetic differences between individuals, and 51 per cent to non-genetic differences.

Individuals of identical genotype are also provided by identical twins in man and cattle, but their use in partitioning the variance is very limited: they will be discussed in a later chapter when the problems that they raise will be better understood. Apart from the severely limited use of identical twins, the partitioning of the variance into genotypic and environmental components depends on the availability of highly inbred lines, and is therefore restricted to experimental populations of plants or small animals.

Three complications arise in connexion with the partitioning of the variance into genotypic and environmental components. They are all things that can usually be neglected or circumvented with little risk of error, but in some circumstances they may be important. The following account of them might well be omitted at a first reading, unless the reader is worried by the logical fallacies introduced by neglecting them.

Dependence of environmental variance on genotype. Experiments of the type illustrated in Example 8.1 rest on the assumption that the environmental variance is the same in all genotypes, and this is certainly not always true. The environmental variance measured in one inbred line or cross is that shown by this one particular genotype, and other genotypes may be more or less sensitive to environmental influences and may therefore show more or less environmental variance. The environmental variance of the mixed population may therefore not be the same as that measured in the genotypically uniform group. Not very much is yet known about this complication except that many characters show more environmental variance among inbred than among outbred individuals, inbreds being more sensitive or less well "buffered." The reality of the complication is therefore not in doubt. Further discussion of the phenomenon will be found under the effects of inbreeding, in Chapter 15, where it more properly belongs. The existence of this complication means that when dealing with genotypically mixed populations we have to define the environmental component of variance as the mean environmental variance of the genotypes in the population, and we have to recognise the possibility that if the frequencies of the genotypes are changed, as by selection, the environmental variance may also be changed in consequence.

Genotype-environment correlation. Hitherto we have tacitly assumed that environmental deviations and genotypic values are independent of each other; in other words that there is no correlation between genotypic value and environmental deviation, such as would arise if the better genotypes were given better environments. Correlation between genotype and environment is seldom an important complication, and can usually be neglected in experimental populations, where randomisation of environment is one of the chief objects of experimental design. There are some situations, however, in which the correlation exists. Milk-yield in dairy cattle provides an example. The normal practice of dairy husbandry is to feed cows according to their yield, the better phenotypes being given more food. This introduces a correlation between phenotypic value and environmental deviation; and, since genotypic and phenotypic values are correlated, there is also a correlation between genotypic value and environmental deviation. The complication of genotype-environment correlation is very simply overcome by regarding the special environment—i.e. the feeding level in the case of cows—as part of the genotype. This situation is covered by the definition of genotypic value, provided genotypic values are taken to refer to genotypes as they occur under the normal conditions of association with specific environments. If genotypic values were not so defined we could not treat the phenotypic variance as simply the sum of the genotypic and environmental variances, but we should have to include a covariance term, thus:

$$V_P = V_G + V_E + 2cov_{GE} \qquad \qquad .....(8.2)$$

where cov_{GE} is the covariance of genotypic values and environmental deviations. If the genotypic variance is estimated, as in Example 8.1, by the comparison of genetically identical with genetically mixed groups, then the covariance would be eliminated with the genotypic variance from the genetically identical group, and the estimate obtained will be of genotypic variance together with twice the covariance. Thus, while on theoretical grounds it is convenient, on practical grounds it is unavoidable, to regard any covariance that may arise from genotype-environment correlation as being part of the genotypic variance.

Genotype-environment interaction. Another assumption that we have made, which is not always justifiable, is that a specific difference of environment has the same effect on different genotypes; or, in other words, that we can associate a certain environmental deviation with a specific difference of environment, irrespective of the genotype on which it acts. When this is not so there is an interaction, in the statistical sense, between genotypes and environments. There are several forms which this interaction may take (Haldane, 1946). For example, a specific difference of environment may have a greater effect on some genotypes than on others; or there may be a change in the order of merit of a series of genotypes when measured under different environments. That is to say, genotype A may be superior to genotype B in environment X, but inferior in environment Y, as in the following example.

EXAMPLE 8.2. The following figures show the growth, between 3 and 6 weeks of age, of two strains of mice reared on two levels of nutrition (original data):

	Good nutrition	*Bad nutrition*
Strain A	17·2 gm.	12·6 gm.
Strain B	16·6 gm.	13·3 gm.

Strain A grows better than strain B under good conditions, but worse under bad conditions.

An interaction between genotype and environment, whatever its nature, gives rise to an additional component of variance. This interaction variance can be isolated and measured only under rather artificial circumstances. We may replicate genotypes by the use of inbred lines or F_1's, and replicate specific environments by the control of such factors as nutrition or temperature. Then an analysis of variance in a two-way classification of genotypes × environments will yield estimates of the genotypic variance (between genotypes), the environmental variance (between environments) and the variance attributable to interaction of genotypes with environments. The specific environments in such an experiment are, however, more in the nature of "treatments" because a population under genetical study would not normally encounter so wide a range of environments as that provided by the different treatments. It is therefore the genotype-environment interaction occurring within one such treatment that is relevant to the genetical study of a population, and this cannot be measured because the separate elements of the environment cannot be isolated and controlled. In an experiment such as that of Example 8.1, which removes the genotypic variance by the use of inbred lines or F_1's, the interaction variance remains with the environmental in the phenotypic variance measured in the genetically uniform individuals. In normal circumstances, therefore, the variance due to genotype-environment interaction, since it cannot be separately measured, is best regarded as part of the environmental variance. When large differences of environment, such as between different habitats, are under consideration, the presence of genotype-environment interaction becomes important in connexion with the specialisation of breeds or varieties to local conditions. This matter will be taken up again later, in Chapter 19, because it can be more profitably discussed from a different viewpoint.

Genetic Components of Variance

The partition into genotypic and environmental variance does not take us far toward an understanding of the genetic properties of a population, and in particular it does not reveal the cause of resemblance between relatives. The genotypic variance must be further divided according to the division of genotypic value into breeding value, dominance deviation, and interaction deviation. Thus we have:

Values $\qquad$ $G \;=\; A \;+\; D \;+\; I$

Variance components $\quad V_G = V_A + V_D + V_I \qquad ......(8.4)$

(genotypic) (additive) (dominance) (interaction)

The *additive variance*, which is the variance of breeding values, is the important component since it is the chief cause of resemblance between relatives and therefore the chief determinant of the observable genetic properties of the population and of the response of the population to selection. Moreover, it is the only component that can be readily estimated from observations made on the population. In practice, therefore, the important partition is into additive genetic variance versus all the rest, the rest being non-additive genetic and environmental variance. This partitioning is most conveniently expressed as the ratio of additive genetic to total phenotypic variance, V_A/V_P, a ratio called the *heritability*.

Estimation of the additive variance rests on observation of the degree of resemblance between relatives and will be described later when we have discussed the causes of resemblance between relatives. Our immediate concern here is to show how the genetic components of variance are influenced by the gene frequency. To do this we have to express the variance in terms of the gene frequency and the assigned genotypic values a and d. We shall consider first a single locus with two alleles, thus excluding interaction variance for the moment.

Additive and dominance variance. The information needed to obtain expressions for the variance of breeding values and the variance of dominance deviations was given in the last chapter in Table 7.3 (p. 125). This table gives the breeding values and dominance deviations of the three genotypes, expressed as deviations from the population mean. It will be remembered that the means of both breeding values and dominance deviations are zero. Therefore no correction for an assumed mean is needed, and the variance is simply the mean of the squared values. The variances are thus obtained by squaring the values in the table, multiplying by the frequency of the genotype concerned, and summing over the three genotypes. (The procedure of multiplying values by frequencies to obtain the mean was explained on p. 114.) The additive variance, which is the variance of breeding values, is obtained as follows:

$$\begin{aligned}
V_A &= \alpha^2[4q^2p^2 + (q-p)^2 . 2pq + 4p^2q^2] \\
&= 2pq\alpha^2(2pq + p^2 + q^2 - 2pq + 2pq) \\
&= 2pq\alpha^2 \qquad\qquad ......(8.5.\,a) \\
&= 2pq[a + d(q-p)]^2 \qquad ......(8.5.\,b)
\end{aligned}$$

K

Similarly the variance of dominance deviations is

$$V_D = d^2(4q^2p^2 + 8p^3q^3 + 4p^2q^2)$$
$$= 4p^2q^2d^2(q^2 + 2pq + p^2)$$
$$= (2pqd)^2 \qquad\qquad\qquad\dots\dots(8.6)$$

It was noted in the last chapter that breeding values and dominance deviations are uncorrelated. From this it follows that the genotypic variance is simply the sum of the additive and dominance variances. Thus

$$V_G = V_A + V_D$$
$$= 2pq[a + d(q - p)]^2 + [2pqd]^2 \qquad\dots\dots(8.7)$$

EXAMPLE 8.3. To illustrate the genetic components of variance arising from a single locus let us return to the pygmy gene in mice, used for several examples in the last chapter. From the values tabulated in Example 7.6 (p. 123) we may compute the components of variance directly. Since the values are expressed as deviations from the population mean, the variance is obtained by multiplying the frequency of each genotype by the square of its value, and summing over the three genotypes. For example, the genotypic variance when $q = 0.1$ is $0.81(0.44)^2 + 0.18(-1.56)^2 + 0.01(-7.56)^2 = 1.1664$. The additive variance is obtained in the same way from the variance of breeding values, and the dominance variance from the variance of dominance deviations. The variances obtained are as follows:

	$q = 0.1$	$q = 0.4$
Genotypic, V_G	1.1664	7.1424
Additive, V_A	1.0368	6.2208
Dominance, V_D	0.1296	0.9216

The variances may be obtained also, and with less trouble, by use of the formulae given above in equations 8.5, 8.6 and 8.7. The values to be substituted were given in Example 7.1; namely, $a = 4$ and $d = 2$. Notice that the dominance variance is quite small in comparison with the additive.

The ways in which the gene frequency and the degree of dominance influence the magnitude of the genetic components of variance can best be appreciated from graphical representations of the relationships derived above, in equations 8.5, 8.6, and 8.7. The graphs in Fig. 8.1 show the amounts of genotypic, additive, and dominance variance arising from a single locus with two alleles, plotted against the gene frequency. Three cases are shown to illustrate the effect of different degrees of dominance: in graph (*a*) there is no dominance ($d=$o); in graph (*b*) there is complete dominance ($d=a$); and in graph (*c*) there is "pure" over-dominance ($a=$o). In the first case the genotypic variance is all additive, and it is greatest when $p=q=$o·5. In the second case the dominance variance is maximal when $p=q=$o·5, and the additive is maximal when the frequency of the recessive allele is $q=$o·75. In the third case the dominance variance is the same as in the second and is maximal

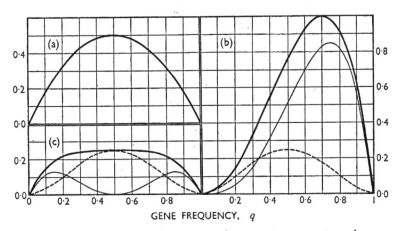

GENE FREQUENCY, q

FIG. 8.1. Magnitude of the genetic components of variance arising from a single locus with two alleles, in relation to the gene frequency. Genotypic variance—thick lines; additive variance—thin lines; dominance variance—broken lines. The gene frequency, q, is that of the recessive allele. The degrees of dominance are: in (*a*) no dominance ($d=$o); in (*b*) complete dominance ($d=a$); and in (*c*) "pure" overdominance ($a=$o). The figures on the vertical scale, showing the amount of variance, are to be multiplied by a^2 in graphs (*a*) and (*b*), and by d^2 in graph (*c*).

when $p=q=$o·5. The additive variance, however, is zero when $p=q=$o·5, and has two maxima, one at $q=$o·15 and the other at $q=$o·85. The genotypic variance, in this case, remains practically constant over a wide range of gene frequency, though its composition changes profoundly. The general conclusion to be drawn from these graphs is that genes contribute much more variance when at intermediate frequencies than when at high or low frequencies: recessives at low frequency, in particular, contribute very little variance.

A possible misunderstanding about the concept of additive genetic variance, to which the terminology may give rise, should be mentioned here. The concept of additive variance does not carry with it the assumption of additive gene action; and the existence of additive variance is not an indication that any of the genes act additively (i.e. show neither dominance nor epistasis). No assumption is made about the mode of action of the genes concerned. Additive variance can arise from genes with any degree of dominance or epistasis, and only if we find that all the genotypic variance is additive can we conclude that the genes show neither dominance nor epistasis.

The existence of more than two alleles at a locus introduces no new principle, though it complicates the theoretical description of the effect of the locus. Expressions for the additive and dominance variances are given by Kempthorne (1955*a*). The locus contributes additive variance arising from the average effects of its several alleles, and dominance variance arising from the several dominance deviations.

To arrive at the variance components expressed in the population the separate effects of all loci that contribute variance have to be combined. The additive variance arising from all loci together is the sum of the additive variances attributable to each locus separately; and the dominance variance is similarly the sum of the separate contributions. But when more than one locus is under consideration then the interaction deviations, if present, give rise to another component of variance, the interaction variance, which is the variance of the interaction deviations.

Interaction variance. We shall treat the interaction variance as a complication, like genotype-environment correlation or interaction, to be circumvented: that is to say, we shall not discuss its properties in detail, but we shall show what happens to it if it is ignored. It is only comparatively recently that the properties of the interaction variance have been worked out (see Cockerham, 1954; Kempthorne, 1954, 1955*a, b*) and little is yet known about its importance in relation to the other components. It seems probable, however, that the amount of variance contributed by it is usually rather small, and that neglect of it is therefore not likely to lead to serious error. Description of the properties of interaction variance rests on its further subdivision into components. It is first subdivided according to the number of loci involved: two-factor interaction arises from the interaction of two loci, three-factor from three loci, etc. Interactions involving larger numbers of loci contribute so little variance that they can be ignored, and we shall confine our attention to two-factor interactions since these suffice to illustrate the principles involved. The next subdivision of the interaction variance is according to whether the interaction involves breeding values or dominance deviations. There are thus three sorts of two-factor interactions. Interaction between the two breeding values gives rise to additive × additive variance, V_{AA}; interaction between the breeding value of one locus and the dominance deviation of the other gives rise to additive × dominance variance, V_{AD}; and interaction between the two dominance deviations gives rise to dominance × dominance variance, V_{DD}. So the interaction variance is broken down into components thus:

$$V_I = V_{AA} + V_{AD} + V_{DD} + \text{etc.} \qquad \dots\dots(8.8)$$

the terms designated "etc." being similar components arising from interactions between more than two loci. At the moment we cannot go further than this in the description of the interaction variance, but we shall show later how it affects the resemblance between relatives and what happens to it when components of variance are estimated from observations on the population.

That completes the description of the nature of the genetic components of variance. The practical value of the partitioning of the variance will not yet be fully apparent because it arises from the causes of resemblance between relatives, which is the subject of the next chapter. The partitioning we have made is essentially a theoretical one, and before we pass on we should consider how much of it can actually be made in practice. When observations of resemblance between relatives are available we can estimate the additive variance and so make the partition $V_A : (V_D + V_I + V_E)$. And if inbred lines are available we can estimate the environmental variance and so make the partition $V_G : V_E$. If both these partitions are made we can separate the additive genetic from the rest of the genetic variance, and so make the three-fold partition into additive genetic, non-additive genetic, and environmental variance, $V_A : (V_D + V_I) : V_E$, the dominance and interaction components being lumped together as non-additive genetic variance. Examples of this partitioning are given in Table 8.2, although at this stage the method by which the additive component is estimated will not be understood. This partitioning is as far as we can go by means of relatively simple experiments. By more elaborate techniques, requiring large numbers of observations, it may be possible to go some way toward separating the dominance from the interaction components, or at least to get an idea of their relative importance. (See, in particular, Robinson and Comstock, 1955; Hayman, 1955, 1958; Cockerham, 1956b.)

TABLE 8.2

Partitioning of the variance of four characters in *Drosophila melanogaster*. Components as percentages of the total, phenotypic, variance.

		Character			
		(1)	(2)	(3)	(4)
		Bristles	Thorax	Ovary	Eggs
Phenotypic	V_P	100	100	100	100
Additive genetic	V_A	52	43	30	18
Non-additive genetic	$V_D + V_I$	9	6	40	44
Environmental	V_E	39	51	30	38

Characters:

(1) Number of bristles on 4th + 5th abdominal segments (Clayton, Morris, and Robertson, 1957; Reeve and Robertson, 1954).
(2) Length of thorax (F. W. Robertson, 1957b).
(3) Size of ovaries, i.e. number of ovarioles in both ovaries. (F. W. Robertson, 1957a).
(4) Number of eggs laid in 4 days (4th to 8th after emergence) (F. W. Robertson, 1957b).

ENVIRONMENTAL VARIANCE

Environmental variance, which by definition embraces all variation of non-genetic origin, can have a great variety of causes and its nature depends very much on the character and the organism studied. Generally speaking, environmental variance is a source of error that reduces precision in genetical studies and the aim of the experimenter or breeder is therefore to reduce it as much as possible by careful management or proper design of experiments. Nutritional and climatic factors are the commonest external causes of environmental variation, and they are at least partly under experimental control. Maternal effects form another source of environmental variation that is sometimes important, particularly in mammals, but is less susceptible to control. Maternal effects are prenatal and postnatal influences, mainly nutritional, of the mother on her young: we shall have more to say about them in the next chapter in connexion with resemblance between relatives. Error of measurement is another source of variation, though it is usually quite trivial. When a character can be measured in units of length or weight it is usually measured so accurately that the variance attributable to measurement is negligible in comparison with the rest of the variance. Some characters, however, cannot strictly speaking be measured, but have to be graded by judgement into classes. Carcass qualities of livestock are an example. With such characters the variance due to measurement may be considerable.

In addition to the variation arising from recognisable causes, such as those mentioned, there is usually also a substantial amount of non-genetic variation whose cause is unknown, and which therefore cannot be eliminated by experimental design. This is generally referred to as "intangible" variation. Some of the intangible variation may be caused by "environmental" circumstances, in the common meaning of the word—that is, by circumstances external to the individual—even though their nature is not known. Some, however, may arise from "developmental" variation: variation, that is, which cannot be attributed to external circumstances, but is attributed, in ignorance of its exact nature, to "accidents" or "errors" of development as a general cause. Characters whose intangible variation is predominantly developmental are those connected with anatomical structure, which do not change after development is complete, such as skeletal form, pigmentation, or bristle number in *Drosophila*. Characters more susceptible to the influences of the external environment, in contrast, are those connected with metabolic processes, such as growth, fertility, and lactation.

EXAMPLE 8.4. Human birth weight provides an example of a character subject to much environmental variation whose nature has been analysed in detail (Penrose, 1954; Robson, 1955). The partitioning of the phenotypic variance given in the table shows the relative importance of all the identified sources of variation, birth weight being regarded as a character of the child. All the environmental variation is "maternal" in the sense that it is connected with the prenatal environment, but several distinct components of the maternal environment are distinguished. "Maternal genotype," which accounts for 20 per cent of the total phenotypic variance, reflects genetic variation (chiefly additive) between mothers in the birth weight of their children; i.e. birth weight regarded as a character of the mother. "Maternal environment, general," which accounts for another 18 per cent, reflects non-genetic variation between mothers in the same way.

These two components, totalling 38 per cent, are maternal causes of variation in birth weight that affect all children of the same mother alike.
"Maternal environment, immediate" means causes attributable to the
mother but differing in successive pregnancies. Two causes of the same
nature—"age of mother" and "parity" (i.e. whether the child is the first,

Partitioning of variance of human birth-weight. Components as percentages of the total, phenotypic, variance.

Cause of variation	% of total	
Genetic		
Additive	15	
Non-additive (approx)	1	
Sex	2	
Total genotypic		18
Environmental		
Maternal genotype	20	
Maternal environment, general	18	
Maternal environment, immediate	6	
Age of mother	1	
Parity	7	
Intangible	30	
Total environmental		82

second, etc.)—are separately identifiable. Finally, the "intangible"
variation is all the remainder, of which the cause cannot be identified. To
explain how these various components were estimated would take too
much space, and could not properly be done until the end of Chapter 10.
It must suffice to say that the estimates all come from comparisons of the
degree of resemblance between identical twins, fraternal twins, full sibs,
children of sisters, and other sorts of cousins.

Multiple measurements. When more than one measurement
of the character can be made on each individual, the phenotypic
variance can be partitioned into variance within individuals and
variance between individuals. This subdivision serves to show how
much is to be gained by the repetition of measurements, and it may
also throw light on the nature of the environmental variation. There
are two ways by which the repetition of a character may provide
multiple measurements: by temporal repetition and by spatial repetition. Milk-yield and litter size are examples of characters repeated
in time. Milk-yield can be measured in successive lactations, and

litter size in successive pregnancies. Several measurements of each individual can thus be obtained. The variance of yield per lactation, or of the number of young per litter, can then be analysed into a component within individuals, measuring the differences between the performances of the same individual, and a component between individuals, measuring the permanent differences between individuals. The within-individual component is entirely environmental in origin, caused by temporary differences of environment between successive performances. The between-individual component is partly environmental and partly genetic, the environmental part being caused by circumstances that affect the individuals permanently. By this analysis, therefore, the variance due to temporary environmental circumstances is separated from the rest, and can be measured.

Characters repeated in space are chiefly structural or anatomical, and are found more often in plants than in animals. For example, plants that bear more than one fruit yield more than one measurement of any character of the fruit, such as its shape or seed content. Spatial repetition in animals is chiefly found in characters that can be measured on the two sides of the body or on serially repeated parts, such as the number of bristles on the abdominal segments of *Drosophila*. With spatially repeated characters the within-individual variance is again entirely environmental in origin but, unlike that of temporally repeated characters, it represents the "developmental" variation arising from localised circumstances operating during development.

In order that we may discuss both temporal and spatial repetition together we shall use the term *special environmental variance*, V_{Es}, to refer to the within-individual variance arising from temporary or localised circumstances; and the term *general environmental variance*, V_{Eg}, to refer to the environmental variance contributing to the between-individual component and arising from permanent or non-localised circumstances. The ratio of the between-individual component to the total phenotypic variance measures the correlation (r) between repeated measurements of the same individual, and is known as the *repeatability* of the character. The partitioning of the phenotypic variance expressed by the repeatability is thus into two components, V_{Es} versus ($V_G + V_{Eg}$), so that the repeatability is

$$r = \frac{V_G + V_{Eg}}{V_P} \qquad \dots (8.9)$$

The repeatability therefore expresses the proportion of the variance of single measurements that is due to permanent, or non-localised, differences between individuals, both genetic and environmental. The repeatability differs very much according to the nature of the character, and also, of course, according to the genetic properties of

TABLE 8.3

SOME EXAMPLES OF REPEATABILITY

Organism and character	*Repeatability*
Drosophila melanogaster:	
Abdominal bristle number (see Example 8.6).	·42
Ovary size (F. W. Robertson, 1957*a*).	·73
Mouse:	
Weight at 6 weeks (repeated on 4 consecutive days. Original data).	·95
Litter size (see Example 8.5).	·45
Sheep:	
Weight of fleece, measured in different years (Morley, 1951).	·74
Cattle:	
Milk-yield (Johansson, 1950).	·40

the population and the environmental conditions under which the individuals are kept. The estimates in Table 8.3 give some idea of the sort of values that may be found with various characters, and two cases are described in more detail in the following examples.

EXAMPLE 8.5. Litter size in mice will serve as an example of a character repeated in time. The number of live young born in first and in second litters was recorded in 296 mice of a genetically heterogeneous stock, and yielded the following components of variance (original data):

$$\text{Between mice} \quad 3\cdot58$$
$$\text{Within mice} \quad 4\cdot44$$

(The procedure for estimating the components of variance from an analysis of variance is described by Snedecor (1956, Section 10.12) and is outlined below, in Chapter 10, p. 173.) The repeatability of litter size is given by the ratio of the between-mice component to the sum of the between-mice and the within-mice components: i.e.

$$r = \frac{3 \cdot 58}{3 \cdot 58 + 4 \cdot 44} = 0 \cdot 45.$$

EXAMPLE 8.6. The number of bristles on the ventral surfaces of the abdominal segments is a character that has been much studied in *Drosophila melanogaster*, because it is technically convenient and its genetic properties are relatively simple. We have already mentioned it several times but have not yet used it as an example. There are about 20 bristles on each of 3 segments in males and each of 4 segments in females. The number of bristles per segment can therefore be treated as a spatially repeated character. The sources of variation in this character have been studied in detail by Reeve and Robertson (1954), and the following components of variance were found:

		♂♂	♀♀
Total phenotypic	V_P	4·24	5·44
Between flies	$V_G + V_{Eg}$	1·82	2·19
Within flies	V_{Es}	2·42	3·25
Repeatability		0·429	0·403

Estimation of the repeatability of a character separates off the component of variance due to special environment, V_{Es}, but it leaves the other component of environmental variance—that due to general environment, V_{Eg}—confounded with the genotypic variance, as shown in the above example. The component due to general environment can be separately estimated only if the genotypic variance (i.e. including the non-additive components) has been estimated, in the manner explained in Example 8.1. This has been done with two characters in *Drosophila*, and the results are given in Table 8.4. The

TABLE 8.4

Partitioning of the environmental variance of two characters in *Drosophila melanogaster* into components due to general, V_{Eg}, and special, V_{Es}, environment. The characters are: abdominal bristle-number (Reeve and Robertson, 1954) as explained in Example 8.6, and ovary size (F. W. Robertson, 1957a), measured in the two ovaries by the number of ovarioles, or "egg strings."

	Bristle number	*Ovary size*
Total environmental, V_E	100	100
General environmental, V_{Eg}	3	9
Special environmental, V_{Es}	97	91

nature of the environmental variation revealed by these results is remarkable. With both characters less than 10 per cent of the environmental variance is general—that is, due to causes influencing the individual as a whole. These characters are therefore very little influenced by the conditions of the external environment: or, perhaps it would be more accurate to say that the experimental technique of rearing the flies has been very successful in eliminating unwanted sources of environmental variation. Yet, fully half the phenotypic variation of one measurement (one segment or ovary) is non-genetic, or environmental in the wide sense, as shown in Table 8.2; and, moreover, is due to strictly localised causes that influence the segments or ovaries independently. Whether this developmental variation represents a real indeterminacy of development, or has material causes still undetected but in principle controllable, is quite unknown. Nor is it known whether the situation revealed in these two characters is at all general. We cannot here pursue further the biological nature of the non-genetic variation: a general discussion of these problems will be found in Waddington (1957).

We must return to the repeatability and consider its uses. Knowledge of the repeatability of a character is useful in two ways. First, it sets upper limits to the values of the two ratios, V_A/V_P and V_G/V_P. The first (additive genetic to total phenotypic variance), is the heritability, which as we shall see in later chapters is of great practical importance. The second (genotypic to phenotypic variance) measures the degree of genetic determination of the character. The repeatability is usually much easier to determine than either of these two ratios, and it may often be known when they are not.

The second way in which knowledge of the repeatability is useful is that it indicates the gain in accuracy to be expected from multiple measurements. Suppose that each individual is measured n times, and that the mean of these n measurements is taken to be the phenotypic value of the individual, say $P_{(n)}$. Then the phenotypic variance is made up of the genotypic variance, the general environmental variance, and one n^{th} of the special environmental variance:

$$V_{P(n)} = V_G + V_{Eg} + \frac{1}{n} V_{Es} \qquad \ldots (8.10)$$

Thus, increasing the number of measurements reduces the amount of variance due to special environment that appears in the phenotypic variance, and this reduction of the phenotypic variance represents the gain in accuracy. The variance of the mean of n measurements as a proportion of the variance of one measurement can be expressed in terms of the repeatability, as follows:

$$\frac{V_{P(n)}}{V_P} = \frac{1 + r(n-1)}{n} \qquad \dots\dots (8.11)$$

where r is the repeatability, or the correlation between the measurements of the same individual. Fig. 8.2 shows how the phenotypic variance is reduced by multiple measurements, with characters of

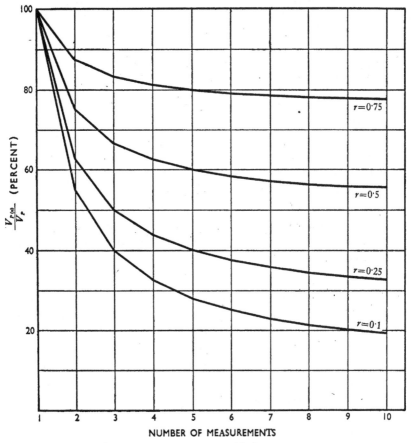

FIG. 8.2. Gain in accuracy from multiple measurements of each individual. The vertical scale gives the variance of the mean of n measurements as a percentage of the variance of one measurement. The horizontal scale gives the number of measurements, up to 10. The four graphs refer to characters of different repeatability as indicated.

different repeatabilities. When the repeatability is high, and there is therefore little special environmental variance, multiple measurements give little gain in accuracy. When the repeatability is low, multiple measurements may lead to a worth-while gain in accuracy. The gain in accuracy, however, falls off rapidly as the number of measurements increases, and it is seldom worth while to make more than two measurements.

EXAMPLE 8.7. Studies of abdominal bristle number in *Drosophila* are generally based on two measurements, i.e. of the fourth and fifth segments, and the phenotypic values are expressed as the sum of the two counts. As an illustration of the nature of the advantage gained by the double measurement we may compare the percentage composition of the phenotypic variance when phenotypic values are based on counts of one or of two segments:

		One segment	*Two segments*
Phenotypic	V_P	100	100
Additive genetic	V_A	34	52
Non-additive genetic	$V_D + V_I$	6	9
Environmental, general	V_{Eg}	2	4
Environmental, special	V_{Es}	58	35

By reducing the amount of environmental variance, the making of two measurements increases the proportionate amount of genetic variance: in practice it is the increase of the proportion of additive variance—in this case from 34 per cent to 52 per cent—that is the important consideration.

There is an important assumption implicit in the idea of repeatability, which we have not yet mentioned. It is the assumption that the multiple measurements are indeed measurements of what is genetically the same character. Consider for example milk-yield in successive lactations. If the assumption were valid it would mean that the genes that influence yield in first lactations are entirely the same, as those that influence yield in second or later lactations; or, to put the matter in another way, that yield in all lactations is dependent on identical developmental and physiological processes. If this assumption is not valid, as it certainly is not for milk-yield in cattle, then the variation within individuals is not purely environmental, and equation *8.11* is erroneous. The variance between the means of individuals will be augmented by additional variance arising from what may formally be regarded as interaction between genotype and "environment," that is between genotype and the time or location of the measurement. And this additional variance may be enough to counteract the reduction of environmental variance which we have described as the chief advantage to be gained from multiple measurements. Consequently an increase in the proportion of additive genetic variance from multiple measurements cannot be relied on until the genetical identity of the character measured has been established. The number of bristles on the abdominal segments of *Drosophila* has been proved to be genetically the same character, as will be explained in Chapter 19, and the conclusions reached in Example 8.7 are valid. Milk-yield in cattle, in contrast, is not the same character in successive lactations, and the proportion of additive variance is actually less for the mean of several lactations than for first lactations only. (See Rendel, *et al.*, 1957.)

CHAPTER 9

RESEMBLANCE BETWEEN RELATIVES

The resemblance between relatives is one of the basic genetic pheno-mena displayed by metric characters, and the degree of resemblance is a property of the character that can be determined by relatively simple measurements made on the population without special experi-mental techniques. The degree of resemblance provides the means of estimating the amount of additive variance, and it is the propor-tionate amount of additive variance (i.e. the heritability) that chiefly determines the best breeding method to be used for improvement. An understanding of the causes of resemblance between relatives is therefore fundamental to the practical study of metric characters and to its application in animal and plant improvement. In this chapter, therefore, we shall examine the causes of resemblance between rela-tives, and show in principle how the amount of additive variance can be estimated from the observed degree of resemblance, leaving the more practical aspects of the estimation of the heritability for con-sideration in the next chapter.

In the last chapter we saw how the phenotypic variance can be partitioned into components attributable to different causes. These components we shall call *causal components* of variance, and denote them as before by the symbol V. The measurement of the degree of resemblance between relatives rests on the partitioning of the pheno-typic variance in a different way, into components corresponding to the grouping of the individuals into families. These components can be estimated directly from the phenotypic values and for this reason we shall call them *observational components* of phenotypic variance, and denote them by the symbol σ^2 in order to keep the distinction clear. Consider, for example, the grouping of individuals into families of full sibs. By the analysis of variance we can partition the total observed variance into two components, within groups and between groups. The within-group component is the variance of individuals about their group means, and the between-group com-ponent is the variance of the "true" means of the groups about the population mean. The true mean of a group is the mean estimated without error from a very large number of individuals. An explanation of the estimation of these two components will be given, with examples, in the next chapter. Now, the resemblance between related individuals, i.e. between full sibs in the case under discussion, can be looked at either as similarity of individuals in the same group, or as difference between individuals in different groups. The greater the similarity within the groups, the greater in proportion will be the difference between the groups. The degree of resemblance can therefore be expressed as the between-group component as a proportion of the total variance. This is the intra-class correlation coefficient and is given by

$$t = \frac{\sigma_B^2}{\sigma_B^2 + \sigma_W^2}$$

where σ_B^2 is the between-group component and σ_W^2 the within-group component. (It is customary to use the symbol t for the intra-class correlation of phenotypic values in order to avoid confusion with other types of correlation for which the symbol r is used.) The between-group component expresses the amount of variation that is common to members of the same group, and it can equally well be referred to as the covariance of members of the groups. In the case of the resemblance between offspring and parents the grouping of the observations is into pairs rather than groups; one parent, or the mean of two parents, paired with one offspring or the mean of several offspring. It is then more convenient to compute the covariance of offspring with parents from the sum of cross-products, rather than from the between-pair component of variance. With offspring-parent relationships, also, it is usually more convenient to express the degree of resemblance as the regression coefficient of offspring on parent, instead of the correlation between them, the regression being given by

$$b_{\text{OP}} = \frac{cov_{\text{OP}}}{\sigma_P^2}$$

where cov_{OP} is the covariance of offspring and parents, and σ_P^2 is the variance of parents.

Thus, the covariance of related individuals is the new property of the population that we have to deduce in seeking the cause of resemblance between relatives, whether sibs or offspring and parents.

The covariance, being simply a portion of the total phenotypic variance, is composed of the causal components described in the last chapter, but in amounts and proportions differing according to the sort of relationship. By finding out how the causal components contribute to the covariance we shall see how an observed covariance can be used to estimate the causal components of which it is composed.

Both genetic and environmental sources of variance contribute to the covariance of relatives. We shall consider the genetic causes of resemblance first, then the environmental causes, and finally, by putting the two causes together, arrive at the phenotypic covariance and the degree of resemblance that can be observed from measurements of phenotypic values. A general description of the covariance, applicable to any sort of relationship, is given by Kempthorne (1955a). Here we shall consider only four sorts of relationship: (1) between offspring and one parent, (2) between half sibs, (3) between offspring and the mean of the two parents, and (4) between full sibs. These are the most important relationships in practice. Identical twins will be considered in the next chapter, because the problems they raise will be better understood then.

GENETIC COVARIANCE

Our object now is to deduce from theoretical considerations the covariance of relatives arising from genetic causes, neglecting for the time being any non-genetic causes of resemblance that there may be. This means that we have to deduce the covariance of the genotypic values of the related individuals. This will be done by reference to two alleles at a locus, but the conclusions are equally valid for loci with any number of alleles. We shall at first omit interaction deviations and the interaction component of variance from consideration, but we shall describe its effects briefly later.

Offspring and one parent. The covariance to be deduced is that of the genotypic values of individuals with the mean genotypic values of their offspring produced by mating at random in the population. If values are expressed as deviations from the population mean, then the mean value of the offspring is by definition half the breeding value of the parent, as explained in Chapter 7. Therefore the covariance to be computed is that of an individual's genotypic value with half its breeding value, i.e. the covariance of G with $\frac{1}{2}A$.

Since $G = A + D$ (D being the dominance deviation) the covariance is that of $(A+D)$ with $\frac{1}{2}A$. Taking the sum of cross-products, we have

$$\text{sum of cross-products} = \Sigma\frac{1}{2}A(A+D)$$
$$= \frac{1}{2}\Sigma A^2 + \frac{1}{2}\Sigma AD$$

Since A and D are uncorrelated (see p. 125), the term $\frac{1}{2}\Sigma AD$ is zero. Then if we divide both sides by the number of paired observations we have

$$cov_{OP} = \frac{1}{2}V_A \qquad\qquad \dots\dots(9.1)$$

since ΣA^2 is the sum of squares of breeding values. The genetic covariance of offspring and one parent is therefore half the additive variance.

The covariance may be derived by another method, which though less concise is perhaps more explicit. Table 9.1 gives the genotypes of the parents, their frequencies in the population, and their genotypic values expressed as deviations from the population mean (from Table 7.3). The right-hand column gives the mean genotypic values

TABLE 9.1

	Parents		Offspring
Genotype	Frequency	Genotypic value	Mean genotypic value
A_1A_1	p^2	$2q(\alpha - qd)$	$q\alpha$
A_1A_2	$2pq$	$(q-p)\alpha + 2pqd$	$\frac{1}{2}(q-p)\alpha$
A_2A_2	q^2	$-2p(\alpha+pd)$	$-p\alpha$

of the offspring, which are half the breeding values of the parents as given in Table 7.3. The covariance of offspring and parent is then the mean cross-product, and is obtained by multiplying together the three columns––frequency × genotypic value of parent × genotypic value of offspring––and summing over the three genotypes of the parents. After collecting together the terms in α^2 and the terms in αd we obtain

$$cov_{OP} = pq\alpha^2(p^2 + 2pq + q^2) + 2p^2q^2\alpha d(-q+q-p+p)$$
$$= pq\alpha^2$$
$$= \frac{1}{2}V_A$$

since from equation 8.5, $V_A = 2pq\alpha^2$. Summing over all loci we again reach the conclusion that the covariance of offspring and one parent is equal to half the additive variance.

Half sibs. Half sibs are individuals that have one parent in common and the other parent different. A group of half sibs is therefore the progeny of one individual mated at random and having one offspring by each mate. Thus the mean genotypic value of the group of half sibs is by definition half the breeding value of the common parent. The covariance is the variance of the means of the half-sib groups, and is therefore the variance of half the breeding values of the parents; this is a quarter of the additive variance:

$$cov_{(HS)} = V_{\frac{1}{2}A} = \tfrac{1}{4}V_A \qquad \ldots(9.2)$$

This covariance also can be demonstrated by the longer method, from the values already given in Table 9.1. The covariance is the variance of the means of the groups of offspring listed in the right-hand column. Squaring the offspring values and multiplying by their frequencies we get

Variance of means of half-sib families

$$= p^2q^2\alpha^2 + 2pq.\tfrac{1}{4}(q-p)^2\alpha^2 + q^2p^2\alpha^2$$
$$= pq\alpha^2[pq + \tfrac{1}{2}(q-p)^2 + pq]$$
$$= pq\alpha^2[\tfrac{1}{2}(p+q)^2]$$
$$= \tfrac{1}{2}pq\alpha^2$$

Therefore, since $2pq\alpha^2 = V_A$ (from equation 8.5),

$$cov_{(HS)} = \tfrac{1}{4}V_A$$

summation being made over all loci.

Offspring and mid-parent. The covariance of the mean of the offspring and the mean of both parents (commonly called the "mid-parent") may be deduced in the following way. Let O be the mean of the offspring, and P and P′ be the values of the two parents. Then we want to find $cov_{O\bar{P}}$; that is, the covariance of O with $\tfrac{1}{2}(P+P')$. This is equal to $\tfrac{1}{2}(cov_{OP} + cov_{OP'})$. If P and P′ have the same variance, then $cov_{OP} = cov_{OP'}$ and $cov_{O\bar{P}} = cov_{OP}$. Thus, provided the two sexes have equal variances, the covariance of offspring and mid-parent is the same as that of offspring with one parent, which we have seen is equal to half the additive variance. This conclusion may be extended to other sorts of relative: the covariance of any individual with the mean value of a number of relatives of the same sort is equal to its covariance with one of those relatives.

The longer method of demonstrating the covariance of offspring with mid-parent is rather laborious, but it must be given since it will

TABLE 9.2

Genotype of parents		Frequencies of matings	Mid-parent value	Progeny A_1A_1 a	Progeny A_1A_2 d	Progeny A_2A_2 $-a$	Mean value of progeny	Progeny mean × Mid-parent	Square of Progeny mean
A_1A_1	A_1A_1	p^4	a	1	—	—	a	a^2	a^2
A_1A_1	A_1A_2	$4p^3q$	$\frac{1}{2}(a+d)$	$\frac{1}{2}$	$\frac{1}{2}$	—	$\frac{1}{2}(a+d)$	$\frac{1}{4}(a^2+2ad+d^2)$	$\frac{1}{4}(a^2+2ad+d^2)$
A_1A_1	A_2A_2	$2p^2q^2$	0	—	1	—	d	0	d^2
A_1A_2	A_1A_2	$4p^2q^2$	d	$\frac{1}{4}$	$\frac{1}{2}$	$\frac{1}{4}$	$\frac{1}{2}d$	$\frac{1}{2}d^2$	$\frac{1}{4}d^2$
A_1A_2	A_2A_2	$4pq^3$	$\frac{1}{2}(-a+d)$	—	$\frac{1}{2}$	$\frac{1}{2}$	$\frac{1}{2}(-a+d)$	$\frac{1}{4}(a^2-2ad+d^2)$	$\frac{1}{4}(a^2-2ad+d^2)$
A_2A_2	A_2A_2	q^4	$-a$	—	—	1	$-a$	a^2	a^2

be needed for arriving at the covariance of full sibs. We shall, how-ever, omit some of the steps of algebraic reduction. A table (Table 9.2) is made in the same manner as for offspring and one parent, but now we have to tabulate types of mating and their frequencies, in-stead of single parents. This was done in Chapter 1 (Table 1.1). Against each type of mating we put the mean genotypic value of the two parents, i.e. the mid-parent value; then the genotypes of the pro-geny and the mean genotypic value of the progeny. The working is made easier by writing the genotypic values in terms of a and d instead of as deviations from the population mean. In the last two columns of the table we put the product of progeny-mean × mid-parent, and the square of the progeny for later use. Now, to get the covariance of progeny-mean and mid-parent value, we take the pro-duct of progeny-mean × mid-parent and multiply it by the frequency of the mating type, and then sum over mating types. This gives the mean product (M.P.) from which we have to deduct a correction for the population mean, since values are not here expressed as deviations from the mean. The correction is simply the square of the population mean (M^2) since the means of parents and of progeny are equal. Both the M.P. and M^2 contain terms in a^2, in ad, and in d^2. By col-lecting together these terms and simplifying a little we obtain

$$\text{M.P.} = a^2[p^3(p+q) + q^3(p+q)] + 2adpq(p^2 - q^2) + d^2pq(p^2 + 2pq + q^2)$$
$$M^2 = a^2(p^2 - 2pq + q^2) \qquad + 4adpq(p-q) \quad + 4d^2p^2q^2$$

Then, $cov_{O\bar{P}} = \text{M.P.} - M^2$

$$= a^2pq - 2adpq(p-q) + d^2pq(p-q)^2$$
$$= pq[a + d(q-p)]^2$$
$$= pq\alpha^2$$
$$= \tfrac{1}{2}V_A \qquad\qquad\qquad \text{.....(9.3)}$$

when summed over all loci.

So the genetic covariance of offspring with the mean of their parents is equal to half the additive genetic variance. That this covariance comes out the same as that of offspring and one parent need cause no surprise when we note that the variance of mid-parent values is half the variance of individual values (see below, p. 162).

Full sibs. The covariance of full sibs is the variance of the means of full-sib families, and is got with little additional work from Table 9.2. The last column shows the squares of progeny means and it will be seen that these squares are all exactly the same as the products of progeny-mean × mid-parent, except for the two entries in the middle involving terms in d^2. The mean square (M.S.) can therefore be got from the mean product (M.P.) already calculated, thus

$$M.S. = M.P. + d^2 . 2p^2q^2 - \tfrac{1}{4}d^2 . 4p^2q^2$$
$$= M.P. + d^2p^2q^2$$

The correction for the mean is the same as before, so we have

$$cov_{(FS)} = cov_{O\bar{P}} + d^2p^2q^2$$
$$= pq\alpha^2 + d^2p^2q^2$$

Since $2pq\alpha^2 = V_A$ (from equation *8.5*) and $4d^2p^2q^2 = V_D$ (from equation *8.6*) the covariance of full sibs is

$$cov_{(FS)} = \tfrac{1}{2}V_A + \tfrac{1}{4}V_D \qquad \qquad \ldots\ldots(9.4)$$

summing over all loci.

So the genetic covariance of full sibs is equal to half the additive genetic variance plus a quarter of the dominance variance. This is the only one of the relationships that we have considered where we find the dominance variance contributing to the resemblance. The reason is that full sibs have both parents in common, and a pair of full sibs have a quarter chance of having the same genotype for any locus.

Covariance due to epistatic interaction. Before we turn to the environmental causes of resemblance between relatives let us briefly examine the role of interaction variance arising from epistasis. In Chapter 8 we noted that the interaction variance, V_I, is subdivided into components according to the number of loci interacting, and according to whether the interaction is between breeding values or dominance deviations. The covariances of relatives, with the contributions of the two-factor interactions included, are shown in Table 9.3

TABLE 9.3

Covariances of relatives including the contributions of two-factor interactions.

Relatives		V_A	V_D	V_{AA}	V_{AD}	V_{DD}
		\multicolumn{5}{c}{}
Offspring-parent:	$cov_{OP} =$	$\tfrac{1}{2}$	—	$\tfrac{1}{4}$	—	—
Half sibs:	$cov_{(HS)} =$	$\tfrac{1}{4}$	—	$\tfrac{1}{16}$	—	—
Full sibs:	$cov_{(FS)} =$	$\tfrac{1}{2}$	$\tfrac{1}{4}$	$\tfrac{1}{4}$	$\tfrac{1}{8}$	$\tfrac{1}{16}$
General:	$cov =$	x	y	x^2	xy	y^2

The header for the Variance components column reads: *Variance components and the coefficients of their contributions* (for details see Kempthorne, 1955a, b). The offspring-parent covariance refers equally to one parent and to mid-parent values. For the sake of clarity the components of variance are shown at the heads of the columns and their coefficients in the covariances are listed below. For example, the offspring-parent covariance is $\frac{1}{2}V_A + \frac{1}{4}V_{AA}$. The contributions of interaction to the covariances are expressible in a simple general form, shown in the bottom line of the table. If the covariance contains xV_A then it contains also x^2V_{AA}; and if it contains yV_D it contains also xyV_{AD} and y^2V_{DD}. Interactions involving more than two loci contribute progressively smaller proportions as the number of loci increases. The effect of the interaction variance on the resemblance between relatives is, in principle, that the offspring-parent covariance is not twice the half-sib covariance, but a little more than twice; and that the excess of the full-sib covariance over the half-sib represents not only dominance variance but also some of the interaction variance.

When the interaction variance was first discussed in Chapter 8 we said we would regard it as a complication to be circumvented, noting only the consequences of neglecting it. These consequences are now apparent. First, only small fractions of it contribute to the covariances and therefore its effect on the resemblance between relatives is small unless the amount of interaction variance is large in comparison with the other components. And second, it appears that there is little we can do in practice except ignore it, because, apart from the special experimental methods mentioned on p. 139, there is no practicable means of separating the interaction from the other components. The consequences of ignoring the interaction variance are thus that any estimate of V_A made from offspring-parent regressions will contain also $\frac{1}{2}V_{AA} + \frac{1}{4}V_{AAA} +$ etc.; any estimate of V_A from half-sib correlations will contain also $\frac{1}{4}V_{AA} + \frac{1}{16}V_{AAA} +$ etc.; and any estimate of V_D obtained from a full-sib correlation will contain also portions of the interaction components. We noted in Chapter 7 that the two definitions of breeding value given there are not equivalent if there is interaction between loci. We can now see how this comes about. Defined in terms of the measured values of progeny—the practical definition—breeding value includes additive × additive interaction deviations in addition to the average effects of the genes carried by the parents; whereas, defined in terms of the average effects of genes—the theoretical definition—it does not.

Effect of linkage. Throughout the discussion of the covariances of relatives we have ignored the effects of linkage, assuming always that the loci concerned segregate independently. The effects of linkage in a random-mating population, where the coupling and repulsion phases are in equilibrium, are as follows (Cockerham, 1956*a*). The covariances of offspring and parents are not affected, but the covariances of half and full sibs are increased; the closer the linkage the greater the increase. The additional covariance due to linkage appears with the interaction component. Therefore what is formally attributed to epistatic interaction may be in part due to linkage.

ENVIRONMENTAL COVARIANCE

Genetic causes are not the only reasons for resemblance between relatives; there are also environmental circumstances that tend to make relatives resemble each other, some sorts of relatives more than others. If members of a family are reared together, as with human families or litters of pigs or mice, they share a common environment. This means that some environmental circumstances that cause differences between unrelated individuals are not a cause of difference between members of the same family. In other words there is a component of environmental variance that contributes to the variance between means of families but not to the variance within the families, and it therefore contributes to the covariance of the related individuals. This between-group environmental component, for which we shall use the symbol V_{Ec}, is usually called the *common environment*, a term that seems more appropriate when we think of the component as a cause of similarity between members of a group than when we think of it as a cause of difference between members of different groups. The remainder of the environmental variance, which we shall denote by V_{Ew}, arises from causes of difference that are unconnected with whether the individuals are related or not. It therefore appears in the within-group component of variance, but does not contribute to the between-group component, which is the variance of the true means of the groups. In considerations of the resemblance between relatives, therefore, the environmental variance must be divided into two components:

$$V_E = V_{Ec} + V_{Ew} \qquad \ldots\ldots(9.5)$$

one of the components, V_{Ec}, contributing to the covariance of the related individuals.

The sources of common environmental variance are many and varied, and only a few examples can be mentioned. Soil conditions may differentiate families of plants when the members of a family are grown together on the same plot: similarly the conditions of the culture medium may differentiate families of *Drosophila* or other small animals. With farm animals, related individuals are likely to have been reared on the same farm, and differences of climate or of management contribute to the resemblance between the relatives. "Maternal effects" are a frequent source of environmental difference between families, especially with mammals. The young are subject to a maternal environment during the first stages of their life, and this influences the phenotypic values of many metric characters even when measured on the adult, causing offspring of the same mother to resemble each other. Finally, members of the same family tend to be contemporaneous, and changes of climatic or nutritional conditions tend to differentiate members of different families. This source of common environmental variation is especially important in animals that produce their young in broods or litters.

These various sources of common environmental variation contribute chiefly to the resemblance between sibs, though some may also cause resemblance between parent and offspring. Maternal effects, in particular, often cause a resemblance between mother and offspring as well as among the offspring themselves. Body size in mice and other mammals provides an example. Large mothers tend to provide better nutrition for their young, both before and after birth, than small mothers. Therefore the young of large mothers tend to grow faster, and the effect of the rapid early growth may persist, so that when adult their body size is larger. Thus mothers and offspring tend to resemble each other in body size.

It will be seen from the examples given that the nature of the component of variance due to common environment differs according to the circumstances. What we designate as the V_{Ec} component depends on the way in which individuals are grouped when we estimate the observational components of phenotypic variance. Whatever the form of the analysis, the part of the variance between the means of groups that can be ascribed to environmental causes is called the V_{Ec} component. The nature of this component thus depends on the form of the analysis applied. If the groups in the analysis are full-sib families then the V_{Ec} component represents environmental causes of similarity between full sibs; if the groups are half sibs it represents causes of similarity between half sibs. And in parent-offspring relationships a comparable covariance term represents environmental causes of resemblance between offspring and parent. Thus, whenever we measure a phenotypic covariance with the object of using it to estimate a causal component of variance we have to decide whether it includes an appreciable component due to common environment, and this is often a matter of judgment based on a biological understanding of the organism and the character. In experiments, much of the V_{Ec} component can often be eliminated by suitable design. For example, members of the same family need not always be reared in the same vial, cage, or plot; they can be randomised over the rearing environments. Or, by replication, the V_{Ec} component can be measured and suitable allowance made for it in the resemblance between the relatives.

Thus relatives of all sorts may in principle be subject to an environmental source of resemblance. In what follows, however, we shall make the simplification of disregarding the V_{Ec} component for all relatives except full sibs, though from time to time we shall put in a reminder of its possible presence. Full sibs are subject to a common maternal environment and this is often the most troublesome source of environmental resemblance to overcome by experimental design. Consequently a V_{Ec} component contributes more often and in greater amount to the covariance of full sibs than to that of any other sort of relative. The simplification of disregarding all other sources of common environmental variance is therefore not entirely unrealistic.

PHENOTYPIC RESEMBLANCE

The covariance of phenotypic values is the sum of the covariances arising from genetic and from environmental causes. Thus by putting together the conclusions of the two preceding sections we arrive at the phenotypic covariances given in Table 9.4. (It will be remembered that some possible sources of environmental covariance are being neglected, particularly in offspring-parent relationships involving the mother.) In all these relationships except that of full sibs the covariance is either a half or a quarter of the additive genetic variance. By observing the phenotypic covariance of relatives we can thus estimate the amount of additive variance in the population and make the partition of the variance into additive versus the rest.

To arrive at the degree of resemblance expressed as a regression or correlation coefficient we have to divide the covariance by the appropriate variance. The resemblance between sibs is expressed as a correlation and the covariance is divided by the total phenotypic variance. The correlation between half sibs, for example, is therefore $\frac{1}{4}V_A/V_P$. The resemblance between offspring and parent is expressed

TABLE 9.4

PHENOTYPIC RESEMBLANCE BETWEEN RELATIVES

Relatives	Covariance	Regression (b) or correlation (t)
Offspring and one parent	$\frac{1}{2}V_A$	$b = \frac{1}{2}\dfrac{V_A}{V_P}$
Offspring and mid-parent	$\frac{1}{2}V_A$	$b = \dfrac{V_A}{V_P}$
Half sibs	$\frac{1}{4}V_A$	$t = \frac{1}{4}\dfrac{V_A}{V_P}$
Full sibs	$\frac{1}{2}V_A + \frac{1}{4}V_D + V_{Ec}$	$t = \dfrac{\frac{1}{2}V_A + \frac{1}{4}V_D + V_{Ec}}{V_P}$

as the regression of offspring on parent, and the covariance is therefore divided by the variance of parents. In the case of single parents this is again the phenotypic variance, and the regression of offspring on one parent is thus $\frac{1}{2}V_A/V_P$. In a random-breeding population the phenotypic variance of parents and offspring is the same, and then the correlation between offspring and one parent is the same as the regression. The case of mid-parent values, however, is a little different. The covariance has to be divided by the variance of mid-parent values, and this is half the phenotypic variance, for the following reason. Let X and Y stand for the phentoypic values of male and female parents respectively. Then $\sigma_X^2 = \sigma_Y^2 = V_P$. The mid-parent value is $\frac{1}{2}X + \frac{1}{2}Y$ and the variance of mid-parent values, assuming X and Y to be uncorrelated, is therefore $\sigma_{\frac{1}{2}X}^2 + \sigma_{\frac{1}{2}Y}^2 = 2\sigma_{\frac{1}{2}X}^2 = 2 \cdot \frac{1}{4}\sigma_X^2 = \frac{1}{2}V_P$. Thus the regression of offspring on mid-parent is $\frac{1}{2}V_A/\frac{1}{2}V_P = V_A/V_P$. The correlalation between offspring and mid-parent values, however, is $\frac{1}{2}V_A/\sigma_{\bar{P}}\sigma_O$, where $\sigma_{\bar{P}}$ and σ_O are the square roots of the phenotypic variances of mid-parents and offspring respectively, and this is not the same as the regression of offspring on mid-parent.

The regressions of offspring on parents and the correlations of sibs are shown in Table 9.4. All except the full-sib correlation are simple fractions of the ratio V_A/V_P. Thus the different degrees of resemblance between different sorts of relatives become apparent. For example, the regression of offspring on one parent is twice the correlation between half sibs, and the correlation between full sibs is twice the correlation between half sibs if there is no dominance and no common environment.

The difference between the full-sib covariance and twice the half-sib covariance can, in principle, be used to estimate the dominance variance, V_D, provided there is no variance due to common environment, though some of the variance due to epistatic interaction would be included, as may be seen from Table 9.3. In practice, however, it is usually very difficult to be certain that there is no variance due to common environment, and estimates of the dominance variance obtained in this way are generally to be regarded as upper limits rather than as precise estimates.

TABLE 9.5

THE RESEMBLANCE BETWEEN RELATIVES FOR SOME CHARACTERS IN MAN

| | | Correlation coefficient | |
| | | Parent- | |
Character	*Reference*	*offspring*	*Full sib*
Stature	(1)	·51	·53
Span	(1)	·45	·54
Length of forearm	(1)	·42	·48
Intelligence	(2)	·49	·49
Birth weight	(3)	—	·50

(1) Pearson and Lee (1903).
(2) Unweighted averages of several estimates, cited by Penrose (1949).
(3) Quoted from Robson (1955).

The chief use of measurements of the degree of resemblance between relatives is to estimate the proportionate amount of additive genetic variance, V_A/V_P, which is the heritability. The meaning of the heritability and the methods of estimating it will be considered more fully in the next chapter. To conclude this chapter we give in Table 9.5 some examples of correlations between relatives in man. These are undoubtedly complicated by covariance due to common environment, and also by assortative mating. The correlation between husband and wife for intelligence, for example, is as high as 0·58 (see Penrose, 1949). For these reasons human correlations cannot easily be used to partition the variation into its components.

CHAPTER 10

HERITABILITY

The heritability of a metric character is one of its most important properties. It expresses, as we have seen, the proportion of the total variance that is attributable to the average effects of genes, and this is what determines the degree of resemblance between relatives. But the most important function of the heritability in the genetic study of metric characters has not yet been mentioned, namely its predictive role, expressing the reliability of the phenotypic value as a guide to the breeding value. Only the phenotypic values of individuals can be directly measured, but it is the breeding value that determines their influence on the next generation. Therefore if the breeder or experimenter chooses individuals to be parents according to their phenotypic values, his success in changing the characteristics of the population can be predicted only from a knowledge of the degree of correspondence between phenotypic values and breeding values. This degree of correspondence is measured by the heritability, as the following considerations will show.

The heritability is defined as the ratio of additive genetic variance to phenotypic variance:

$$h^2 = \frac{V_A}{V_P} \qquad \ldots\ldots(10.1)$$

(The customary symbol h^2 stands for the heritability itself and not for its square. The symbol derives from Wright's (1921) terminology, where h stands for the corresponding ratio of standard deviations.) An equivalent meaning of the heritability is the regression of breeding value on phenotypic value:

$$h^2 = b_{AP} \qquad \ldots\ldots(10.2)$$

The equivalence of these meanings can be seen from reasoning similar to that by which we derived the genetic covariance of offspring and one parent on p. 153. If we split the phenotypic value into breeding value and a remainder (R) consisting of the environmental, dominance, and interaction deviations, then $P = A + R$. Since A and R are uncorrelated, $cov_{AP} = V_A$ and so $b_{AP} = V_A/V_P$.

We may note also that the correlation between breeding values and phenotypic values, r_{AP}, is equal to the square root of the heritability. This follows from the general relationship between correlation and regression coefficients, which gives

$$r_{AP} = b_{AP} \frac{\sigma_P}{\sigma_A}$$

$$= h^2 \frac{1}{h}$$

$$= h \qquad \qquad \ldots\ldots(10.3)$$

By regarding the heritability as the regression of breeding value on phenotypic value we see that the best estimate of an individual's breeding value is the product of its phenotypic value and the heritability:

$$A_{(\text{expected})} = h^2 P \qquad \qquad \ldots\ldots(10.4)$$

breeding values and phenotypic values both being reckoned as deviations from the population mean. In other words the heritability expresses the reliability of the phenotypic value as a guide to the breeding value, or the degree of correspondence between phenotypic value and breeding value. For this reason the heritability enters into almost every formula connected with breeding methods, and many practical decisions about procedure depend on its magnitude. These matters, however, will be considered in the next chapters; here we are concerned only to point out that the determination of the heritability is one of the first objectives in the genetic study of a metric character.

It is important to realise that the heritability is a property not only of a character but also of the population and of the environmental circumstance to which the individuals are subjected. Since the value of the heritability depends on the magnitude of all the components of variance, a change in any one of these will affect it. All the genetic components are influenced by gene frequencies and may therefore differ from one population to another, according to the past history of the population. In particular, small populations maintained long enough for an appreciable amount of fixation to have taken place are expected to show lower heritabilities than large populations. The environmental variance is dependent on the conditions of culture or management: more variable conditions reduce the heritability, more uniform conditions increase it. So, whenever a value is stated for the heritability of a given character it must be understood to refer to a particular population under particular conditions. Values found in other populations under other circumstances will be more or less the same according to whether the structure of the population and the environmental conditions are more or less alike.

Very many determinations of heritabilities have been made for a variety of characters, chiefly in farm animals. Some representative examples are given in Table 10.1. Different determinations of the heritability of the same character show a considerable range of variation. This is partly due to statistical sampling, but some of the variation reflects real differences between the populations or the conditions under which they are studied. For these reasons, and because estimations of heritabilities can seldom be very precise, the figures quoted in the table are rounded to the nearest 5 per cent. From Table 10.1 it can be seen that the magnitude of the heritability shows some connexion with the nature of the character. On the whole, the characters with the lowest heritabilities are those most closely connected with reproductive fitness, while the characters with the highest heritabilities are those that might be judged on biological grounds to be the least important as determinants of natural fitness. This is well seen in the gradation of the four characters of *Drosophila*.

TABLE 10.1

Approximate values of the heritability of various characters in domestic and laboratory animals.

Cattle

Amount of white spotting in Friesians (Briquet and Lush, 1947)	·95
Butterfat % (Johansson, 1950)	·6
Milk-yield (Johansson, 1950)	·3
Conception rate (in 1st service) (A. Robertson, 1957a)	·01

Pigs

Thickness of back fat (Fredeen and Jonsson, 1957)	·55
Body length (Fredeen and Jonsson, 1957)	·5
Weight at 180 days (Whatley, 1942)	·3
Litter size (Lush and Molln, 1942)	·15

(*Continued overleaf*)

Sheep (Australian Merino)

Length of wool (Morley, 1955)	·55
Weight of fleece (Morley, 1955)	·4
Body weight (Morley, 1955)	·35

Poultry (White Leghorn)

Egg weight (Lerner and Cruden, 1951)	·6
Age at laying of first egg (King and Henderson, 1954b)	·5
Egg-production (annual, of surviving birds) (King and Henderson, 1954b)	·3
Egg-production (annual, of all birds) (King and Henderson, 1954b)	·2
Body weight (Lerner and Cruden, 1951)	·2
Viability (Robertson and Lerner, 1949)	·1

Rats

Expression of hooded gene (amount of white) (from data of Castle and Wright, 1916)	·4
Ovary response to gonadotrophic hormone (Chapman, 1946)	·35
Age at puberty in females (Warren and Bogart, 1952)	·15

Mice

Tail length at 6 weeks (Falconer, 1954b)	·6
Body weight at 6 weeks (Falconer, 1953)	·35
Litter size (1st litters) (Falconer, 1955)	·15

Drosophila melanogaster

Abdominal bristle number (Clayton, Morris, and Robertson, 1957)	·5
Body size (thorax length) (F. W. Robertson, 1957b)	·4
Ovary size (F. W. Robertson, 1957a)	·3
Egg production (F. W. Robertson, 1957b)	·2

ESTIMATION OF HERITABILITY

Let us first compare the merits of the different sorts of relatives for estimating either the additive genetic variance from the covariance, or the heritability from the regression or correlation coefficient. Table 10.2 shows again the composition of the phenotypic covariances, and shows also the regression or correlation expressed in terms of the heritability. The choice depends on the circumstances. In addition

TABLE 10.2

Relatives	Covariance	Regression (b) or correlation (t)
Offspring and one parent	$\frac{1}{2}V_A$	$b = \frac{1}{2}h^2$
Offspring and mid-parent	$\frac{1}{2}V_A$	$b = h^2$
Half sibs	$\frac{1}{4}V_A$	$t = \frac{1}{4}h^2$
Full sibs	$\frac{1}{2}V_A + \frac{1}{4}V_D + V_{Ec}$	$t > \frac{1}{2}h^2$

to the practical matter of which sorts of relatives are in fact obtainable, there are two points to consider—sampling error and environmental sources of covariance. The statistical precision of the estimate depends on the experimental design and also on the magnitude of the heritability being estimated, and so no hard and fast rule can be made. The matter of statistical precision will be further considered in a later section of this chapter. The question of environmental sources of covariance is generally more important than the statistical precision of the estimate, because it may introduce a bias which cannot be overcome by statistical procedure. From considerations of the biology of the character and the experimental design we have to decide which covariance is least likely to be augmented by an environmental component, a matter already discussed in the last chapter. Generally speaking the half-sib correlation and the regression of offspring on father are the most reliable from this point of view. The regression of offspring on mother is sometimes liable to give too high an estimate on account of maternal effects, as it would, for example, with body size in most mammals. The full-sib correlation, which is the only relationship for which an environmental component of covariance is shown in the table, is the least reliable of all. The component due to common environment is often present in large amount and is difficult to overcome by experimental design; and the full-sib covariance is further augmented by the dominance variance. The full-sib correlation can therefore seldom do more than set an upper limit to the heritability.

EXAMPLE 10.1. The heritability of abdominal bristle number in *Drosophila melanogaster* has been determined by three different methods, applied to the same population (Clayton, Morris, and Robertson, 1957), with the following results:

Method of estimation	Heritability
Offspring-parent regression	·51 ± ·07
Half-sib correlation	·48 ± ·11
Full-sib correlation	·53 ± ·07
Combined estimate	·52

The estimates obtained by the three methods are in very satisfactory agreement. In this case, the character—bristle number—is free of complications arising from maternal effects and common environment.

Let us now consider briefly some technical matters concerning the translation of observational data into estimates of heritability. We shall deal first with the estimation of the heritability; and we shall later discuss the standard error of the estimate, and the design that gives an experiment its greatest precision.

Selection of parents and assortative mating. In the treatment of resemblance between relatives we have supposed the parents to be a random sample of their generation and to be mated at random. Quite often, however, one or other of these conditions does not hold, and the choice of which sort of relative to use in the estimation of heritability is then somewhat restricted. In experimental and domesticated populations the parents are often a selected group and consequently the phenotypic variance among the parents is less than that of the population as a whole and less than that of the offspring. The regression of offspring on parents, however, is not affected by the selection of parents because the covariance is reduced to the same extent as the the variance of the parents, so that the slope of the regression line is unaltered. Thus the regression of offspring on one parent is a valid measure of $\frac{1}{2}h^2$, and that of offspring on mid-parent is a valid measure of h^2. But the covariance is not a valid measure of V_A, nor the variance of parents of V_P; moreover, the correlation and regression coefficients are not equal.

Sometimes the mating of parents is not made at random but according to their phenotypic resemblance, a system known as assortative mating. There is then a correlation between the phenotypic values of the mated pairs. The consequences of assortative mating are described by Reeve (1955b) but they are too complicated to explain in detail here. They can be deduced by modification of Table 9.2, the frequencies of the different types of mating being altered according to the correlation between the mated pairs. The variance of mid-parent values is increased and consequently also the covariance of full sibs. The regression of offspring on mid-parent, however, is very little affected and it can be taken as a valid measure of h^2. The increased variance of mid-parent values under assortative mating has the practical advantage of reducing the sampling error of the regression coefficient and thus of the estimate of heritability.

Offspring-parent relationship. The estimation of heritability from the regression of offspring on parent is comparatively straightforward and needs little comment apart from the points mentioned in the preceding paragraphs. The data are obtained in the form of measurements of parents and the mean values of their offspring. The covariance is then computed in the usual way from the cross-products of the paired values. The mean values of offspring may be weighted according to the number of offspring in each family, if the numbers differ. The appropriate weighting is discussed by Kempthorne and Tandon (1953) and by Reeve (1955c).

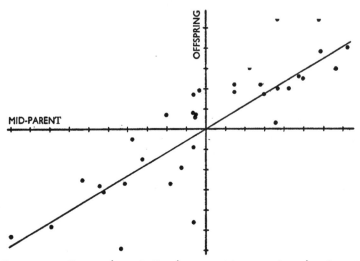

FIG. 10.1. Regression of offspring on mid-parent for wing-length in *Drosophila*, as explained in Example 10.2. Mid-parent values are shown along the horizontal axis, and mean value of offspring along the vertical axis. (Drawn from data kindly supplied by Dr E. C. R. Reeve.)

EXAMPLE 10.2. Fig. 10.1 illustrates the regression of offspring on mid-parent values for wing length in *Drosophila melanogaster* (Reeve and Robertson, 1953). There are 37 pairs of parents and a mean of 2·73 offspring were measured from each pair of parents. The parents were mated assortatively, with the result that the variance of mid-parent values is greater than it would be if mating had been at random. Each point on the graph represents the mean value of one pair of parents (measured along the horizontal axis), and the mean value of their offspring (measured along the vertical axis). The axes are marked at intervals of 1/100 mm., and they intersect at the mean value of all parents and all offspring. The sloping line is the linear regression of offspring on mid-parent. The slope of this line estimates the heritability, and has the value ($\pm$ standard error):

$$h^2 = b_{O\bar{P}} = 0.577 \pm 0.07$$

A complication in the use of the regression of offspring on mid-parent arises if the variance is not equal in the two sexes. We noted in the previous chapter that the genetic covariance of offspring and mid-parent is equal to half the additive variance on condition that the sexes are equal in variance. If this is not so, the regression on mid-parent cannot, strictly speaking, be used, and the heritability must be estimated separately for each sex from the regression of daughters on mothers and of sons on fathers. If the heritabilities are found to be equal in the two sexes, then a joint estimate can be made from the regression on mid-parent, by taking the mean value of the offspring as the unweighted mean of males and females.

Sib analysis. The estimation of heritability from half sibs is more complicated than appears at first sight and needs more detailed comment. A common form in which data are obtained with animals is the following. A number of males (sires) are each mated to several females (dams), and a number of offspring from each female are measured to provide the data. The individuals measured thus form a population of half-sib and full-sib families. An analysis of variance is then made by which the phenotypic variance is divided into observational components attributable to differences between the progeny of different males (the between-sire component, σ_S^2); to differences between the progeny of females mated to the same male (between-dam, within-sires, component, σ_D^2); and to differences between individual offspring of the same female (within-progenies component, σ_W^2). The form of the analysis is shown in Table 10.3. There are supposed to be s sires, each mated to d dams, which produce k offspring each. The values of the mean squares are denoted by MS_S, MS_D, and MS_W. The mean square within progenies is itself the estimate of the within-progeny variance component, σ_W^2; but the other mean squares are not the variance components. The compositions of the mean squares in terms of the observational components of variance are shown in the right-hand column of the table, consideration of which will show how the variance components are to be estimated. The between-dam mean square, for example, is made up of the within-progeny component together with k times the between-dam component; so the between-dam component is estimated as $\sigma_D^2 = (1/k)(MS_D - MS_W)$, i.e. we deduct the mean square for progenies from the mean square for dams and divide by the number of offspring per dam. Similarly the between-sire component is estimated as $\sigma_S^2 = (1/dk)(MS_S - MS_D)$, where dk is the number of off-

TABLE 10.3

FORM OF ANALYSIS OF HALF-SIB AND FULL-SIB FAMILIES

Source	d.f.	Mean Square	Composition of Mean Square
Between sires	$s-1$	MS_S	$= \sigma_W^2 + k\sigma_D^2 + dk\sigma_S^2$
Between dams (within sires)	$s(d-1)$	MS_D	$= \sigma_W^2 + k\sigma_D^2$
Within progenies	$sd(k-1)$	MS_W	$= \sigma_W^2$

s = number of sires
d = number of dams per sire
k = number of offspring per dam

spring per sire. If there are unequal numbers of offspring from the dams, or of dams in the sire groups, the exact solution, which is described by King and Henderson (1954a), Williams (1954), and Snedecor (1956, section 10.17) becomes too complicated for description here. We can, however, use the mean values of d and k with little error, provided the inequality of numbers is not very great.

The next step is to deduce the connexions between the observational components that have been estimated from the data and the causal components, in particular the additive genetic variance, the estimation of which is the main purpose of the analysis. Though all the information needed has already been given, the interpretation of the observational components, which is given in Table 10.4, is not immediately apparent without explanation. The first point to note is that the estimate of the phenotypic variance is given by the sum (σ_T^2) of the three observational components: $V_P = \sigma_T^2 = \sigma_S^2 + \sigma_D^2 + \sigma_W^2$. This is not necessarily equal to the observed variance as estimated from the total sum of squares, though the two seldom differ by much. Now consider the interpretation of the between-sire component,

σ_S^2. This is the variance between the means of half-sib families and it therefore estimates the phenotypic covariance of half sibs, $cov_{(HS)}$, which is $\frac{1}{4}V_A$. Thus $\sigma_S^2 = \frac{1}{4}V_A$. Next consider the within-progeny component, σ_W^2. Since any between-group variance component is equal to the covariance of the members of the groups, it follows that a within-group component is equal to the total variance minus the covariance of members of the groups. The progenies of the dams are

TABLE 10.4

Interpretation of the observational components of variance
in a sib analysis

Observational component		*Covariance and causal components estimated*	
Sires:	$\sigma_S^2 =$	$cov_{(HS)}$	$= \frac{1}{4}V_A$
Dams:	$\sigma_D^2 =$	$cov_{(FS)} - cov_{(HS)}$	$= \frac{1}{4}V_A + \frac{1}{4}V_D + V_{Ec}$
Progenies:	$\sigma_W^2 =$	$V_P - cov_{(FS)}$	$= \frac{1}{2}V_A + \frac{3}{4}V_D + V_{Ew}$
Total: $\sigma_T^2 = \sigma_S^2 + \sigma_D^2 + \sigma_W^2 =$		V_P	$= V_A + V_D + V_{Ec} + V_{Ew}$
Sires + Dams: $\sigma_S^2 + \sigma_D^2 =$		$cov_{(FS)}$	$= \frac{1}{2}V_A + \frac{1}{4}V_D + V_{Ec}$

full-sib families and so the within-progeny variance estimates $V_P - cov_{(FS)}$. This leads to the interpretation $\sigma_W^2 = \frac{1}{2}V_A + \frac{3}{4}V_D + V_{Ew}$. Finally, there remains the between-dam component, and what it estimates can be found by subtraction as follows:

$$\sigma_D^2 = \sigma_T^2 - \sigma_S^2 - \sigma_W^2 = cov_{(FS)} - cov_{(HS)} = \frac{1}{4}V_A + \frac{1}{4}V_D + V_{Ec}$$

Consideration of the between-sire and between-dam components will show that their sum gives an estimate of the full-sib covariance, $cov_{(FS)}$, but this provides no new information for estimating the causal components. These conclusions about the connexion between observational and causal components of variance are summarised in Table 10.4. The contributions of the interaction variance to the observational components is given by Kempthorne (1955a), and can be deduced from the contributions to the covariances given in Table 9.3.

EXAMPLE 10.3. As an illustration of the estimation of heritability from a sib analysis we refer to the study of Danish Landrace pigs based on the records of the Danish Pig Progeny Testing Stations (Fredeen and Jonsson, 1957). The data came from 468 sires each mated to 2 dams, the analysis being made on the records of 2 male and 2 female offspring from each dam. Only one such analysis is given here: that of body length in the male offspring. The analysis, shown in the table, was made within stations and within years, and this accounts for the degrees of freedom being fewer than would appear appropriate from the numbers stated above. The interpretation of the analysis, shown at the foot of the table, has been slightly

Sib analysis of body length in Danish Landrace pigs; data
for male offspring only (from Fredeen and Jonsson, 1957).

Source	d.f.	Mean Square	Component of variance	
Between sires	432	6·03	$\sigma_S^2 = \frac{1}{4}(6·03 - 3·81) = 0·555$	
Between dams, within sires	468	3·81	$\sigma_D^2 = \frac{1}{2}(3·81 - 2·87) = 0·47$	
Within progenies	936	2·87	$\sigma_W^2 =$	2·87
			$\sigma_T^2 =$	3·895

Interpretation of analysis

Sib correlations   Estimates of heritability

Half sibs: $t_{(HS)} = \dfrac{\sigma_S^2}{\sigma_T^2} = 0·142$ Sire-component: $h^2 = \dfrac{4\sigma_S^2}{\sigma_T^2} = 0·57$

Dam-component: $h^2 = \dfrac{4\sigma_D^2}{\sigma_T^2} = 0·48$

Full sibs: $t_{(FS)} = \dfrac{\sigma_S^2 + \sigma_D^2}{\sigma_T^2} = 0·263$ Sire + Dam: $h^2 = \dfrac{2(\sigma_S^2 + \sigma_D^2)}{\sigma_T^2} = 0·53$

simplified by the omission of some minor adjustments not relevant for us at this stage. The between-dam component is not greater than the between-sire component, so there cannot be much non-additive genetic variance or variance due to common environment. The two estimates of the heritability, from the sire and dam components respectively, can therefore be regarded as equally reliable, and their combination based on the resemblance between full sibs may be taken as the best estimate.

EXAMPLE 10.4. We have not yet had an example to illustrate the effect of common environment in augmenting the full-sib correlation. This is provided by body size in mice. The analysis given in table (i) refers to the

Table (i)

Source	d.f.	Mean Square	Composition of M.S.	Components
Sires	70	17·10	$\sigma_W^2 + k'\sigma_D^2 + dk'\sigma_S^2$	$\sigma_S^2 = 0·48$
Dams	118	10·79	$\sigma_W^2 + k\sigma_D^2 +$	$\sigma_D^2 = 2·47$
Progenies	527	2·19	σ_W^2	$\sigma_W^2 = 2·19$

$k = 3·48; \quad k' = 4·16; \quad d = 2·33$ $\sigma_T^2 = 5·14$

weight of female mice at 6 weeks of age (J. C. Bowman, unpublished). There were 719 offspring from 74 sires and 192 dams, each with one litter. These were spread over 4 generations and the analysis was made within generations. The analysis is complicated by the inequality of the number of offspring per dam and of dams per sire. We shall not attempt to explain the adjustments made for these inequalities, but simply give the compositions of the mean squares from which the components are estimated. The dam component is much greater than the sire component, indicating a substantial amount of variance due to common environment. Therefore only the sire component can be used to estimate the heritability. The estimate obtained is $h^2 = 4 \times 0.48/5.14 = 0.37$. Let us now use the analysis to estimate the causal components according to the interpretation given in Table 10.4, but with the assumption that non-additive genetic variance is negligible in amount. Table (ii) gives the estimates and shows how they

Table (ii)

$$V_P = \sigma_T^2 \qquad\qquad = 5.14 = 100\%$$
$$V_A = 4\sigma_S^2 \qquad\qquad = 1.92 = 37\%$$
$$V_{Ec} = \sigma_D^2 - \sigma_S^2 \quad = 1.99 = 39\%$$
$$V_{Ew} = \sigma_W^2 - 2\sigma_S^2 = 1.23 = 24\%$$

are derived. The percentage contribution of each component to the total variance is given in the right-hand column. It will be seen that the variance due to common environment (V_{Ec}) amounts to 39 per cent of the total, and is greater than the environmental variance within full-sib families (V_{Ew}) which amounts to only 24 per cent of the total.

Intra-sire regression of offspring on dam. The heritability can be estimated from the offspring-parent relationship in a population with the structure described in the foregoing section, but a slight modification is necessary. Since each male is mated to several females, the regression of offspring on mid-parent is inappropriate; and, since there are usually rather few male parents, the simple regressions on one or other parent are both unsuitable. The heritability can, however, be satisfactorily estimated from the average regression of offspring on dams, calculated within sire groups. That is to say, the regression of offspring on dam is calculated separately for each set of dams mated to one sire, and the regressions from each set pooled in a weighted average. This method is commonly used for the estimation of heritabilities in farm animals. The intra-sire regression of offspring on dam estimates half the heritability, as the following consideration will show. The progeny of one sire has a mean deviation from the population mean equal to half the breeding value of the sire, provided the females he is mated to are a random sample from the population. The progeny of one dam deviates from the mean of the sire group by half the breeding value of the dam. Therefore the within-sire covariance of offspring and dam is equal to half the additive variance of the population as a whole; and the within-sire regression of offspring on dam is equal to half the heritability, just like the simple regression of offspring on one parent. The validity of the estimate is, of course, dependent on the absence of maternal effects contributing to the resemblance between daughters and dams. Inequality of the variance of males and females calls for an adjustment if the heritability is to be estimated from the intra-sire regression of male offspring on dams. The regression coefficient should then be multiplied by the ratio of the phenotypic standard deviation of females to that of males.

EXAMPLE 10.5. The heritability of abdominal bristle-number in *Drosophila melanogaster*, estimated from the offspring-parent regression, was cited in Example 10.1. This was in fact a joint estimate based on intra-sire regressions of daughters on dams and of sons on dams, the latter being corrected for inequality of variance in the two sexes (Clayton, Morris, and Robertson, 1957). The separate regression coefficients, with the correction for inequality of variances, and the estimates of the heritability are given in the table.

		Estimate of heritability
Standard deviation: females	3·54	
Standard deviation: males	3·03	
Standard deviation: female/male	1·17	
Regression coefficient: daughter-dam	0·269	0·54
Regression coefficient: son-dam	0·206	
Regression coefficient: son-dam corrected		
0·206 × 1·17 =	0·241	0·48
Joint estimate, as given in Example 10.1, ·		0·51

THE PRECISION OF ESTIMATES OF HERITABILITY

It is of the greatest importance to know the precision of any estimate of heritability. When an estimate has been obtained one wants to be able to indicate its precision by the standard error. And when an experiment aimed at estimating a heritability is being planned one wants to choose the method and design the experiment so that the estimate will have the greatest possible precision within the limitations imposed by the scale of the experiment. The precision of an estimate depends on its sampling variance, the lower the sampling variance the greater the precision; and the standard error is the square root of the sampling variance. Estimates of heritability are derived from estimates of either a regression coefficient or an intra-class correlation coefficient, and the sampling variances of these are given in textbooks of statistics. We shall therefore present the necessary formulae without explanation of their derivation. The information on the design of experiments given here is derived from the paper by A. Robertson (1959a) on this subject.

The problems of experimental design are, first, the choice of method and, second, the decision of how many individuals in each family are to be measured. Since the total number of individuals measured cannot be increased indefinitely, an increase of the number of individuals per family necessarily entails a reduction of the number of families. The problem is therefore to find the best compromise between large families and many families. In assessing the relative efficiencies of different methods and designs we have to compare experiments made on the same scale; that is to say, with the same total expenditure in labour or cost. We must therefore decide first what are the circumstances that limit the scale of the experiment. If the labour of measurement is the limiting factor, as for example in experiments with *Drosophila*, then the limitation is in the total number of individuals measured, including the parents if they are measured. If, on the other hand, breeding and rearing space is the limiting factor, as it generally is with larger animals, the limitation may be either in the number of families or in the total number of offspring that can be produced for measurement, and measurements of the parents may be included without additional cost. We cannot here take account of all the possible ways in which the scale of the experiment may be limited. Therefore for the sake of illustration we shall consider only a limitation of the total number of individuals measured. That is to say, we shall assume the total number of individuals measured to be the same for all methods and all experimental designs. What we have to do, then, is to consider each method on this basis and see what design and which method will give an estimate of the heritability with the lowest sampling variance.

Offspring-parent regression. Consider first estimates based on the regression of offspring on parents. Let X be the independent variate, which may be either the value of a single parent or the mid-parent value. Let Y be the dependent variate, which may be either a single offspring of each parent or the mean of n offspring. Let σ_X^2 and σ_Y^2 be the variances of X and Y respectively; let b be the regression of Y on X, and N the number of paired observations of X and Y, which is equivalent to the number of families in the experiment. Let T be the total number of individuals measured, which is fixed by the scale of the experiment. The number of offspring measured is nN, and the number of parents N or $2N$ according to whether the regression is on one parent or on the mid-parent value. So, with one parent measured, $T = N(n+1)$, and with both parents measured $T = N(n+2)$. With these symbols, the variance of the estimate of the regression coefficient is

$$\sigma_b^2 = \frac{1}{N-2}\left[\frac{\sigma_Y^2}{\sigma_X^2} - b^2\right] \qquad \ldots\ldots(10.5)$$

For use as a guide to design this formula is more convenient if put in a simplified and approximate form. The regression coefficient is usually small enough that b^2 can be ignored; and we may suppose that N is fairly large, so that the variance of the estimate may be put, approximately, as

$$\sigma_b^2 = \frac{1}{N} \cdot \frac{\sigma_Y^2}{\sigma_X^2} \qquad \text{(approx.)} \quad \ldots\ldots(10.6)$$

When only one parent is measured the variance of parental values is equal to the phenotypic variance, i.e. $\sigma_X^2 = V_P$. When both parents are measured (provided they were not mated assortatively) the variance of mid-parent values is half the phenotypic variance, i.e. $\sigma_X^2 = \frac{1}{2}V_P$. The variance of the offspring values, σ_Y^2, is the variance of the means of families of n individuals. This depends on the phenotypic correlation, t, between members of families, in a manner that will be explained in Chapter 13, (see Table 13.2), where it will be shown that

$$\sigma_Y^2 = \frac{1 + (n-1)t}{n} V_P$$

Therefore by substitution for σ_X^2 and σ_Y^2 in equation *10.6* the sampling variance of the regression on one parent becomes

$$\sigma_b^2 = \frac{1 + (n-1)t}{nN} \quad \text{(approx.)} \qquad \ldots\ldots(10.7)$$

and that of the regression on mid-parent is twice as great. Since the phenotypic correlation, t, depends on the heritability it will not generally be known at the time an experiment is being planned. Therefore the best design cannot be exactly determined in advance. We can, however, get an approximate idea of how many offspring of each parent should be measured. On the assumption already stated, that the total number of individuals measured including the parents is fixed, it can be shown that the sampling variance given in equation 10.7 is minimal when $n = \sqrt{(1-t)/t}$ if one parent is measured and when $n = \sqrt{2(1-t)/t}$ if both parents are measured. Consider, for example, a character with a heritability of 20 per cent and no variance due to common environment, so that the phenotypic correlation in full-sib families is $t = 0.1$. Then the optimal family size works out to be $n = 3$ when only one parent is measured and $n = 4$ when both parents are measured. If we had taken a higher heritability the optimal family size would have been lower. Large families are advantageous only for the estimation of very low heritabilities. For example, full-sib families of about 10 or 14 would be optimal for estimating a heritability of 2 per cent.

So far we have considered only the sampling variance of the regression coefficient, and how this can be reduced by the design of the experiment. Now let us consider the sampling variance of the estimate of heritability, so that we can compare methods, i.e. the use of one parent or of mid-parent values. A just comparison can only be made on the assumption of the optimal design for each method, and therefore we can only illustrate the comparison by reference to a particular case. We shall consider the particular case mentioned above where the phenotypic correlation is $t = 0.1$, which would be found in full-sib families when the heritability is 20 per cent. The optimal family sizes are 3 or 4 as stated above. For the purpose of comparison we have to express the sampling variance of the regression coefficient given in equation 10.7 in terms of the total number of individuals measured, T, since this is assumed to be the same for all methods. We therefore substitute in equation 10.7 as follows. When one parent is measured $N = T/(n+1)$, and $n = 3$. When both parents are measured $N = T/(n+2)$, and $n = 4$. Substitution in equation 10.7 then yields $\sigma_b^2 = 4.8/3T$ when one parent is measured, and $\sigma_b^2 = 3.9/T$ when both are measured. The regression on one parent must be doubled to give the estimate of heritability, but the regression on mid-parent is itself the estimate. So the sampling variances of the estimates of heritability, in the special case under consideration, are:

By regression on one parent: $\sigma_{h^2}^2 = 4\sigma_b^2 = 6\cdot4/T$ (approx.)

By regression on mid-parent: $\sigma_{h^2}^2 = \sigma_b^2 = 3\cdot9/T$ (approx.)

Thus the estimate based on mid-parent values has considerably less sampling variance. A regression on mid-parent values, in general, yields a more precise estimate of heritability for a given total number of individuals measured.

Sib analyses. Now let us consider estimates obtained from the intra-class correlation of full-sib or half-sib families. We shall at first suppose for simplicity that half-sib families are not subdivided into full-sib families; i.e. that only one offspring from each dam is measured in paternal half-sib families. In the case of full-sib families we shall assume that there is no variance due to common environment so that the estimate of heritability is a valid one. Let N be the number of families, and n the number of individuals per family, so that the total number of individuals measured is $T=nN$. Let the intra-class correlation be t. The sampling variance of the intra-class correlation is then

$$\sigma_t^2 = \frac{2[1 + (n-1)t]^2(1-t)^2}{n(n-1)(N-1)} \qquad \ldots\ldots(10.8)$$

When the value of $T=nN$ is limited by the size of the experiment it can be shown that the sampling variance of the intra-class correlation is minimal when $n=1/t$, approximately. Therefore the optimal family size depends on the heritability. In the case of full-sib families $h^2=2t$, and in the case of half-sib families, $h^2=4t$. So the most efficient design has the following family sizes:

With full-sib families: $n = \dfrac{2}{h^2}$

With half-sib families: $n = \dfrac{4}{h^2}$

Since prior knowledge of the heritability will be at the best only approximate, the optimal family size cannot be exactly determined before-hand. The loss of efficiency, however, is much greater if the family size is below the optimum than if it is above. It is therefore better to err on the side of having too large families. A. Robertson (1959*a*) shows that, in the absence of prior knowledge of the heritability, half-sib analyses should generally be designed with families of between 20 and 30.

If the experiment has the most efficient design, with $n = 1/t$, then the sampling variance of the intra-class correlation is approximately

$$\sigma_t^2 = \frac{8t}{T} \qquad\qquad .....(10.9)$$

Therefore under optimal design the sampling variances of the estimates of heritability are as follows:

From full-sib families: $\sigma_{h^2}^2 = 4\sigma_t^2 = \dfrac{16h^2}{T}$ (approx.)

From half-sib families: $\sigma_{h^2}^2 = 16\sigma_t^2 = \dfrac{32h^2}{T}$ (approx.)

Thus, other things being equal, an estimate from full-sib families is twice as precise as one from half-sib families.

At this point let us compare the precision of estimates from sib analyses with those from offspring-parent regressions, assuming optimal design in each case. Again we have to choose a specific case for illustration of the comparison. Let us for simplicity suppose as we did before that the heritability to be estimated is 20 per cent. And, though perhaps not very representative of situations likely to arise in practice, let us compare an estimate obtained from a half-sib analysis with one obtained from the regression of offspring on one parent when the offspring consist of full-sib families. The variance of the estimate of heritability from the half-sib analysis would then be $6.4/T$ by substitution in the formula given above, and from the regression of offspring on one parent it would also be $6.4/T$ as we found previously. In this case, therefore, these two methods would give equally precise estimates for a given total number of individuals measured. If we had considered a higher heritability, then the regression method would have had the lower sampling variance. The comparison we have made, though referring to a particular case, illustrates the general conclusion, which is that the regression method is preferable for estimating moderately high heritabilities and the sib correlation method is preferable for low heritabilities, the critical heritability being, very roughly, about 20 per cent when the comparison is made on the basis of an equal total number of individuals measured.

Finally let us consider briefly a sib analysis where the half-sib families are subdivided into full-sib families. The situation is then more complicated, and for details the reader should consult the papers of Osborne and Paterson (1952) and A. Robertson (1959a). The conclusions are as follows. In many cases the estimation of heritability will be based only on the between-sire component, i.e. the half-sib correlation. This will arise when common environment renders the full-sib correlation unsuitable. The most efficient design then has only one offspring per dam, and is exactly the same as the half-sib analysis discussed above. If there is no common environment and it is desired to estimate the correlations from sire and from dam components with equal precision, then the optimal design has 3 or 4 dams per sire with the number of offspring per dam equal to $2/h^2$. In the absence of prior knowledge of the heritability the analysis should be planned with 3 or 4 dams per sire, and 10 offspring per dam.

IDENTICAL TWINS

Identical twins seem at first sight to provide, for man and cattle, a means of estimating the genotypic variance. They provide individuals of identical genotype, just as inbred lines, or crosses between lines, do for laboratory animals or for plants. The phenotypic variance within pairs of identical twins should, therefore, estimate the environmental variance and so allow the partition of the phenotypic variance into genotypic and environmental components to be made. (This would not estimate the heritability, but the use of identical twins seems nevertheless most appropriately discussed at this point.) Many studies of human twins have been made, and have shown the members of the pairs to be extremely alike in most characters, even when reared apart from childhood (see Stern, 1949, Ch. 23, for review and references). Studies of cattle twins, though on a much smaller scale, show the same thing (see Hancock, 1954; Brumby, 1958). Taken at their face value these studies seem to indicate a very high degree of genetic determination—up to 90 per cent or even more—for many characters. The use of identical twins in this way is, however, vitiated by the additional similarity due to common environment. Twins share a common environment from conception to birth, and over the period during which they are reared together, so that the within-pair variance contains only a part, and perhaps only a small part, of the total environmental variance. This difficulty may be partially overcome by the comparison of identical with fraternal twins. Fraternal twins are full sibs which share a common environment to approximately the same extent as identical twins. Let us therefore consider how the causal components of variance contribute to the observational components between pairs and within pairs for the two sorts of twins. The composition of the observational components are given in Table 10.5, the between-pair component being the phenotypic covariance. The environmental components are shown as being the same for fraternal as for identical twins. This is not necessarily true, but one can proceed only on the assumption that it is.

TABLE 10.5

Composition of the components of variance between and within pairs of twins.

	Between pairs	Within pairs
Identicals	$V_A + V_D + V_{Ec}$	V_{Ew}
Fraternals	$\frac{1}{2}V_A + \frac{1}{4}V_D + V_{Ec}$	$\frac{1}{2}V_A + \frac{3}{4}V_D + V_{Ew}$
Difference	$\frac{1}{2}V_A + \frac{3}{4}V_D$	$\frac{1}{2}V_A + \frac{3}{4}V_D$

The contributions of the interaction variance, which for simplicity are omitted, can be added from Table 9.3 (p. 157). If the environmental components are the same for the two sorts of twins, then the difference between identicals and fraternals in either of the two components estimates half the additive variance together with three-quarters of the dominance variance (and more than three-quarters of the interaction variance). To take the partitioning further it is necessary to have an estimate of the additive variance, reliably free from admixture with variance due to common environment. By subtraction of half the additive variance we may then obtain an estimate of three-quarters of the dominance variance together with more than three-quarters of the interaction variance. This would give at least an approximate idea of the amount of non-additive genetic variance. There is, however, a difficulty with cattle in comparisons between identical and fraternal twins, connected again with the environmental components of variance. Vascular anastomoses frequently occur in the placentae of both sorts of twins, so that the blood of the two twins is mixed. This will not make identicals any more alike, but it may make fraternals more alike than they would otherwise be.

Some results of twin-studies are quoted in Table 10.6, in order to illustrate the degree of resemblance between identical and between fraternal twins in both man and cattle. The difference between the correlation coefficients of identicals and fraternals, given in the right-hand column, could be taken as an estimate of half the heritability if there were no non-additive genetic variance and if there were no complications arising from a common circulation. But since non-additive variance cannot reasonably be assumed to be absent, the difference can only be regarded as setting an upper limit to half the heritability. The vascular anastomoses in cattle twins may, however, render the estimates of the heritability, or of its upper limit, too low.

TABLE 10.6

RESEMBLANCE BETWEEN TWINS

Character	Reference	Correlation coefficients		
		Identicals	Fraternals	Difference
Man				
Height	(1)	·93	·64	·29
Weight	(1)	·92	·63	·29
Intelligence	(1)	·88	·63	·25
Birth weight	(2)	·67	·58	·09
Cattle	(3)			
Milk-yield, 1st lactation		·91	·65	·26
Butterfat-yield, 1st lactation		·90	·51	·39
Fat % in milk, 1st lactation		·95	·86	·09
Weight at 96 weeks		·83	·78	·05
Body length at 96 weeks		·75	·62	·13

(1) Newman, Freeman, and Holzinger (1937). Based on 50 pairs of identicals and 50 pairs of fraternals, corrected for age differences.
(2) Quoted from Robson (1955).
(3) Brumby and Hancock (1956). Based on 10 pairs of identicals and 11 pairs of fraternals.

CHAPTER 11

SELECTION:

I. THE RESPONSE AND ITS PREDICTION

Up to this point in our treatment of metric characters we have been concerned with the description of the genetic properties of a population as it exists under random mating, with no influences tending to change its properties; now we have to consider the changes brought about by the action of breeder or experimenter. There are two ways, as we noted in Chapter 6, in which the action of the breeder can change the genetic properties of the population; the first by the choice of individuals to be used as parents, which constitutes selection, and the second by control of the way in which the parents are mated, which embraces inbreeding and cross breeding. We shall consider selection first, and in doing so we shall ignore the effects of inbreeding, even though we cannot realistically suppose that we are always dealing with a population large enough for its effects to be negligible.

The basic effect of selection is to change the array of gene frequencies in the manner described in Chapter 2. The changes of gene frequency themselves, however, are now almost completely hidden from us because we cannot deal with the individual loci concerned with a metric character. We therefore have to describe the effects of selection in a different manner, in terms of the observable properties —means, variances and covariances—though without losing sight of the fact that the underlying cause of the changes we describe is the change of gene frequencies. Before we come to details let us consider the change of gene frequencies a little further in general terms.

To describe the change of the genetic properties from one generation to the next we have to compare successive generations at the same point in the life cycle of the individuals, and this point is fixed by the age at which the character under study is measured. Most often the character is measured at about the age of sexual maturity or on the young adult individuals. The selection of parents is made after the measurements, and the gene frequencies among these selected individuals are different from what they were in the whole population before selection. If there are no differences of fertility among the selected individuals or of viability among their progeny, then the gene frequencies are the same in the offspring generation as in the selected parents. Thus artificial selection—that is, selection resulting from the action of the breeder in the choice of parents—produces its change of gene frequency by separating the adult individuals of the parent generation into two groups, the selected and the discarded, that differ in gene frequencies. Natural selection, operating through differences of fertility among the parent individuals or of viability among their progeny, may cause further changes of gene frequency between the parent individuals and the individuals on which measurements are made in the offspring generation. Thus there are three stages at which a change of gene frequency may result from selection: the first through artificial selection among the adults of the parent generation; the second through natural differences of fertility, also among the adults of the parent generation; and the third through natural differences of viability among the individuals of the offspring generation. Though natural differences of fertility and viability are always present they are not necessarily always relevant, because they are not necessarily connected with the genes concerned with the metric character.

RESPONSE TO SELECTION

The change produced by selection that chiefly interests us is the change of the population mean. This is the *response* to selection, which we shall symbolise by R; it is the difference of mean phenotypic value between the offspring of the selected parents and the whole of the parental generation before selection. The measure of the selection applied is the average superiority of the selected parents, which is called the *selection differential*, and will be symbolised by S. It is the mean phenotypic value of the individuals selected as parents expressed as a deviation from the population mean, that is from the mean phenotypic value of all the individuals in the parental generation before selection was made. To deduce the connexion between response and selection differential let us imagine two successive generations of a population mating at random, as represented diagrammatically in Fig. 11.1. Each point represents a pair of parents and their progeny, and is positioned according to the mid-parent value measured along the horizontal axis and the mean value of the

progeny measured along the vertical axis. The origin represents the
population mean, which is assumed to be the same in both generations.
The sloping line is the regression line of offspring on mid-parent.
(A diagram of this sort, plotted from actual data was given in Fig.
10.1.) Now let us regard a group of individuals in the parental
generation as having been selected—say those with the highest
values. These pairs of parents and their offspring are indicated by
solid dots in the figure. The parents have been selected on the basis

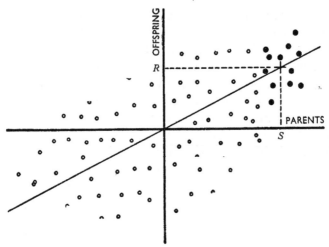

FIG. 11.1. Diagrammatic representation of the mean values of
progeny plotted against the mid-parent values, to illustrate the
response to selection, as explained in the text.

of their own phenotypic values, without regard to the values of their
progeny or of any other relatives. (This chapter deals exclusively
with selection made in this way: other methods will be described in
Chapter 13.) Let S be the mean phenotypic value of these selected
parents, expressed as a deviation from the population mean. And
similarly let R be the mean deviation of their offspring from the
population mean. Then S is the selection differential and R is the
response. The point marked by the cross represents the mean value
of the selected parents and of their progeny, and it lies on the regression line. The regression coefficient of offspring on parents is thus
equal to R/S. Therefore the connexion between response and selection
differential is

$$R = b_{0\bar{P}}S \qquad \qquad .....(11.1)$$

We saw in the last chapter that the regression of offspring on mid-parent is equal to the heritability, provided there is no non-genetic cause of resemblance between offspring and parents. To this we must add the further condition that there should be no natural selection: that is to say, that fertility and viability are not correlated with the phenotypic value of the character under study. Provided these conditions hold, therefore, the ratio of response to selection differential is equal to the heritability, and the response is given by

$$R = h^2 S \qquad \qquad \dots \dots (11.2)$$

The connexion between the response and the selection differential, expressed in equation *11.2*, follows directly from the meaning of the heritability. We noted in the last chapter (equation *10.2*) that the heritability is equivalent to the regression of an individual's breeding value on its phenotypic value. The deviation of the progeny from the population mean is, by definition, the breeding value of the parents, and so the response is equivalent to the breeding value of the parents. Thus it follows that the expected value of the progeny is given by $R = h^2 S$.

There is one point at which the situation envisaged in deducing the equations of response does not coincide with what is actually done in selection. We supposed the individuals of the parent generation to have mated at random and the selection to have been applied subsequently. In practice, however, the selection is usually made before mating, on the basis of the individuals' values and not the mid-parent values. The effect of this is that the individuals, when regarded as part of the whole parental population, have been mated assortatively. Assortative mating, however, has very little effect on the offspring-parent regression, as we noted in the last chapter, and this feature of selection procedure can therefore be disregarded.

Prediction of response. The chief use of these equations of response is for predicting the response to selection. Let us consider a little further the nature of the prediction that can be made. First, it is clear that equation *11.1* is not a prediction but simply a description, because the regression of offspring on parent cannot be measured until the offspring generation has been reared. We could, however, measure the regression, $b_{O\bar{P}}$, in a previous generation, and then use the equation $R = b_{O\bar{P}}S$ to predict the response to selection. There is no genetics involved in this; it is simply an extrapolation of direct observation, and the only conditions on which it depends are the absence of environmental change and the absence of genetic change between the generations from which the regression was estimated and the generation to which selection is applied. The equation $R=h^2S$, however, provides a means of prediction based on observations made only on the individuals of the parent generation before selection. Its validity rests on obtaining a reliable estimate of h^2 from the resemblance between relatives, such as half sibs; and on the truth of the identity $b_{O\bar{P}}=h^2$.

EXAMPLE 11.1. The selection for abdominal bristle number in *Drosophila melanogaster*, by Clayton, Morris, and Robertson (1957), will provide an illustration of the prediction of the response, and will serve also to indicate the extent of the agreement between observation and prediction. (The data for this example were kindly supplied by Dr G. A. Clayton.) The heritability of bristle number was first estimated from the base population before selection, and the value found was 0·52, as stated in Example 10.1. Five samples of 100 males and 100 females were taken from the base population, and selection for high and for low bristle number was made in each of the five samples, the 20 most extreme individuals of each sex being selected as parents. The mean deviations of these selected individuals from the mean of the sample out of which they were selected are given in the table in the columns headed *S*, the negative signs under downward selection being omitted. These are the selection differentials. The expected responses are obtained by multiplying the selection differentials by the heritability, according to equation *11.2*. The observed responses

| | *Upward selection* | | | *Downward selection* | | |
| | | *Response* | | | *Response* | |
Line	*S*	*Exp.*	*Obs.*	*S*	*Exp.*	*Obs.*
1	5·29	2·75	2·60	4·63	2·41	2·44
2	5·12	2·66	2·23	4·58	2·38	2·29
3	4·44	2·31	2·43	4·36	2·27	0·67
4	4·32	2·25	3·12	5·60	2·91	1·13
5	4·88	2·54	2·68	4·12	2·14	2·68
Mean	4·81	2·50	2·61	4·66	2·42	1·84

are the differences between the progeny means and the sample means out of which the parents were selected. The expected and observed responses are also given in the table, negative signs being again omitted. Comparison of the observed with the expected responses shows that on the whole there is fairly good agreement, though in some lines—particularly lines 3 and 4 selected downward—there are quite serious discrepancies. These discrepancies, which are typical of selection experiments, illustrate the fact that a single generation of selection in only one line cannot be relied on to follow the prediction at all closely.

The prediction of response is valid, in principle, for only one generation of selection. The response depends on the heritability of the character in the generation from which the parents are selected. The basic effect of the selection is to change the gene frequencies, so the genetic properties of the offspring generation, in particular the heritability, are not the same as in the parent generation. Since the changes of gene frequency are unknown we cannot strictly speaking predict the response to a second generation of selection without re-determining the heritability. Experiments have shown, however, that the response is usually maintained with little change over several generations—up to five, ten, or even more. This will be seen in the graphs of responses to selection given later in this chapter and in the next. In practice, therefore, the prediction may be expected to hold good over several generations. The effects of selection over longer periods, and also its effects on properties other than the mean, will be discussed in a later section.

The selection differential. We have seen that the change of the population mean brought about by selection—i.e. the response—depends on the heritability of the character and on the amount of selection applied as measured by the selection differential. The selection differential will not be known, however, until the selection among the parental generation has actually been made. So the equations of response in the form given above are only of limited usefulness for predicting the response. To be able to predict further ahead we need to know what determines the magnitude of the selection differential. Consideration of the factors that influence the selection differential will also enable us to see more clearly the means by which the breeder may improve the response to selection.

The magnitude of the selection differential depends on two factors: the proportion of the population included among the selected group, and the phenotypic standard deviation of the character. The dependence of the selection differential on these two factors is illustrated diagrammatically in Fig. 11.2. The graphs show the distribution of phenotypic values, which is assumed to be normal. The individuals with the highest values are supposed to be selected, so that the distribution is sharply divided at a point of truncation, all individuals above this value being selected and all below rejected.

The arrow in each figure marks the mean value of the selected group, and S is the selection differential. In graph (a) half the population is selected, and the selection differential is rather small: in graph (b) only 20 per cent of the population is selected, and the selection differential is much larger. In graph (c) 20 per cent is again selected, but

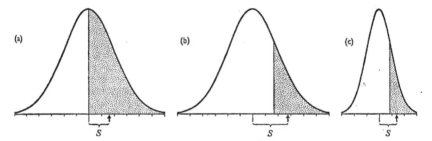

FIG. 11.2. Diagrams to show how the selection differential, S, depends on the proportion of the population selected, and on the variability of the character. All the individuals in the stippled areas, beyond the points of truncation, are selected. The axes are marked in hypothetical units of measurement.

(a) 50% selected; standard deviation 2 units: $S = 1.6$ units
(b) 20% selected; standard deviation 2 units: $S = 2.8$ units
(c) 20% selected; standard deviation 1 unit: $S = 1.4$ units

the character represented is less variable and the selection differential is consequently smaller. The standard deviation in (c) is half as great as in (b) and the selection differential is also half as great.

The standard deviation, which measures the variability, is a property of the character and the population, and it sets the units in which the response is expressed—i.e. so many pounds, millimetres, bristles, etc. The response to selection may be generalised if both response and selection differential are expressed in terms of the phenotypic standard deviation, σ_P. Then R/σ_P is a generalised measure of the response, by means of which we can compare different characters and different populations; and S/σ_P is a generalised measure of the selection differential, by means of which we can compare different methods or procedures for carrying out the selection. The "standardised" selection differential, S/σ_P, will be called the *intensity of selection*, symbolised by i. The equation of response (11.2) then becomes

$$\frac{R}{\sigma_P} = \frac{S}{\sigma_P} h^2$$

or $$R = i\sigma_P h^2 \qquad \ldots\ldots(\mathit{11.3})$$

By noting that $h = \sigma_A/\sigma_P$, where σ_A is the standard deviation of breeding values (square root of the additive genetic variance), we may write this equation in the form

$$R = ih\sigma_A \qquad \ldots\ldots(\mathit{11.4})$$

which is sometimes used in comparisons of different methods of selection.

The intensity of selection, i, depends only on the proportion of the population included in the selected group, and, provided the

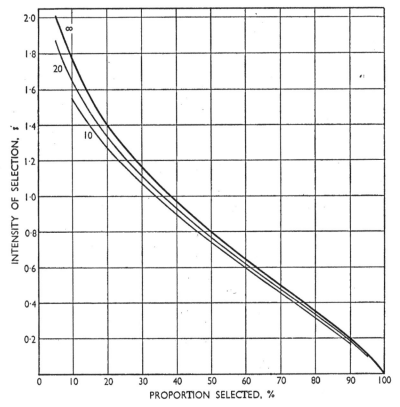

FIG. 11.3. Intensity of selection in relation to proportion selected. The intensity of selection is the mean deviation of the selected individuals, in units of phenotypic standard deviations. The upper graph refers to selection out of a large total number of individuals measured: the lower two graphs refer to selection out of totals of 20 and 10 individuals respectively.

distribution of phenotypic values is normal, it can be determined
from tables of the properties of the normal distribution. If p is the
proportion selected—i.e. the proportion of the population falling
beyond the point of truncation—and z is the height of the ordinate at
the point of truncation, then it follows from the mathematical
properties of the normal distribution that

$$\frac{S}{\sigma_P} = i = \frac{z}{p} \qquad \qquad .....(11.5)$$

Thus, given only the proportion selected, p, we can find out by how
many standard deviations the mean of the selected individuals will
exceed the mean of the population before selection: that is to say, the
intensity of selection, i. The graphs in Fig. 11.3 show the relationship between i and p; the value of i for any given value of p can be
read from the graphs with sufficient accuracy for most purposes. The
relationship between i and p given in equation 11.5 applies, strictly
speaking, only to a large sample: that is to say, when a large number of
individuals have been measured, among which the selection is to be
made. When selection is made out of a small number of measured
individuals, the mean deviation of the selected group is a little less.
The intensity of selection can be found from tables of deviations of
ranked data (Table XX of Fisher and Yates, 1943). The two lower

TABLE 11.1

Intensities of selection when selection is made out of a small
number of individuals measured. The figures in the table
are values of $i = S/\sigma_P$ = mean deviation in standard measure.

Number selected	Size of sample							
	9	8	7	6	5	4	3	2
1	1·49	1·42	1·35	1·27	1·16	1·03	0·85	0·56
2	1·21	1·14	1·06	0·96	0·83	0·67	0·42	—
3	1·00	0·91	0·82	0·70	0·55	0·34	—	—
4	0·82	0·72	0·62	0·48	0·29	—	—	—
5	0·66	0·55	0·42	0·25	—	—	—	—
6	0·50	0·38	0·23	—	—	—	—	—
7	0·35	0·20	—	—	—	—	—	—
8	0·19	—	—	—	—	—	—	—

curves in Fig. 11.3 show the intensity of selection for samples of 10
and 20. Selection intensities for samples smaller than 10 are given
in Table 11.1.

EXAMPLE 11.2. A comparison of the expected and observed responses under different intensities of selection was made by Clayton, Morris, and Robertson (1957), studying abdominal bristle number in *Drosophila*. The heritability was first determined by three methods which yielded a combined estimate of 0·52 (see Example 10.1). The standard deviation of bristle number (average of the two sexes) was 3·35. Selection at four different intensities was carried on for five generations, both upward and downward (i.e. both for increased and for decreased bristle number). In each case 20 males and 20 females were selected as parents, the intensity being varied by the number out of which these were selected, as shown in the first column of the table. The intensities of selection corresponding to these proportions selected may be read off the graphs in Fig. 11.3. They are given in the second column of the table. The expected responses are

| | | *Mean response per generation* | | |
| *Proportion* | *Intensity of* | *Exp-* | *Observed* | |
selected, p	*selection, i*	*ected*	*Up*	*Down*
20/100 = 0·20	1·40	2·44	2·62	1·48
20/75 = 0·267	1·23	2·14	2·20	1·26
20/50 = 0·40	0·97	1·65	1·46	0·79
20/25 = 0·80	0·34	0·59	0·28	− 0·08

then found from equation *11.3*. Under the most intense selection, for example, it is $R = 1·4 \times 3·35 \times 0·52 = 2·44$. There were five replicate lines in both directions under the most intense selection, and three replicates under the other intensities. The observed responses are quoted in the last two columns of the table. Although they do not agree very precisely with expectation, they show how the change made by selection falls off as the intensity of selection is reduced, and the data serve to illustrate the computation of the expected response.

It will now be clear that there are two methods open to the breeder for improving the rate of response to selection: one by increasing the heritability and the other by reducing the proportion selected and so increasing the intensity of selection. The heritability can be increased only by reducing the environmental variation through attention to the technique of rearing and management. Reducing the proportion selected seems at first sight to be a straightforward means of improving the response, but there are several factors to be considered which set a limit to what the breeder can do in this way. First is the matter of population size and inbreeding. This sets a lower limit to the number of individuals to be used as parents. In experimental work, for example, one might decide to use not less than 10 or even 20 pairs of parents; and in livestock improvement, particularly if artificial insemination came into general use as a means of intense selection on males, care would have to be taken not to restrict the number of males too much. For this reason the intensity of selection can be increased above a certain point only by increasing the total number of individuals measured, out of which the selection is made. With organisms that have a high reproductive rate, such as *Drosophila* and plants, very large numbers can, in principle, be measured; but in practice a limit is set to the intensity of selection by the time and labour required for the measurement. With organisms that have a low reproductive rate the limit to the intensity of selection is set by the reproductive rate, since the proportion saved can never be less than the proportion needed for replacement; that is to say, two individuals are needed on the average to replace each pair of parents. Usually fewer males are needed than females, because each male can mate with several females, and so the males leave more offspring than the females. A higher intensity of selection can then be made on males than on females. Suppose, for example, that females leave on the average 5 offspring, and each male mates with 10 females, so that males leave on the average 50 offspring. Then the proportion of females selected cannot be less than 1/5, but only 1/50 of the males need be selected. The upper limits of the intensity of selection in this case would be 1·40 for females, and 2·64 for males.

The number of offspring produced by a pair of parents depends not only on their reproductive rate but also on how long the breeder is willing to wait before he makes the selection. This introduces a new factor—the interval of time between generations—which we have not yet taken into account in the treatment of the response to selection, and which we must now consider.

Generation interval. The progress per unit of time is usually more important in practice than the progress per generation, so the interval between generations is an important factor in reckoning the response to selection. The generation interval is the interval of time between corresponding stages of the life cycle in successive generations, and it is most conveniently reckoned as the average age of the parents when the offspring are born that are destined to become parents in the next generation. By waiting until more offspring have been reared before he makes the selection the breeder can increase the intensity of selection and the response per generation; but in doing so he inevitably increases the generation interval and may thereby reduce the response per unit of time. There is thus a conflict of interest between intensity of selection and generation interval, and the best compromise must be found between the two. Increasing the number of offspring will pay up to a certain point, and beyond this point it will not. The optimal number of offspring cannot be stated in general terms, and each case must be worked out according to its special circumstances. The procedure is explained in the following example, referring to mice.

EXAMPLE 11.3. Let us suppose that selection is to be applied to some character in mice, and that speed of progress per unit of time is the aim. The question is: how many litters should be raised? To find the number of litters that will give the maximum speed of progress we have to find the intensity of selection and the generation interval. The ratio of the two will then give the relative speed. The actual speed could be obtained by multiplying by the heritability and the standard deviation, but these factors will be assumed to be independent of the number of litters raised. A comparison of the expected rates of progress per week is made in the table. The comparison is made for three different average sizes of litter, meaning the number of young reared per litter. It is assumed that the character to be selected can be measured before sexual maturity, and that first litters are born when the parents are 9 weeks old, subsequent litters following at intervals of 4 weeks. It is assumed also that the population is large enough to be treated as a large sample in reckoning the intensity of selection; and that equal numbers of males and females are selected. The optimal number of litters differs according to the number reared per litter. If 6

		$N=6$			$N=4$			$N=2$		
L	t	p	i	i/t	p	i	i/t	p	i	i/t
1	9	·333	1·10	·122	·50	0·80	·089	1·0	0·00	·000
2	13	·167	1·50	·115	·25	1·27	·098	·50	0·80	·062
3	17	·111	1·71	·101	·167	1·50	·088	·333	1·10	·065
4	21	·083	1·85	·088	·125	1·65	·079	·25	1·27	·060

Column headings: L = number of litters raised.
t = generation interval in weeks.
p = proportion selected.
i = intensity of selection.
i/t = relative speed of progress.
N = number of young reared per litter.

young are reared the maximum speed is attained by rearing only one litter. If 4 young are reared it is worth while to wait for second litters before making the selection, but not for third litters. If only 2 young are reared per litter, raising three litters gives the maximum speed of progress.

Most mouse stocks are able to rear 6 young per litter, so under most circumstances it is best to make the selection from the first litters, and not to wait for second litters. This conclusion could hardly have been guessed at without the computations shown in the table.

MEASUREMENT OF RESPONSE

When one or more generations of selection have been made the measurement of the response actually obtained introduces several problems. These are matters of procedure rather than of principle and will be only briefly discussed.

Variability of generation means. The first problem to be solved arises from the variability of generation means. Inspection of any of the graphs of selection given in the examples shows that the generation means do not progress in a simple regular fashion, but fluctuate erratically and more or less violently. There are two main causes of this variation between the generation means: sampling variation, depending on the number of individuals measured; and environmental change, which is usually the more important of the two. The consequence of this variation between generation means is that the response can seldom be measured with any pretence of accuracy until several generations of selection have been made. The best measure of the average response per generation is then obtained from the slope of a regression line fitted to the generation means, the assumption being made that the true response is constant over the period. The variation between generation means appears as error variation about the regression line, and the standard error of the estimate of response is based on it. Variation due to changes of environment can, of course, be overcome, or at least reduced, by the use of a control population. The measurement of the response can, however, be improved in accuracy if the "control" is not an unselected population but is selected in the opposite direction. This is known as a "two-way" selection experiment. The response measured from the divergence of the two lines is then about twice as great as that of the lines separately, and the variation between generations is reduced to the extent that the environmental changes affect both lines alike. An unselected control is, however, preferable if for practical reasons one is interested only in the change in one direction, because the response is not always equal in the two directions. This point will be discussed in the next chapter.

EXAMPLE 11.4. Fig. 11.4 shows the results of 11 generations of two-way selection for body weight in mice (Falconer, 1953). On the left the "up" and "down" lines are shown separately, and on the right the divergence between the two is shown. Linear regression lines are fitted to the observed

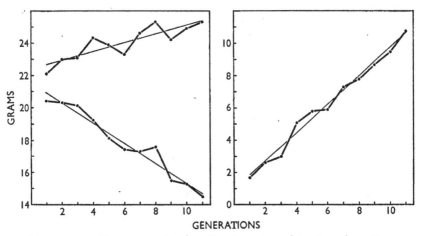

FIG. 11.4. Two-way selection for 6-week weight in mice. Explanation in Example 11.4. (Redrawn from Falconer, 1953.)

generation means. (The first generation of selection is disregarded because the method of selection was different.) The estimates of the average response per generation, with their standard errors, are as follows:

*Response ± standard error
in grams per generation.*

Up	0·27 ± 0·050
Down	0·62 ± 0·046
Divergence	0·88 ± 0·036

The difference between the upward and downward responses will be discussed in the next chapter.

The foregoing example shows how the variation of the generation means can be reduced when the response is measured from the difference between two lines, each acting in the manner of a control for the other. Controls, however, are not always available, and then a more serious difficulty may arise from progressive changes of environment. This makes it difficult to assess the effectiveness of selection in the improvement of domesticated animals, and to a lesser extent of plants, because in the absence of a control there is no sure way of deciding how much of the improvement is due to selection and how much to a progressive change in the conditions of management.

EXAMPLE 11.5. Lush (1950) has assembled a number of graphs showing the improvement of farm animals that has taken place during the present century. Instead of reproducing any of these graphs we give in the table an indication of the increase of yield per individual over a period of years, as a percentage of the initial yield. It is difficult to avoid the conclusion that much of the improvement of these characters is the result of selection, but in the absence of any standard of comparison it is very difficult to decide how much is due to selection and how much to improved methods of feeding and management.

	Character	Country	Period	Improvement, %
Cows:	Milk-yield	Sweden	1920–1944	21
	Butterfat-yield	New Zealand	1910–1940	47
	Fat % in milk	Netherlands	1906–1945	22
Pigs:	Efficiency of growth	Denmark	1922–1949	16
	Body length	Denmark	1926–1949	5
Sheep:	Fleece weight	Australia	1881–1945	71
Hens:	Egg production	U.S.A.	1909–1950	64

Weighting the selection differential. In experimental selection the selection differential as well as the response has to be measured because it is the relationship between the two, and not the response alone, that is of interest from the genetic point of view. We have to distinguish between the expected and the effective selection differential, because in practice the individual parents do not contribute equally to the offspring generation. Differences of fertility are always present so that some parents contribute more offspring than others. To obtain a measure of the selection differential that is relevant to the response observed in the mean of the offspring generation we therefore have to weight the deviations of the parents according to the number of their offspring that are measured. The expected selection differential is the simple mean phenotypic deviation of the parents as defined at the beginning of this chapter; the effective selection differential is the weighted mean deviation of the parents, the weight given to each parent, or pair of parents, being their proportionate contribution to the individuals that are measured in the next generation.

The weighting of the selection differential takes account of a good part of the effects of natural selection. If the differences of fertility

re related to the parents' phenotypic values for the character being
elected, then this natural selection will either help or hinder the
rtificial selection. If, for example, the more extreme phenotypes are
ess fertile or more frequently sterile, then natural selection is working
gainst artificial selection. By weighting the selection differential we
easure the joint effects of natural and artificial selection together.
comparison of the effective (i.e. weighted) with the expected selec-
ion differential may thus be used to discover whether natural selec-
ion is operative.

EXAMPLE 11.6. In an experiment with mice, selection for body size
(weight at 6 weeks) was carried through 30 generations in the upward
direction and 24 generations in the downward direction (see Falconer,
1955). Comparisons are made in the table between the effective (weighted)
and the expected (unweighted) selection differentials in the two lines. The
period of selection is divided into two parts and the comparisons are made
separately in each. Throughout the whole of the upward selection there
was virtually no difference between the effective and expected selection
differential, and we can conclude that natural selection was unimportant
as a factor influencing the response. The situation in the downward
selected line, however, is different, the effective selection differential being
less than the expected, especially in the second part. From this we can
conclude that natural selection was operating in favour of large size, thus
hindering the artificial selection and reducing the response obtained,
particularly in the latter part of the experiment. The cause of the natural
selection and the reason why it operated only in the downward selected
line were as follows. Large mice produce larger litters than small mice; but
for the purpose of standardisation, litters were artificially reduced to 8
young at birth. At the beginning, and throughout the whole period in the
upward selected line, there were few litters with less than 8 young, and so

Direction of selection	Generation numbers	Selection differential per generation (gms.)		
		Expected	Effective	$\dfrac{Effective}{Expected}$
Upwards	1–22	1·39	1·36	0·98
	23–30	1·08	1·09	1·01
Downwards	1–18	1·03	0·96	0·93
	19–24	0·82	0·70	0·86

the differential fertility had no consequence in the upward selected line.
In the downward selected line, however, there was soon no standardisation
because there were few litters with as many as 8 young. Thus the smaller

mice produced fewer young and this reduced the effective selection differential. In the second part of the experiment the smallest mice did not breed at all and this reduced the effective selection differential still further.

The weighting of the selection differential does not take account of the whole effect of natural selection. We noted at the beginning of the chapter that natural selection may operate at two stages, through differences of fertility among the parents and through differences of viability among the offspring. The effect of differences of viability among the offspring are not accounted for in the effective selection differential. For further examples and a fuller account of the interaction of natural and artificial selection see Lerner (1954, 1958).

Realised heritability. The equation of response, $R = h^2 S$ (*11.2*), which we discussed earlier from the point of view of predicting the response, can be looked at the other way round, as a means of estimating the heritability from the result of selection already carried out, the heritability being estimated as the ratio of response to selection differential:

$$h^2 = \frac{R}{S} \qquad\qquad\qquad .....(11.5)$$

The same conditions are necessary for the valid use of the equation for estimating heritability as for predicting response, except that now by weighting the selection differential a good part of the effects of natural selection can be taken account of. There is also the condition that the observed response should not be confounded with systematic changes of generation mean due to the environment or the effects of inbreeding. This, and the absence of maternal effects, are the important conditions for the valid estimation of heritability from the response to selection.

The ratio of response to selection differential, however, has an intrinsic interest of its own, quite apart from whether it provides a valid estimate of the heritability. It provides the most useful empirical description of the effectiveness of selection, which allows comparison of different experiments to be made even when the intensity of selection is not the same. The term *realised heritability* will be used to denote the ratio R/S, irrespective of its validity as a measure of the true heritability. The realised heritability is estimated as follows. The generation means are plotted against the cumulated selection differential. That is to say, the selection differentials, appropriately weighted, are summed over successive generations so as to give the total selection applied up to the generation in question. A regression line is then fitted to the points and the slope of this line measures the average value of R/S, the realised heritability.

EXAMPLE 11.7. Fig. 11.5 shows the results of 21 and 18 generations of two-way selection for 6-week weight in mice (Falconer, 1954 *a*). The

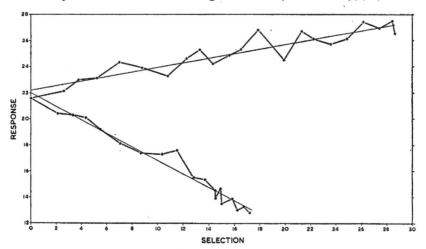

FIG. 11.5. Two-way selection for 6-week weight in mice. Response plotted against cumulated selection differential, as explained in Example 11.7. (From Falconer, 1954*a*; reproduced by courtesy of the editor of the *International Union of Biological Sciences*.)

generation means are plotted against the cumulated selection differential and linear regression lines are fitted to the points. The realised heritabilities, estimated from the slopes of these lines, are:

Upward selection: 0·175 ± 0·0161
Downward selection: 0·518 ± 0·0231

The difference between the upward and downward selection is referred to in the next chapter.

CHANGE OF GENE FREQUENCY UNDER ARTIFICIAL SELECTION

It was pointed out at the beginning of this chapter that the change of the population mean resulting from selection is brought about through changes of the gene frequencies at the loci which influence the character selected. But since the effects of the loci cannot be individually identified, the changes of gene frequency cannot in practice be followed. Consequently the process of selection for a metric character had to be described in terms of the selection differential, or the intensity of selection, and of the change of the population mean, representing the combined effects of all the loci. This leaves unanswered the fundamental question: How great are the changes of gene frequency underlying the response of a metric character to selection? To answer this question, and so to bridge the gap between the treatment of selection given in this chapter and that given earlier in Chapter 2, we have to find the connexion between the intensity of selection (i) and the coefficient of selection (s) operating on a particular locus.

The effect of selection for a metric character on one of the loci concerned may best be pictured in the manner illustrated in Fig. 11.6. This refers to a locus with two alleles of which one (A_1) is com-

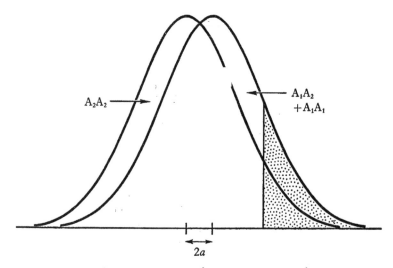

FIG. 11.6. Selection for a metric character operating on one of the loci concerned. The frequency of A_2A_2 as depicted is $q^2 = \frac{1}{2}$.

pletely dominant. With respect to this locus, therefore, the population is divided into two portions which differ in their mean phenotypic values by an amount $2a$, this being the difference between the two homozygotes in the notation of earlier chapters (see Fig. 7.1, p. 113). It is assumed that the residual variance within each portion is the same, this residual variance arising from all the other loci as well as from environmental causes. The proportion of individuals in the two portions depends on the gene frequency at the locus, q^2 being in the portion consisting of A_2A_2 genotypes, and $1-q^2$ in the portion containing A_1A_1 and A_1A_2 genotypes. When artificial selection is applied, a proportion of the whole population lying beyond the point of truncation is cut off, and the proportion of A_2A_2 genotypes is lower among this selected group than in the population as a whole, selection acting in the case illustrated against the A_2 allele. Now, the new gene frequency, q_1, is the frequency of A_2 genes among the selected group of individuals. This may be found by deducing the regression of gene frequency on phenotypic value, b_{qP}. The selected group deviates in mean phenotypic value from the population mean by an amount S, which is the selection differential. The gene frequency among the selected group will then be given by the regression equation

$$q_1 = q + b_{qP}S \qquad \dots\dots(11.6)$$

The regression of gene frequency on phenotypic value is found as follows. The three genotypes are listed in Table 11.2 with their

TABLE 11.2

		q	$G=P$
A_1A_1	p^2	0	a
A_1A_2	$2pq$	$\frac{1}{2}$	d
A_2A_2	q^2	1	$-a$

frequencies in the whole population. The third column of the table gives the frequency of the A_2 allele among each of the three genotypes, which is simply 0, $\frac{1}{2}$, and 1. The last column gives the genotypic values. Provided there is no correlation between genotype and environment, these are also the mean phenotypic values of each genotype. There is now no assumption of complete dominance. The covariance of gene frequency with phenotypic value is obtained from the sum of the products of q and P, each multiplied by the frequency of the genotype. From this sum of products must be deducted the product of the means of the gene frequency and the phenotypic value. Thus the covariance is $cov_{qP} = pqd - q^2a - qM$, where M is the population mean. Substituting the value of M from equation 7.2, the covariance reduces to $-pq[a+d(q-p)] = -pq\alpha$, where α is the average effect of the gene substitution (see equation 7.5). The regression of gene frequency on phenotypic value is therefore

$$b_{qP} = -\frac{pq\alpha}{\sigma_P^2} \qquad\qquad .....(11.7)$$

where σ_P^2 is the phenotypic variance.

Next, we substitute this regression coefficient in equation 11.6, putting also $S = i\sigma_P$ from equation 11.5. This gives the gene frequency among the selected parents as

$$q_1 = q - \frac{pq\alpha}{\sigma_P'^2} i\sigma_P$$

and the change of gene frequency resulting from the selection reduces to

$$\Delta q = -ipq\frac{\alpha}{\sigma_P} \qquad\qquad .....(11.8)$$

The change is negative because selection is acting against the allele A_2 whose frequency is q. This formula enables us to translate the intensity of selection, i, into the coefficient of selection, s, against A_2, because equations for the change of gene frequency in terms of s were given in Chapter 2. We shall take the approximate equations given in 2.7 and 2.8. If dominance is complete, $d = a$ and $\alpha = 2qa$. Then equating 11.8 with 2.8 gives

$$ipq\frac{2qa}{\sigma_P} = sq^2(1-q).$$

If there is no dominance $d = 0$ and $\alpha = a$. Then equating 11.8 with 2.7 gives

$$ipq\frac{a}{\sigma_P} = \tfrac{1}{2}sq(1-q)$$

Both these equations, on simplification, reduce to

$$s = i\frac{2a}{\sigma_P} \qquad\qquad .....(11.9)$$

Thus we find that the two ways of expressing the "force" of selection —by the intensity and the coefficient of selection—are very simply related to each other. The coefficient of selection operating on any locus is directly proportional to the intensity of selection and to the quantity $2a/\sigma_P$. This quantity is the difference of value between the two homozygotes expressed in terms of the phenotypic standard deviation.

11.7)
For want of a more suitable term we shall refer to this, rather loosely, as the "proportionate effect" of the locus. There is nothing more that we can do with the relationship expressed in equation *11.9* at the moment, but we shall use it in the next chapter to draw some tentative conclusions about the "proportionate effects" of loci concerned with metric characters.

ency

ction

'11.8)

allele
e the
of A.

CHAPTER 12

SELECTION:

II. The Results of Experiments

In the last chapter we saw that the theoretical deductions about the effects of artificial selection are limited to the change of the population mean, and strictly speaking over only one generation. By changing the gene frequencies selection changes the genetic properties of the population upon which the effects of further selection depend. And, because the effects of the individual loci are unknown, the changes of gene frequency cannot be predicted, and so the response to selection can be predicted only for as long as the genetic properties remain substantially unchanged. Thus there are many consequences of selection that can be discovered only by experiment. The object of this chapter is to describe briefly what seem to be the most general conclusions about these consequences that have emerged from experimental studies of selection. It should be noted, however, that the drawing of conclusions from the results of experiments in the field of quantitative genetics is to some extent a matter of personal judgement. Many of the conclusions put forward in this chapter therefore represent a personal viewpoint, and are not necessarily accepted generally. The most important questions to be answered by experiment concern the long-term effects of selection. For how long does the response continue? By how much can the population mean ultimately be changed? What is the genetic nature of the limit to further progress? These questions will be dealt with in the latter part of the chapter. First we shall consider two questions raised by the examples in the last chapter.

Repeatability of Response

In Example 11.1 we saw that the response in one generation of selection was very variable when the selection was replicated in a number of lines. Though the average response agreed fairly well with the prediction, the responses of the individual lines did not. This raises the question: How consistent, or repeatable, are the results of selection? If selection is applied to different samples drawn from the same population, how closely will the results agree? Part of the problem here concerns sampling variation—the extent to which the samples differ in gene frequencies, both initially and during the course of the continued selection. This depends, of course, on the size of the populations, or lines, during the course of the selection; but it depends also on the initial gene frequencies in the base population from which the samples were drawn. If most of the loci concerned with the character have genes at more or less intermediate frequencies then the response to selection is not likely to be much influenced by sampling variation. On the other hand, if there are loci with genes at low frequency then these will be included in some samples drawn from the initial population but will be absent from others. Then, if any of these low-frequency genes have a fairly large effect on the character their presence or absence may appreciably influence the outcome of selection. The experiment on abdominal bristle-number in *Drosophila* whose first generation was quoted in Example 11.1, provides the only evidence on this point (Clayton, Morris, and Robertson, 1957). Fig. 12.1 shows the responses in the five up and the five down lines over 20 generations. The responses are reasonably consistent over the first 5 generations in the up lines and over about 10 generations in the down lines. Thereafter the lines begin to differentiate, and by the twentieth generation there are substantial differences between them. The conclusion suggested by the early similarity and the later divergence between the replicate lines is that the early response is governed chiefly by genes at more or less intermediate frequencies, but in the later stages genes at initially low frequencies begin to come into play, the initial sampling having caused differences between the lines in respect of these genes.

The question of repeatability of the response to selection may be extended to differences between populations. This is not a matter of sampling variation but of the differences in the genetic properties of populations. We noted in Chapter 10 that heritabilities frequently differ between populations, and consequently we should not expect the responses to selection to be the same. It is of interest nevertheless to compare the results of selection applied to different populations and to see how they do actually differ. Fig. 12.2 shows the results of selection for thorax length in *Drosophila melanogaster* applied to three

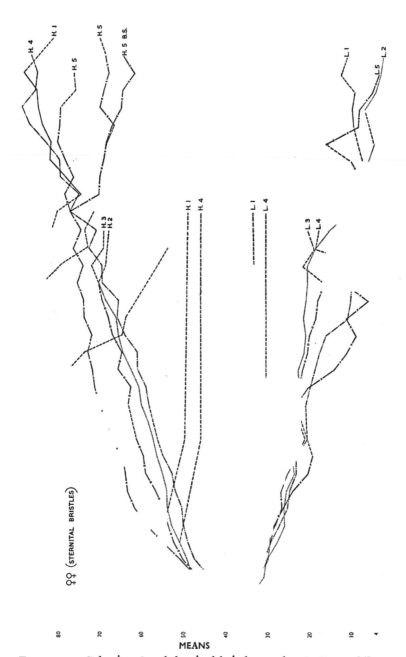

QQ (STERNITAL BRISTLES)

MEANS

FIG. 12.1. Selection for abdominal bristle number in *Drosophila melanogaster*, replicated in 5 lines in each direction. The broken lines refer to suspended selection and the thin continuous lines to inbreeding without selection. (From Clayton, Morris, and Robertson, 1957; reproduced by courtesy of the authors and the editor of the *Journal of Genetics*.)

MASS SELECTION ——— THORAX

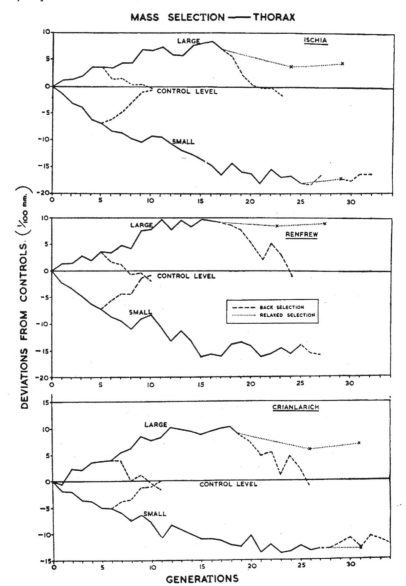

FIG. 12.2.   Selection for thorax length in *Drosophila melanogaster* from three different base populations. The broken lines refer to reversed selection and the dotted lines to suspended selection. (From F. W. Robertson, 1955; reproduced by courtesy of the author and the editor of the *Cold Spring Harbor Symposia on Quantitative Biology*.)

different wild populations, (F. W. Robertson, 1955). The responses of the three populations, both upward and downward, are fairly alike.

It is not possible to discuss further the degree of repeatability between the responses found in these two experiments, because there is no objective criterion for deciding how closely the responses ought to agree. One can therefore only regard them as empirical evidence of what in practice does occur.

ASYMMETRY OF RESPONSE

A surprising feature of the experimental results illustrated in the last chapter is the inequality of the responses to selection in opposite directions, seen particularly well in Fig. 11.5. This asymmetry of response has been found in many two-way selection experiments, but its cause is not yet known. For this reason we shall not discuss the phenomenon in detail, but shall merely note the possible causes, of which there are several. These possible causes are, briefly, as follows.

1. Selection differential. The selection differential may differ between the upward and downward selected lines, for several reasons. (i) Natural selection may aid artificial selection in one direction or hinder it in the other. (ii) The fertility may change so that a higher intensity of selection is achieved in one direction than in the other. (iii) The variance may change as a result of the change of mean: the selection differential will increase as the variance increases and decrease as it decreases. This is a "scale-effect," to be discussed more fully in Chapter 17. These three causes operating through the selection differential were all found in the experiment with mice cited in the last chapter, but they operated in the direction opposite to that of the asymmetry found. The selection differential was greater in the upward selection but the response was greater in the downward selection. Differences of the selection differential influence the response per generation, but they affect the realised heritability only a little. Therefore if the response is plotted against the cumulated selection differential and there is still much asymmetry, as in Fig. 11.5, it cannot be attributed to any cause operating through the selection differential.

2. "Genetic asymmetry." There are two sorts of asymmetry in the genetic properties of the initial population that could give rise

to asymmetry of the responses to selection (Falconer, 1954*a*). These concern the dominance and the gene frequencies of the loci concerned with the character. The dominant alleles at each locus may be mostly those that affect the character in one direction, instead of being more or less equally distributed between those that increase and those that decrease it. We shall refer to this situation as *directional dominance*. If the initial gene frequencies were about 0·5, the response would be expected to be greater in the direction in which the alleles tend to be recessive. It will be shown in Chapter 14 that this is also the direction in which the mean is expected to change on inbreeding. Therefore we should, in general, expect characters that show inbreeding depression to respond more rapidly to downward selection than to upward selection. There may also be asymmetry in the distribution of gene frequencies. The more frequent alleles at each locus may be mostly those that affect the character in one direction—a situation that we shall refer to as *directional gene frequencies*. In the absence of directional dominance this would be expected to cause a more rapid response to selection in the direction of the less frequent alleles. Under natural selection the less favourable alleles, in respect to fitness, will have been brought to lower frequencies. Therefore if selection in one direction reduces fitness more than selection in the other, we should expect a more rapid response in the direction of the greater loss of fitness. The asymmetry of the response to selection theoretically expected from these two causes may be seen by consideration of Fig. 2.3, which shows the expected response arising from one locus. Neither of these two causes—directional dominance and directional gene frequencies—would, however, be expected to give rise to immediate asymmetry; that is, in the first few generations of selection. The asymmetry would appear only as the gene frequencies in the upward and downward selected lines become differentiated. The asymmetry found in some experiments undoubtedly appears sooner than would be expected from these causes.

3. Selection for heterozygotes. If selection in one direction favours heterozygotes at many loci, or at a few loci with important effects, the response would become slow as the gene frequencies approach their equilibrium values. But the response in the other direction would be rapid until the favoured alleles approach fixation. This situation, which is a form of directional dominance, would also be expected to give rise to an asymmetrical response (Lerner, 1954); but, again, not immediately.

4. Inbreeding depression. Most experiments on selection are made with populations not very large in size, and there is usually therefore an appreciable amount of inbreeding during the progress the selection. If the character selected is one subject to inbreeding depression, there will be a tendency for the mean to decline through inbreeding. This will reduce the rate of response in the upward direction and increase it in the downward direction, thus giving rise to asymmetry. An unselected control population will reveal how much asymmetry can be attributed to this cause. Inbreeding depression has been shown to be an insufficient cause of the asymmetry in the experiments cited in the last chapter.

5. Maternal effects. Characters complicated by a maternal effect may show an asymmetry of response associated with the maternal component of the character. The situation envisaged may best be explained by reference to the selection for body weight in mice (Falconer, 1955), which showed the strong asymmetry illustrated in Fig. 11.5. The character selected—6-week weight—may be divided into two components, weaning weight and post-weaning growth, the former being maternally determined. It was found that all the asymmetry resided in the weaning weight and none in the post-weaning growth. The weaning weight increased hardly at all in the large line but decreased very much in the small line. Thus it was the mothering ability that changed asymmetrically under selection and not the growth of the young themselves. To attribute an asymmetrical response to maternal effects does not, however, solve the problem, because the asymmetry has merely been shifted from the character selected to another, and is still just as much in need of an explanation.

These, then, are the possible causes of asymmetry that may be suggested. There are probably others. Until the causes of asymmetry are better understood it is clear that predictions of the rate of response to selection must be made with caution. Where there is asymmetry of response the mean of the realised heritabilities in the two directions will presumably correspond with the heritability estimated from the resemblance between relatives. Therefore the response predicted will presumably be about the mean of the two-way responses actually obtained. If the asymmetry found in the mouse experiment should prove to be characteristic of selection for economically desirable characters in mammals, it means that we must expect actual progress to fall short of the predicted progress. In this experiment the mean realised heritability was 35 per cent, but the upward progress was only at the rate of 18 per cent. In other words the progress made was only about half as rapid as would, presumably, have been predicted.

LONG-TERM RESULTS OF SELECTION

The response to selection cannot be expected to continue indefinitely. Sooner or later it is to be expected that all the favourable alleles originally segregating will be brought to fixation. As they approach fixation the genetic variance should decline and the rate of response diminish, till, when fixation is complete, the response should cease. The population should then fail also to respond to selection in the opposite direction, and further response to selection in either direction will depend on the origin of new genetic variation by mutation. But how many generations must elapse before the response ceases, and how great will be the total response are questions that can be answered only by experiment. Let us first see what evidence is available on these points, and then see how far the long-term effects of selection conform to the simple theoretical picture outlined above.

Total response and duration of response. When the response to selection has ceased, the population is said to be at the *selection limit*. It is usually impossible to decide exactly at what point the limit is reached, because the limit is approached gradually, the response becoming progressively slower. The total response, and particularly the duration of the response, can therefore be estimated only approximately. Bearing this in mind, we may examine the results of four two-way selection experiments, two with *Drosophila* and two with mice, given in Table 12.1. The asymmetry of the responses is disregarded, and the total response is taken as the sum of the total responses in the two directions. This is the difference between the upper and lower selection limits, and may be called the *total range*. In the table the total range is expressed in three ways: as a percentage of the initial population mean, M_0; in terms of the phenotypic standard deviation, σ_P, in the initial population; and in terms of the standard deviation of breeding values, σ_A, (i.e. the square root of the additive variance) in the initial population. To draw general conclusions from these four experiments would be rash, because the experiments differed in several ways—in the intensity of selection, the population

P　　　　　　　　　　　　　　　　　　　　　　　　　　F.Q.G.

size, and the nature of the initial population—all of which would be expected to affect the duration of response and the total range. Despite these differences, however, the picture they give is fairly consistent. The response continues for about 20 to 30 generations;

TABLE 12.1

TOTAL RESPONSES IN FOUR SELECTION EXPERIMENTS

Experiment	Duration	Total range		
	(generations)	$/M_0$ (%)	$/\sigma_P$	$/\sigma_A$
Drosophila:				
(1) abdominal bristles	30	189	20	28
(2) thorax length	20	24	12	22
Mice:				
(3) 6-week weight	25	69	8	16
(4) 60-day weight	20	122	10	21

References:

(1) Clayton and Robertson (1957).

(2) F. W. Robertson (1955).

(3) Falconer (1955).

(4) MacArthur (1949); Butler (1952).

and the total range is between 15 and 30 times the square root of the additive variance, or about 10 to 20 times the phenotypic standard deviation in the initial population. The relationship between the total range and the original population mean, however, is quite irregular.

The total response produced by selection in these experiments, though it may be impressive when reckoned in terms of the variation present in the original population, is not at all spectacular when compared with the achievements of the breeders of domestic animals. For example, the upper limits of body weight of the mice in the experiments quoted are 2 to 3 times the lower limits; but the weights of the largest breeds of dog are about 75 times greater than those of the smallest (Sierts-Roth, 1953). The reason for the disappointing results of experimental selection when viewed against the differences between the breeds of domestic animals is that experiments are carried out with closed populations of not very large size. The limits are set by the gene content of the foundation individuals, since no genes are brought in after selection has been started. The breeder of domestic animals, in contrast, by intermittent crossing casts his net far wider in the search for genes favourable to his purposes.

The effects of inbreeding during the selection have been ignored in this account of selection limits. It is clear on theoretical grounds that inbreeding will tend to cause fixation of unfavourable alleles at some loci. Both the total response and the duration of the response must therefore be expected to be reduced if the selection is carried out in a small population with a fairly high rate of inbreeding. There is, however, little experimental evidence on the magnitude of this effect of inbreeding. The four experiments discussed above were all carried out on fairly large populations, so that the rate of inbreeding was fairly low.

Number of "loci." When the total range has been determined by experiment it is possible, in principle, to deduce the number of loci that gave rise to the response, and the magnitude of their effects. The estimates that can be made in practice, however, are only rough ones, because the properties of the individual loci are unknown and have to be guessed at. But even though we can do no more than establish the order of magnitude of the number and effects of the loci, this is better than no estimate at all; so let us see how these estimates may be obtained. The limitations will become apparent as we proceed.

The estimates come from a comparison of the total range with the amount of additive genetic variance in the original population. In principle it is clear that with a given amount of initial variation a small number of genes will produce less total response than a larger number; and that if a given amount of variation is produced by few genes the magnitude of their effects must be greater than if it is produced by many. It is clear, also, that linkage is an important factor in the relationship between variance and total response. Some segments of chromosome that segregate as units in the initial population will recombine during the selection and appear as many genes contributing to the total response. Other segments may fail to recombine and will be counted as single genes. In order to emphasise this limitation, the estimate of the number of loci may be referred to as the number of "effective factors" or as the "segregation index." There are, however, other uncertainties, and we shall simply refer to it as the number of "loci," letting the inverted commas serve to remind us of the unavoidable limitations and qualifications.

We must first suppose that there has been no inbreeding and when the selection limits have been reached all loci are fixed for the favourable allele. The total range is then $2\Sigma a$, where $2a$ is the difference of genotypic value between the two most extreme homozygotes at a particular locus, and is the precise meaning of what we have loosely called the "effect" of the locus. If R is the total range and n is the number of loci that have contributed to the response, then

$$R = 2n\bar{a} \qquad \qquad \ldots\ldots(12.1)$$

where $\bar{a}$ is the mean value of a. Next we must suppose that each locus has only two alleles. The additive variance arising from one locus is then $\sigma_A^2 = 2pq[a + d(q - p)]^2$, from equation 8.5. (We shall use σ^2 here to denote variance instead of V, because it simplifies the formulation when standard deviations are involved.) The gene frequencies at the individual loci thus enter the picture. Unless the initial population was made from crosses between inbred lines, the gene frequencies are not known and we shall therefore have to insert hypothetical values. We shall suppose that all segregating genes are at frequencies of 0·5, as they would be if the initial population were made from a cross between two inbred lines. The additive variance contributed by one locus then becomes $\sigma_A^2 = \frac{1}{2}a^2$, and the degree of dominance becomes irrelevant. Next we have to suppose there is no linkage between the loci, so that the additive variance due to all n loci together is

$$\sigma_A^2 = \frac{1}{2}n(\overline{a^2}) \qquad \qquad \ldots(12.2)$$

where $(\overline{a^2})$ is the mean of the squares of a for each locus. Finally we shall suppose that all loci have equal effects, so that equations 12.1 and 12.2 become

$$R = 2na \qquad \qquad \ldots\ldots(12.3)$$

and

$$\sigma_A^2 = \frac{1}{2}na^2 \qquad \qquad \ldots\ldots(12.4)$$

Squaring equation 12.3 and substituting $a^2 = (2/n)\sigma_A^2$ from equation 12.4 gives $R^2 = 8n\sigma_A^2$, whence

$$n = \frac{1}{8} \cdot \frac{R^2}{\sigma_A^2} \qquad \qquad \ldots\ldots(12.5)$$

This equation gives the basis for estimating the number of "loci." Their effects may then be estimated from equation 12.4. The most meaningful measure of the "effect" of a locus, however, is what we have earlier called the "proportionate effect," $2a/\sigma_P$, which is the difference between the homozygotes expressed in terms of the phenotypic standard deviation. By rearrangement of equation *12.4* this becomes

$$\frac{2a}{\sigma_P} = 2h \sqrt{\left(\frac{2}{n}\right)} \qquad \ldots\ldots(12.6)$$

where h is the square root of the heritability.

Let us see what results these theoretical deductions yield when applied to the experiments quoted in Table 12.1. The estimates of the number of "loci" and of the proportionate effects of the genes are

TABLE 12.2

Experiment	Number of "loci"	Proportionate effect $(2a/\sigma_P)$
Drosophila:		
(1) abdominal bristles	99	0·21
(2) thorax length	59	0·20
Mice:		
(3) 6-week weight	35	0·23
(4) 60-day weight	53	0·19

(For references to experiments see Table 12.1)

given in Table 12.2. Since the estimation of the number of "loci" is necessarily so imprecise it does not seem worth while to discuss in detail its limitations or the errors that may have been introduced by the assumptions that were made. These matters are discussed by Wright (1952*b*). The results given in Table 12.2, then, suggest that the responses to selection in these experiments have resulted from about 100 loci (i.e. more nearly 100 than 10 or 1,000); and that on the average the difference in value between homozygotes at one locus amounts to about one-fifth of the phenotypic standard deviation.

Nature of the selection limit. The deductions made in the last section from the observed total response were based on the assumption that the selection limit represents fixation of all favourable alleles. The simple theoretical expectation is that selection should lead to fixation with the consequent loss of genetic variance. Let us now consider the evidence from experiments about the nature of the selection limit and see how far it conforms to this simple theoretical picture. If the genetic variance declines as the limit is approached this ought to be apparent in a decline of phenotypic variance. In many experiments, however, the phenotypic variance has been found not to decline, even when the selection limit has been reached, and when due allowance for "scale effects" has been made as will be explained in Chapter 17. A fairly typical example is provided by the experiment with mice which was described in the last chapter (Fig. 11.5). The phenotypic variance is shown in Fig. 12.3, expressed in the form of the coefficient of variation in order to eliminate scale

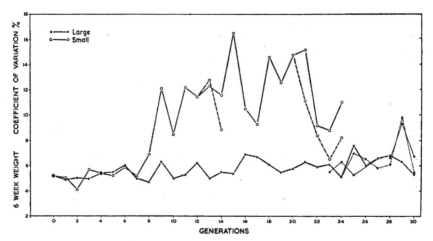

FIG. 12.3. Coefficient of variation of 6-week weight in mice. The thin continuous line starting at generation 23 refers to the un-selected control. The broken lines refer to reversed selection and the dotted lines to suspended selection. (From Falconer, 1955; reproduced by courtesy of the editor of the *Cold Spring Harbor Symposia on Quantitative Biology.*)

effects. The variance in the large line remains at the same level throughout the experiment, and after the limit has been reached at about the twenty-fifth generation a comparison with the unselected control shows the variance not to have declined at all. The variance in the small line shows a sudden and large increase, but we shall return to this point later. An example from *Drosophila* is provided by the experiment on abdominal bristle-number illustrated in Fig. 12.1. The phenotypic variance in the base population and in the most extreme of the high and of the low lines after 35 and 34 generations respectively is illustrated by frequency distributions in Fig. 12.4. In this case the variance not only failed to decline but increased very much during the selection in both directions. Before we consider the reasons for this behaviour of the variance we shall mention another fact often found in selection experiments. It is that when the response to continued selection has ceased the population will often respond to selection in the reverse direction and will often respond rapidly. This is well illustrated in Fig. 12.2, where the three lines selected for

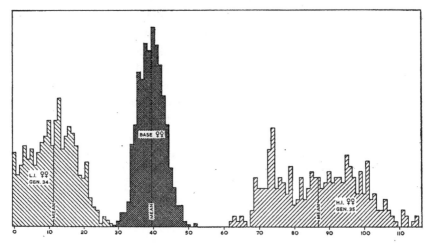

FIG. 12.4. Frequency distributions of abdominal bristle number in *Drosophila melanogaster* (females), in the base population and in the most extreme high and low lines after 35 and 34 generations of selection. (From Clayton, Morris, and Robertson, 1957; reproduced by courtesy of the authors and the editor of the *Journal of Genetics*.)

increased thorax length returned rapidly to the unselected level when the direction of selection was reversed after the upward responses had ceased. The lines selected for reduced thorax length, however, did not respond to reversed selection. From this brief outline of the evidence it is clear that the simple theoretical picture of the selection limit is not substantiated by experiment. Instead, we find—not always but often—no loss of phenotypic variance and the ability to respond rapidly to reversed selection. Let us now consider what may be the possible reasons for these facts, and what conclusions about the genetic nature of the selection limit can be drawn from them.

1. The failure of the phenotypic variance to decline may be due to an increase of non-genetic variance compensating for the expected reduction of genetic variance. With the approach to fixation of the

loci concerned, and of others linked to them, the frequency of homozygotes will increase. There is evidence, mentioned in Chapter 8
and to be discussed more fully in Chapter 15, that homozygotes are
sometimes more variable from environmental causes than heterozygotes. This could cause an increase of environmental variance which
might counterbalance a reduction of genetic variance; but there is
little experimental evidence concerning the matter.

2. If the population, after the selection limit has been reached,
responds to reversed selection we can only conclude that genetic
variance of some sort remains. The continued presence of genetic
variance could result from the following causes:

(i) We saw in Example 11.6 how natural selection opposed the
artificial selection for small size in mice, partly because small mice
are less fertile than large ones and partly because the smallest mice
were sterile. Natural selection acting in this sort of way may increase
as the population mean changes further from the original level, until
it becomes strong enough to counteract completely the artificial
selection. The response would then cease, but reversed selection
would be aided by natural selection and the population would respond.

(ii) Selection may favour heterozygotes at some loci. At the
selection limit the genes would be in equilibrium at more or less
intermediate frequencies, and they would give rise to genetic variance. But the variance would be non-additive, and there would be
no immediate response to reversed selection. If reversed selection
were continued a response would slowly develop and become more
rapid as the gene frequencies changed away from the equilibrium
values. The behaviour of populations at the selection limit, however,
does not seem commonly to be of this sort.

(iii) If there is superiority of heterozygotes arising from the combined action of artificial and natural selection then the situation is
quite different. Consider a locus at which the heterozygote A_1A_2 is
superior in the character selected to the homozygote A_1A_1, and the
homozygote A_2A_2 is inviable or sterile. Artificial selection will choose
A_1A_2, or perhaps A_2A_2 if it is viable, but natural selection will reject
A_2A_2, so that under the combined effect of artificial and natural
selection the heterozygote is superior. The pygmy gene in mice
which was used for several examples in Chapter 7 provides just such a
case, when artificial selection is in the direction of small size. Heterozygotes are favoured because they are smaller than normal homozy-

gotes; homozygous pygmies are smaller still but are sterile. When the selection limit is reached under this situation there will be genetic variance due to the gene, but no further response. When selection is reversed, however, it is only the artificial selection that is reversed in direction, and one homozygote will be favoured. The population will therefore respond immediately. This may be regarded as an extreme form of asymmetrical response to selection. It leads to the anomaly of a high heritability—about 50 per cent—estimated from the offspring-parent regression, but a realised heritability of zero in one direction and up to 100 per cent in the other direction. The anomaly, however, is only apparent because the estimation of heritability and the prediction of the response to selection are valid only if natural selection does not interfere with the appearance of the genotypes in their proper Mendelian ratios.

The situation described above was proved to exist in one of the lines of *Drosophila* selected for high bristle number in the experiment illustrated in Fig. 12.1. There was a gene present which was lethal in the homozygote and which in the heterozygote increased bristle number by 22, which is 5·8 times the original phenotypic standard deviation (Clayton and Robertson, 1957). The line carrying this gene was the one whose distribution is shown in Fig. 12.4, and the bimodality of the distribution can be seen. It seems probable that in cases like this the gene does not have so large an effect in the original population, but that the effect of the heterozygote is enhanced during the selection, either by "modifying" genes or by a cross-over which separates a linked gene whose effect is in the opposite direction. A mechanism of this sort seems to be required to account for the very great increase of variance often found in selected lines (F. W. Robertson and Reeve, 1952a; Clayton and Robertson, 1957).

The selection of heterozygotes at one or a few loci with major effects through the combined action of artificial and natural selection in the manner explained above seems to be a common situation in *Drosophila* populations at the selection limit. Whether it occurs as frequently in other organisms is not known because the genetic analyses required to detect it are more difficult to make. The increase of variance in the mice selected for small size shown in Fig. 12.3 may well have been due to this cause.

The deleterious effect on fitness is an essential part of the situation, so genes of this sort will always be at low frequencies in the initial population. The appearance of any particular gene in a selected line will therefore depend very much on the chances of sampling, or on its occurring later by mutation. Consequently such genes will be a cause of differences between replicated lines, such as we noted at the beginning of this chapter in the experiment on *Drosophila* bristle number, and they will render the selection limit to a large extent unpredictable in its level and its precise genetic nature.

Relevance of selection limits to animal and plant improvement. It may be thought that experimental studies of long continued selection are of little relevance to the practice of selection in animal and plant improvement, because the breeder is concerned only with the first five or ten generations. This, however, is not necessarily so. The breeds of animals and varieties of plants which he seeks to improve have already been under selection for more or less the same characters over a long time. They may therefore by now be approaching, if they are not already at, the selection limits. An understanding of the nature of the selection limit and of the behaviour of populations at the selection limit may therefore be very relevant in the field of practice.

CHAPTER 13

SELECTION:

III. Information from Relatives

In our consideration of selection we have up to now supposed that individuals are measured for the character to be selected and that the best are chosen to be parents in accordance with the individual phenotypic values. An individual's own phenotypic value, however, is not the only source of information about its breeding value; additional information is provided by the phenotypic values of relatives, particularly by those of full or half sibs. With some characters, indeed, the values of relatives provide the only available information. Milk-yield, to take an obvious example, cannot be measured in males, so the breeding value of a male can only be judged from the phenotypic values of its female relatives. Ovarian response to gonadotropic hormone, a character for which selection has been applied in rats (Kyle and Chapman, 1953), cannot be measured on the living animal, so selection can only be based on the phenotypic values of female relatives. The use of information from relatives is of great importance in the application of selection to animal breeding, for two reasons. First, the characters to be selected are often ones of low heritability, and with these the mean value of a number of relatives often provides a more reliable guide to breeding value than the individual's own phenotypic value. And, second, when the outcome of selection is a matter of economic gain even quite a small improvement of the response will repay the extra effort of applying the best technique. In this chapter we shall outline the principles underlying the use of information from relatives and the choice of the best method of selection, but we shall not discuss the technical details of procedure in the application of selection to animal breeding.

Methods of Selection

If the family structure of the population is taken into account we can compute the mean phenotypic value of each family; this is known as the *family mean*. Suppose, then, that we have a population in which the individuals are grouped in families, which may be full or half sibs, and we have measurements of each individual and of the means of every family. A choice of procedure for applying selection to this population is then open, according to the use we make of the family means. Let us first look at the problem from the point of view of the additional information provided by the values of relatives. Suppose, for example, that we have an individual whose own value puts it on the border-line between selection and rejection, and it has a number of sibs with high values, so that the family to which it belongs has a high mean. We may interpret the situation in one of two ways. Either we may say that the individual's own rather poor value has been due to poor environmental circumstances, and that the high family mean suggests that its breeding value is likely to be a good deal better than its phenotypic value. Or we may say that the high family mean has been due to a favourable common environment, provided perhaps by a good mother, from which the individual in question must also have benefited; on this interpretation, therefore, the individual's breeding value is likely to be less good than its phenotypic value. In the first case we should regard the information from the relatives as favourable and we should select the individual in question, while in the second case we should regard it as unfavourable and should reject the individual. Here then is the problem: how do we decide which is the correct interpretation? It turns out that only three things need be known: the kind of family (whether full or half sibs), the number of individuals in the families (i.e. the family size), and the phenotypic correlation between members of the families with respect to the character. The choice of method is thus a relatively simple matter in practice. But the explanation of the principles underlying the choice is more complicated. Before embarking on this explanation we shall therefore give a brief general account of the different methods of selection according to the use made of the information from relatives, indicating the circumstances to which each method is specially suited. Then we shall explain how the response expected under each method is deduced; and finally we shall compare the relative merits of the methods under different circumstances.

The phenotypic value of an individual, P, measured as a deviation from the population mean, is the sum of two parts: the deviation of its family mean from the population mean, P_f, and the deviation of the individual from the family mean, P_w (the within-family deviation);

so that

$$P = P_f + P_w \qquad \ldots\ldots(13.1)$$

The procedure of selection, then, varies according to the attention paid, or the weight given, to these two parts. If we select on the basis of individual values only, as assumed in the last two chapters, we give equal weight to the two components P_f and P_w of the individual's value P. This is known as *individual selection*. We may, alternatively, select on the basis of the family mean P_f alone, disregarding the within-family deviation P_w entirely. This is known as *family selection* and it corresponds to the procedure adopted in the first case discussed above. Again, we may select on the basis of the within-family deviation P_w alone, disregarding the family mean P_f entirely. This is known as *within-family selection* and it corresponds to the second case discussed above. Finally, we may take account of both components P_f and P_w but give them different weights chosen so as to make the best use of the two sources of information. This is known as selection by optimum combination, or *combined selection*. It represents the general solution for obtaining the maximum rate of response, and the other three simpler methods are special cases in which the weights given to the two sources of information are either 1 or 0. It is therefore in principle always the best method. But its advantage over one or other of the simpler methods is never very great, and it is a refinement that is not often worth while in practice. Beyond showing why this is so, we shall therefore not give very much attention to combined selection.

The salient features of the three simpler methods are as follows, the differences of procedure between them being illustrated diagrammatically in Fig. 13.1.

Individual selection. Individuals are selected solely in accordance with their own phenotypic values. This method is usually the simplest to operate and in many circumstances it yields the most rapid response. It should therefore be used unless there are good reasons for preferring another method. *Mass selection* is a term often used for individual selection, especially when the selected individuals are put together *en masse* for mating, as for example *Drosophila* in a bottle. The term individual selection is used more specifically when the matings are controlled or recorded, as with mice or larger animals.

Family selection. Whole families are selected or rejected as units according to the mean phenotypic value of the family. Individual values are thus not acted on except in so far as they determine the family mean. In other words the within-family deviations are given zero weight. The families may be of full sibs or half sibs, families of more remote relationship being of little practical significance. The use of full-sib families is dependent on a high reproductive rate and with slow-breeding organisms half sibs must generally be used.

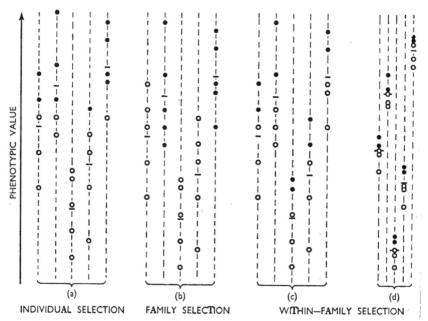

FIG. 13.1. Diagram to illustrate the different methods of selection. The dots and circles represent individuals plotted on a vertical scale of merit, those with the best measurements being at the top. The individuals to be selected are those shown as dots. There are 5 families each with 5 individuals; (a), (b), and (c) show identical arrangements of the same 25 individuals. The families are separated laterally, with the individuals of each family placed one above the other. The mean of each family is shown by a crossbar. The situation in which within-family selection is most useful is shown in (d), where the variation between families is very great in comparison with the variation within families. (Redrawn from Falconer, 1957a.)

The chief circumstance under which family selection is to be preferred is when the character selected has a low heritability. The efficacy of family selection rests on the fact that the environmental deviations of the individuals tend to cancel each other out in the mean value of the family. So the phenotypic mean of the family comes close to being a measure of its genotypic mean, and the advantage gained is greater when environmental deviations constitute a large part of the phenotypic variance, or in other words when the heritability is low. On the other hand, environmental variation common to members of a family impairs the efficacy of family selection. If this component is large, as illustrated in Fig. 13.1 (*d*), it will tend to swamp the genetic differences between families and family selection will be correspondingly ineffective. Another important factor in the efficacy of family selection is the number of individuals in the families, or the family size. The larger the family the closer is the correspondence between mean phenotypic value and mean genotypic value. So the conditions that favour family selection are low heritability, little variation due to common environment, and large families.

There are practical difficulties in the application of family selection, particularly in laboratory populations. They arise from the conflict between the intensity of selection and the avoidance of inbreeding. It is generally desirable to keep the rate of inbreeding as low as possible. If the minimum number of parents is fixed by considerations of inbreeding—say at ten pairs—then under family selection ten families must be selected, since each family represents only one pair of parents in the previous generation. And, if a reasonably high intensity of selection is to be achieved, the number of families bred and measured must be perhaps twice to four times this number. Family selection is thus costly of space, and if breeding space is limited the intensity of selection that can be achieved under family selection may be quite small. The two following methods are variants of family selection.

Sib selection. Some characters, we have already noted, cannot be measured on the individuals that are to be used as parents, and selection can only be based on the values of relatives. This amounts to family selection but with the difference that now the selected individuals have not contributed to the estimate of their family mean. The difference affects the way in which the response is influenced by family size. Where the distinction is of consequence we shall use the term *sib selection* when the selected individuals are not measured and family selection when they are measured and included in the family mean. When families are very large the two methods are equivalent, and the term family selection is then to be understood to cover both.

Progeny testing is a method of selection widely applied in animal breeding. We shall not discuss it in detail, except in so far as it can be treated as a form of family selection. The criterion of selection, as the name implies, is the mean value of an individual's progeny. At first sight this might seem to be the ideal method of selection and the easiest to evaluate because, as we saw in Chapter 7, the mean value of an individual's offspring comes as near as we can get to a direct measure of its breeding value, and is in fact the operational definition of breeding value. In practice, however, it suffers from the serious drawback of a much lengthened generation interval, because the selection of the parents cannot be carried out until the offspring have been measured. The evaluation of selection by progeny testing is apt to be rather confusing because of the inevitable overlapping of generations, and because of a possible ambiguity about which generation is being selected, the parents or the progeny. The progeny, whose mean is used to judge the parents, are ready to be used as parents just when the parents have been tested and await selection. Thus both the selected parents and their progeny are used concurrently as parents. The difficulty of interpretation may be partially overcome by regarding progeny testing as a modified form of family selection. The progenies are families, usually of half sibs, and selection is made between them on the basis of the family means in the manner described above. The only difference is that the selected families are increased in size by allowing their parents to go on breeding. The additional, younger, members of the families do not contribute to the estimates of the family means and are therefore selected by sib selection. Increasing the size of the selected families by unmeasured individuals does not improve the accuracy of the selection, but it reduces the replacement rate and so increases the intensity of selection that can be achieved. This is the principal advantage of progeny testing, but it can only be realised in operations on a large scale, when the danger of inbreeding is not introduced by limitation of space.

Within-family selection. The criterion of selection is the deviation of each individual from the mean value of the family to which it belongs, those that exceed their family mean by the greatest amount being regarded as the most desirable. This is the reverse of family selection, the family means being given zero weight. The chief condition under which this method has an advantage over the others is a large component of environmental variance common to members of a family. Fig 13.1 (*d*) shows how within-family selection would be applied in this situation. Pre-weaning growth of pigs or mice might be cited as examples of such a character. A large part of the variation of individuals' weaning weights is attributable to the mother and is therefore common to members of a family. Selection within families would eliminate this large non-genetic component from the variation operated on by selection. An important practical advantage of selection within families, especially in laboratory experiments, is that it economises breeding space, for the same reason that family selection is costly of space. If single-pair matings are to be made, then two members of every family must be selected in order to replace the parents. This means that every family contributes equally to the parents of the next generation, a system that we saw in Chapter 4 renders the effective population size twice the actual. Thus when selection within families is practised, the breeding space required to keep the rate of inbreeding below a certain value is only half as great as would be required under individual selection.

EXPECTED RESPONSE

To evaluate the relative merits of the different methods of selection we have to deduce the response expected from each. There is nothing to be added here about individual selection to what was said in Chapter 11. The expected response was given in equation 11.3 as $R = i\sigma_P h^2$, where i is the intensity of selection (i.e. the selection differential in standard deviations), σ_P is the standard deviation, and h^2 the heritability, of the phenotypic values of individuals. The response expected under family selection or within-family selection is arrived at in an analogous manner. Under family selection, the criterion of selection is the mean phenotypic value of the members of a family, so the expected response to family selection is

$$R_f = i\sigma_f h_f^2 \qquad \qquad \ldots\ldots(13.2)$$

where i is the intensity of selection, σ_f is the observed standard deviation of family means, and h_f^2 is the heritability of family means. In the same way the expected reponse to within-family selection is

$$R_w = i\sigma_w h_w^2 \qquad \qquad \ldots\ldots(13.3)$$

where σ_w is the standard deviation, and h_n^2 the heritability of within-family deviations.

Q   F.Q.G.

The concept of heritability applied to family means or to within-family deviations introduces no new principle. It is simply the proportion of the phentoypic variance of these quantities that is made up of additive genetic variance. These heritabilities can be expressed in terms of the heritability of individual values (which we shall continue to refer to simply as the heritability, with symbol h^2), the phenotypic correlation between members of families, and the number of individuals in the families, all of which can be estimated by observation. To arrive at the appropriate expressions we have to consider again how the observational components of variance are made up of the causal components, as explained in Chapters 9 and 10 (see in particular Tables 9.4 and 10.4). First let us simplify matters by supposing that all families contain a large number of individuals, so that the means of all families are estimated without error. Consider first the phenotypic variance. The intra-class correlation, t, between members of families is the between-group component divided by the total variance: $t = \sigma_B^2/\sigma_T^2$. Therefore the between-group component can be expressed as $\sigma_B^2 = t\sigma_T^2$, and the within-group component as $\sigma_W^2 = (1-t)\sigma_T^2$. This expresses the partitioning of the phenotypic variance into its observational components. The total variance, written here as σ_T^2, is the phenotypic variance which we shall write as V_P in the context of causal components. Now, the partitioning of the additive variance between and within families can be expressed in the same way, in terms of the correlation of breeding values, for which we shall use the symbol r. (The meaning of this correlation will be explained in a moment.) Thus the additive variance between families is rV_A and the additive variance within families is $(1-r)V_A$. The dual partitioning is summarised in Table 13.1.

TABLE 13.1

Partitioning of the variance between and within families of large size.

Observational component	Additive variance	Phenotypic variance
Between families, σ_B^2	rV_A	tV_P
Within families, σ_W^2	$(1-r)V_A$	$(1-t)V_P$

This partitioning of both the additive and the phenotypic variance leads at once to the heritabilities of family means and of within-family deviations, since these heritabilities are simply the ratios of the additive variance to the phenotypic variance. Thus, when the families are large, the heritability of family means is rV_A/tV_P, or $(r/t)h^2$, since V_A/V_P is the heritability of individual values, h^2.

The correlation of breeding values between members of families is a measure of the degree of relationship, usually called the "coefficient of relationship." The correlation between the breeding values of relatives in a random-mating population is twice their coancestry

$$r = 2f \qquad \qquad \ldots\ldots(13.4)$$

that is to say, twice the inbreeding coefficient of their progeny if the relatives were mated together. Its values in full-sib and half-sib families can be seen from Table 9.4; for full sibs it is $\frac{1}{2}$ and for half sibs it is $\frac{1}{4}$. In order to be able to discuss full-sib and half-sib families at the same time in what follows, we shall retain the symbol r in the formulae instead of inserting the appropriate values of $\frac{1}{2}$ or $\frac{1}{4}$.

The foregoing account of the heritabilities of family means and within-family deviations was simplified by the supposition of large families. This simplification is not justified in practice and we must now remove it by considering families of finite size. We shall, however, suppose that all families are of equal size. The number of individuals in a family—called the family size—has to be taken into consideration for the following reason. If selection is based on the family mean, or on the deviations from the family mean, then it is the observed mean that we are concerned with and not the true mean. In other words we are not concerned with the observational components of variance which we have hitherto discussed, but with the variance of the observed means and of the observed within-family deviations. The observed means of groups are subject to sampling variance which comes from the within-group variance. If there are n individuals in a group then the sampling variance of the group-mean is $(1/n)\,\sigma_W^2$, where σ_W^2 is the component of variance within the group. Thus the variance of observed group-means is augmented by $(1/n)\,\sigma_W^2$, and the variance of

TABLE 13.2

Composition of observed variances with families of size n.

Observed variance	Observational components	Causal components Additive	Causal components Phenotypic
of family means	$\sigma_B^2 + \dfrac{1}{n}\sigma_W^2$	$\dfrac{1+(n-1)r}{n}\,V_A$	$\dfrac{1+(n-1)t}{n}\,V_P$
of within-family deviations	$\sigma_W^2 - \dfrac{1}{n}\sigma_W^2$	$\dfrac{(n-1)(1-r)}{n}\,V_A$	$\dfrac{(n-1)(1-t)}{n}\,V_P$

observed deviations within groups is correspondingly diminished by the same amount. The observed variances, with family size n, are therefore made up of the observational components as shown in Table 13.2. The causal components entering into the observed variances can now be found by translating the observational components into causal components from Table 13.1. They are shown in the two right-hand columns of Table 13.2.

To find the heritabilities of family means and of within-family deviations we have only to divide the additive component by the phenotypic component of the observed variances. Thus the heritability of family means is

$$h_f^2 = \frac{1 + (n-1)r}{1 + (n-1)t} h^2$$

and the heritability of within-family deviations is

$$h_w^2 = \frac{1 - r}{1 - t} h^2$$

At this point sib selection has to be distinguished from family selection. The foregoing account referred to family selection where the individuals to be selected were themselves measured and contributed to the observed family mean. Sib selection differs in that the individuals selected are not measured. This does not affect the phenotypic component, because this is simply the observed variance of what is measured. But it does affect the additive component, because the mean breeding value with which we are concerned is not that of the individuals whose phenotypic values have been measured, but of others that have not been measured. Therefore the appropriate variance of mean breeding values is simply the between-family component of additive variance, rV_A, irrespective of the number of other individuals that have been measured. The heritability of family means appropriate to sib selection is therefore

$$h_s^2 = \frac{nr}{1 + (n-1)t} h^2$$

The heritabilities of the different methods of selection, whose derivations have now been explained, are listed in Table 13.3.

To deduce the expected response is now a simple matter. Let us take family selection for illustration. The expected response was given in equation *13.2* as

$$R_f = i\sigma_f h_f^2$$

where σ_f is the standard deviation of observed family means. This expression, however, is not much use as it stands, because it does not readily allow a comparison to be made with the other methods. It will be most convenient to cast it into a form that facilitates comparison with individual selection. This can be done by substituting the

TABLE 13.3

Heritability and expected response under different methods of selection.

Method of selection	Heritability	Expected response
Individual	h^2	$R = i\sigma_P h^2$
Family	$h_f^2 = h^2 \cdot \dfrac{1 + (n-1)r}{1 + (n-1)t}$	$R_f = i\sigma_P h^2 \cdot \dfrac{1 + (n-1)r}{\sqrt{n\{1 + (n-1)t\}}}$
Sib	$h_s^2 = h^2 \cdot \dfrac{nr}{1 + (n-1)t}$	$R_s = i\sigma_P h^2 \cdot \dfrac{nr}{\sqrt{n\{1 + (n-1)t\}}}$
Within-family	$h_w^2 = h^2 \cdot \dfrac{(1-r)}{(1-t)}$	$R_w = i\sigma_P h^2 \cdot (1-r)\sqrt{\left[\dfrac{n-1}{n(1-t)}\right]}$
Combined	—	$R_c = i\sigma_P h^2 \sqrt{\left[1 + \dfrac{(r-t)^2}{(1-t)} \cdot \dfrac{(n-1)}{1 + (n-1)t}\right]}$

    i = intensity of selection (selection differential in standard measure): assumed to be equal for all methods, but not necessarily so.

    σ_P = standard deviation in phenotypic values of individuals.

    h^2 = heritability of individual values.

    r: with full-sib families, $r = \frac{1}{2}$

         with half-sib families, $r = \frac{1}{4}$

    t = correlation of phenotypic values of members of the families.

    n = number of individuals in the families.

expression for the heritability of family means, h_f^2, given above, and by putting the standard deviation of observed family means, σ_f, in terms of the standard deviation of individual phenotypic values, $\sigma_P(=\sqrt{V_P})$ from the right-hand column of Table 13.2. The expected response then becomes

$$R_f = i \sqrt{\frac{1 + (n-1)t}{n}} \cdot \sigma_P \cdot \frac{1 + (n-1)r}{1 + (n-1)t} \cdot h^2$$

which reduces to

$$R_f = i\sigma_P h^2 \left[\frac{1 + (n-1)r}{\sqrt{[n\{1 + (n-1)t\}]}} \right]$$

The term $i\sigma_P h^2$ is equivalent to the expected response under individual selection, so the expression within the square brackets is the factor that compares family selection with individual selection. The expression looks very complicated but it contains only three simple quantities: n, which is the family size; r, which is $\frac{1}{2}$ for full-sib and $\frac{1}{4}$ for half-sib families; and t, which is the phenotypic intra-class correlation.

The expected responses under the different methods of selection are listed in Table 13.3, all expressed in this manner which allows the comparisons to be made with individual selection. The relative merits of the different methods will be discussed in the next section: first we must deal with combined selection.

Combined selection. We shall deal very briefly with combined selection, referring the reader to Lush (1947), Lerner (1950) and. A. Robertson (1955a) for details. First we have to find what are the appropriate weighting factors to be used in its application. We saw before that the phenotypic value of an individual is made up of two parts, the family mean and the within-family deviation, $P = P_f + P_w$, and that each part gives some information about the individual's breeding value. In Chapter 10 we saw that the heritability is equivalent to the regression of breeding value on phenotypic value (equation 10.2), so that the best estimate of an individual's breeding value to be derived from its phenotypic value is h^2P. This idea can be applied separately to the two parts of the phenotypic value, since these are uncorrelated and supply independent information about the breeding value. Therefore, taking both parts of the phenotypic value into account, the best estimate of an individual's breeding value is given by the multiple regression equation

$$\text{expected breeding value} = h_f^2 P_f + h_w^2 P_w$$

(P_f being measured as a deviation from the population mean, and P_w as a deviation from the family mean). The weighting factors that make the most efficient use of the two sources of information are therefore the two heritabilities, appropriate to family means and to within-family deviations respectively. The criterion of selection under combined selection is thus an index, I, in the form

$$I = h_f^2 P_f + h_w^2 P_w \qquad \qquad \dots\dots(13.5)$$

If the values of the heritabilities are inserted from Table 13.3 it will be seen that the term h^2 is common to both weighting factors, and this term may therefore be omitted without affecting the relative weighting. We then have an index for the computation of which only n, r, and t need be known. In practice it is more convenient to work with the individual values in place of the within-family deviations, and to assign them a weight of 1. The family mean is thus used in the manner of a correction, supplementing the information provided by the individual itself. Rearrangement of the appropriate weighting factor for the family mean leads to an index made up as follows (Lush, 1947):

$$I = P + \left[\frac{r-t}{1-r} \cdot \frac{n}{1+(n-1)t} \right] P_f \qquad \dots\dots(13.6)$$

where P is the individual value and P_f the family mean, in which the individual itself is included.

This solution of the problem of how we can best make use of the information provided by relatives is now cast in precisely the form in which the problem was introduced at the beginning of this chapter. The expression in the square brackets in equation *13.6*, which contains nothing but easily measurable quantities, shows how we can best use the family mean to supplement the individual values in making the selection.

The expected response to combined selection, cast in a form suitable for comparison with individual selection, is given at the foot of Table 13.3. For its derivation see Lush (1947).

Relative Merits of the Methods

The formulae for the expected responses that we have derived enable us to compare one method of selection with another and discover what are the conditions that determine the choice of the best method. Before making detailed comparisons let us note the reason for individual selection being usually better than either family selection or within-family selection. The reason is that the standard deviations of family means and of within-family deviations are both bound to be less than the standard deviation of individual values; and the standard deviation of the criterion of selection is one of the factors governing the response. If we compare, for example, family selection with individual selection by writing the expected responses in the form

$$R=i\sigma_P h^2 \quad \text{(for individual selection)}$$
and
$$R_f=i\sigma_f h_f^2 \quad \text{(for family selection)}$$

then it is clear that family selection cannot be better than individual selection unless the heritability of family means, h_f^2, is greater than the heritability of individual values, h^2, by an amount great enough to counterbalance the lower standard deviation of family means. And the same applies to within-family selection.

A general picture of the circumstances that make one method better than another can best be obtained from graphical representations of the relative responses: that is, the response expected from one method expressed as a proportion of the response expected from another, the expected responses being taken from Table 13.3. In making these comparisons we shall assume that the intensity of selection is the same for all methods. Though not necessarily true, this simplification is unavoidable because no generalisation can be made about the proportions selected under the different methods. We shall make the comparisons separately for full-sib families ($r=\frac{1}{2}$) and for half-sib families ($r=\frac{1}{4}$). Then the relative responses depend only on two factors, the family size, n, and the intra-class correlation of phenotypic values, t. If there is no variance due to common environment contributing to the variance of family means, then the correlation in full-sib families is equal to half the heritability, and that in half-sib families to one quarter of the heritability. This lets us see in a general way how the heritability of the character influences the relative response. It is, however, the correlation and not the heritability that is the determining factor, so only the correlation need be known when a choice of method is to be made.

Fig. 13.2 gives a general picture of all the methods, showing how their relative merits depend on the phenotypic correlation. The graphs refer only to full-sib families and only to the two extremes of family size: infinitely large families in (a) and families of 2 in (b). The comparisons are made here with combined selection since this is necessarily the method that gives the greatest response. The graphs therefore show the ratio of the response for each method to that for combined selection: e.g. for family selection, the ratio R_f/R_c. The general picture indicated by the graphs is as follows. The relative merit of individual selection is greatest when the correlation is 0·5 and falls off as the correlation drops below or rises above this value. The relative merit of family selection is greatest when the correlation is low, and that of within-family selection when the correlation is

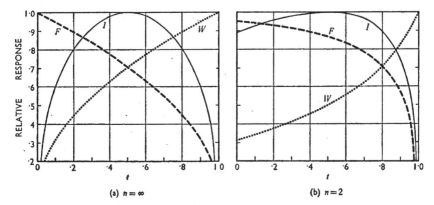

(a) $n = \infty$ (b) $n = 2$

FIG. 13.2. Relative merits of the different methods of selection, with full-sib families. Responses relative to that for combined selection plotted against the phenotypic intra-class correlation, t. I = individual selection; F = family selection; W = within-family selection.

high. Now, a low correlation between sibs can only result from a character of low heritability, and with very little variance due to common environment. These therefore are the circumstances that favour family selection. A high correlation can only result from a large amount of variance due to common environment. Even if the heritability were 100 per cent the correlation between full sibs could not exceed 0·5 without augmentation by common environment. A large amount of variation due to common environment is therefore the circumstance that favours within-family selection. We shall examine the three simpler methods in more detail in a moment. First let us look at what may be gained from combined selection. Though combined selection is always as good as or superior to any other method, its superiority is never very great. With large families its superiority is greatest when the correlation is close to 0·25 or 0·75, but even then its superiority is not much more than 10 per cent.

With families of 2 its superiority reaches 20 per cent when the correlation is 0·875. Thus the range of circumstances under which combined selection is more than a few per cent better than one or other of the simpler methods is very narrow. In general, therefore, there is little to be gained from the extra trouble of applying combined selection, and we shall not give it any further consideration.

Let us now examine the simpler methods in more detail. The most useful comparison to make now is with individual selection. The expected responses will therefore be expressed as a proportion of the response to individual selection. We shall examine each method in turn, commenting on the special questions that arise in connexion with each.

Family selection. Fig. 13.3 shows the relative response R_f/R plotted against the family size, n, for full-sib families in (*a*) and for half-sib families in (*b*). These graphs therefore show primarily the effect of family size on the relative merit of family selection, but the magnitude of the correlation, t, is taken into account by separate curves for different correlations. Only the circumstances when family selection is superior to individual selection are shown on the graphs. The chief points made clear by the graphs are these. (i) As we saw from Fig. 13.2, there is a critical value of the correlation, above which family selection cannot be superior to individual selection. From the expected responses in Table 13.3 it is easy to show that when the families are large the relative response expected is $R_f/R = r/\sqrt{t}$. So, with large families, family selection becomes superior to individual selection when r exceeds $\sqrt{t}$. The critical value of the correlation, t, depends a little on the family size and differs between full-sib and half-sib families. Family selection with full sibs is very little better than individual selection unless the correlation is below 0·2; and with half sibs unless it is below 0·05. (ii) The effect of family size is greatest when the correlation is low. Therefore there is little to be gained from very large families unless the correlation is well below the critical value. There is, however, another consideration in connexion with the family size which will be explained later. (iii) Finally, there is the question whether full sibs or half sibs are to be preferred for family selection. This depends so much on the special circumstances that general conclusions cannot be drawn. From the graphs it would appear that full sibs must always be better than half sibs. But the full-sib correlation is more likely to be increased by common environment, and full-sib families are likely to be a good deal smaller

^{cor-} than half-sib families. Both these factors work in favour of half-sib
families. It has been shown that in selection for egg-production in
poultry the factor of family size makes half-sib families superior to
^{:fore,} full sibs (Osborne, 1957*a*).
^{)ined}

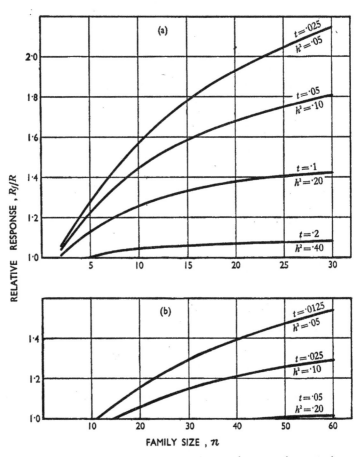

^{:tion.}
^{)rtion}

^{:exion}

FIG. 13.3. Responses expected under family selection relative to
that for individual selection, plotted against family size. The
separate curves refer to different values of the phenotypic correlation, *t*, as indicated. The corresponding values of the heritability, h^2, in the absence of variation due to common environment,
are also given. (*a*) full-sib families; (*b*) half-sib families.

Sib selection. The use of this method is usually dictated by
necessity rather than by choice, and comparisons with other methods
are of less interest. The chief practical question that arises concerns

the family size: how many sibs should be measured? Or, how far is it worth while increasing family size? The effect of family size on the response to sib selection is shown in Fig. 13.4. The graphs show the response with family size n, as a percentage of the response with infinitely large families, which would be the maximum possible

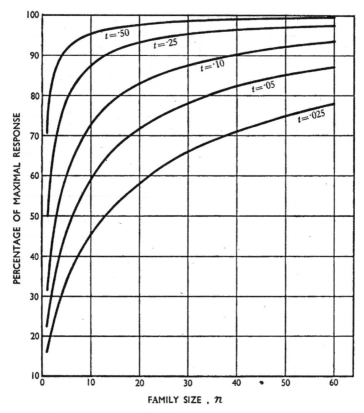

FIG. 13.4. Effect of family size on the response to sib selection, with either full- or half-sib families. The expected response is shown as a percentage of the response with infinitely large families. The separate curves refer to different values of the phenotypic correlation, t, as indicated.

response. The graphs are valid for both full and half sibs. Again the effect of increasing family size is greatest when the correlation is low. But with sib selection as with family selection there is another consideration to be taken into account in connexion with the family size, which will now be explained.

Optimal family size. Though the graphs suggest that the larger the family size the greater will be the response, under both family selection and sib selection, this is not so in practice because the intensity of selection is involved as a factor in the following way. In practice there is always a limitation on the amount of breeding space or facilities for measurement. The total available space can be filled with a large number of small families, or with a small number of large families. Considerations of inbreeding set a lower limit to the number of families that will be selected, so the larger the number of families measured the greater will be the intensity of selection. Therefore there is a conflict of advantage between the size of the families and the intensity of selection: large families lead to a lower intensity of selection. When the intensity of selection is taken into consideration it turns out that there is an optimal family size which gives the greatest expected response. The optimal family size with half-sib families can be found approximately from the following simple formula (A. Robertson, 1957 b):

$$n = 0.56 \sqrt{\frac{T}{Nh^2}} \qquad \qquad \ldots\ldots(13.7)$$

where n is the otpimal family size, T is the total number of individuals that can be accommodated and measured, N is the number of families to be selected, and h^2 is the heritability of the character.

Within-family selection. Fig. 13.5 shows the relative response, R_w/R, for within-family selection applied to full-sib families. Half-sib families need not be considered since the method is unlikely to be applied to them. The graphs show primarily the effect of the phenotypic correlation, t, on the response. Four graphs are given representing family sizes between 2 and 30, and it can be seen that the family size does not have a great effect. The relative response when the families are very large can be shown from the expected responses given in Table 13.3 to be $R_w/R = (1 - r)/\sqrt{(1 - t)}$. So, with large families, within-family selection will be superior to individual selection if $(1 - r)$ exceeds $\sqrt{(1 - t)}$. The graphs in Fig. 13.5 show that the correlation, t, in full-sib families would have to exceed about 0.75 to 0.85, according to the family size. Correlations as high as this cannot arise without a large amount of variation due to common environment. Correlations high enough to make within-family selection superior to individual selection are, however, not commonly found, and the advantage of within-family selection therefore comes chiefly from the reduced rate of inbreeding which was mentioned earlier. Fig. 13.5 shows how much will be sacrificed in the rate of response if within-family selection is applied. Most characters have full-sib correlations below about 0·5, and within-family selection is then only about half as effective as individual selection.

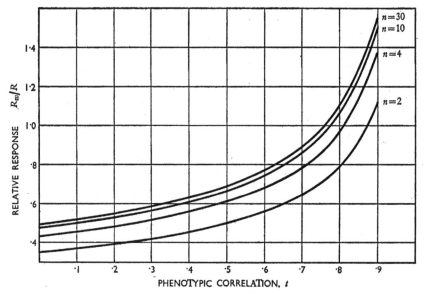

FIG. 13.5. Response expected under within-family selection relative to that of individual selection, plotted against the phenotypic correlation, t. The separate curves refer to different family sizes, as indicated.

Weights to be attached to families of different size. Throughout this chapter we have assumed that all families whose mean values are to be used in selection have equal numbers of individuals in them; i.e. n is the same for all families. This is a reasonable enough assumption to make when we are considering the expected response from the point of view of the planner who has to decide on the method of selection to be applied. But, in practice, families are very seldom of equal size and if we are to apply any method of selection based on family means we are immediately faced with the problem of how to make allownace for different numbers in the families. Obviously the mean of a large family is more reliable than that of a small one, and should be given more weight when the selection is being made. The solution of the problem comes from a consideration of the heritability as the regression of breeding value on phenotypic value. The best estimate of the breeding value of a family is obtained by multiplying the family mean (measured as a deviation from the population mean) by the heritability of family means. The appropriate weighting factor for family means is therefore the heritability of family means, calculated separately for each family according to its size. Quantities that are constant for all families may be omitted without altering the relative weights. Thus, in the application of family selection, each family mean, calculated as a deviation from the population mean, should be weighted by $[1 + (n - 1)r]/[1 + (n - 1)t]$, and in sib selection by $nr/[1 + (n - 1)t]$. The heritability of within-family deviations does not contain the term n, and is therefore unaffected by family size. Thus no weighting is required in the application of within-family selection. The weighting factor to be used in combined selection has already been given in equation *13.6*.

We conclude this chapter with an example from a laboratory experiment which compared the responses actually obtained under different methods of selection.

EXAMPLE 13.1. In an experiment with *Drosophila melanogaster* selection for abdominal bristle-number was made by three methods (Clayton, Morris, and Robertson, 1957). The responses to individual selection at different intensities were quoted in Example 11.2. Sib selection was also applied in both full-sib and half-sib families and the responses compared with expectation. Here we shall compare the responses under sib selection with the response under individual selection, according to the formula in Table 13.3. The same proportion of the population was selected in each case, namely 20 per cent, but the intensities of selection under sib selection

	Data			Relative response, R_s/R	
	Full sibs	*Half sibs*		*Full sibs*	*Half sibs*
i	1·33	1·27	Exp.	0·832	0·614
n	12	20	Obs. up	0·618	0·527
r	0·50	0·275	Obs. down	0·919	0·635
t	0·265	0·121			

were lower than under individual selection because there was a smaller total number of families than of individuals—10 half-sib families, 20 full-sib families, and 100 individuals. The intensity of selection under individual selection was 1·40. Those under sib selection are given in the table, together with the other data needed for calculating the expected responses under sib selection relative to that under individual selection.

In applying the formula from Table 13.3 we have to take account of the intensity of selection, multiplying by the ratio of the intensity under sib selection to the intensity under individual selection. It will be seen that the correlation of breeding values, r, between half sibs is a little greater than $\frac{1}{4}$. This is because the females mated to a male were not entirely unrelated to each other. The ratios of the responses expected and observed are given in the right-hand half of the table. The expectation is that individual selection should be the best method, and so it proved to be. There is, however, some discrepancy between the upward and downward responses, of which the reason is not known.

CHAPTER 14

INBREEDING AND CROSSBREEDING:

I. Changes of Mean Value

We turn our attention now to inbreeding, the second of the two ways open to the breeder for changing the genetic constitution of a population. The harmful effects of inbreeding on reproductive rate and general vigour are well known to breeders and biologists, and were mentioned in Chapter 6 as one of the two basic genetic phenomena displayed by metric characters. The opposite, or complementary, phenomenon of hybrid vigour resulting from crosses between inbred lines or between different races or varieties is equally well known, and forms an important means of animal and plant improvement. The production of lines for subsequent crossing in the utilisation of hybrid vigour is one of two main purposes for which inbreeding may be carried out. The other is the production of genetically uniform strains, particularly of laboratory animals, for use in bioassay and in research in a variety of fields. Inbreeding in itself, however, is almost universally harmful and the breeder or experimenter normally seeks to avoid it as far as possible, unless for some specific purpose. Mention should be made here of naturally self-fertilising plants, to which much of the discussion in this chapter is inapplicable. Since inbreeding is their normal mating system they cannot be further inbred: they can, however, be crossed, but they do not regularly show hybrid vigour.

In the treatment of inbreeding given in Chapter 3 the consequences were described in terms of the expected changes of gene frequencies and of genotype frequencies. Here we have to show how the changes of gene and genotype frequencies are expected to affect metric characters. And at the same time we have to consider the observed consequences of inbreeding and crossing, and see what light they throw on the properties of the genes concerned with metric characters. We shall first consider the changes of mean value and then, in the next chapter, the changes of variance resulting from inbreeding and crossbreeding. Finally, in Chapter 16, we shall consider the combination of selection with inbreeding and crossbreeding by means of which hybrid vigour may be utilised in animal and plant improvement.

Inbreeding Depression

The most striking observed consequence of inbreeding is the reduction of the mean phenotypic value shown by characters connected with reproductive capacity or physiological efficiency, the phenomenon known as *inbreeding depression*. Some examples of inbreeding depression are given in Table 14.1, from which one can see what sort of characters are subject to inbreeding depression, and—very roughly—the magnitude of the effect. From the results of these and many other studies we can make the generalisation that inbreeding tends to reduce fitness. Thus, characters that form an important component of fitness, such as litter size or lactation in mammals, show a reduction on inbreeding; whereas characters that contribute little to fitness, such as bristle number in *Drosophila*, show little or no change.

In saying that a certain character shows inbreeding depression, we refer to the average change of mean value in a number of lines. The separate lines are commonly found to differ to a greater or lesser extent in the change they show, as, indeed, we should expect in consequence of random drift of gene frequencies. This matter of differentiation of lines will be discussed later when we deal with changes of variance. It is mentioned here only to emphasise the fact that the changes of mean value now to be discussed refer to changes of the mean value of a number of lines derived from one base population. As in our earlier account of inbreeding we have to picture the "whole population" consisting of many lines. The population mean then refers to the whole population and inbreeding depression refers to a reduction of this population mean. Let us now consider the theoretical basis of the change of population mean on inbreeding.

First, we may recall and extend some of the conclusions from Chapter 3, supposing at first that selection does not in any way interfere with the dispersion of gene frequencies. Since the gene frequencies in the population as a whole do not change on inbreeding, any change of the population mean must be atributed to the changes of genotype frequencies. Inbreeding causes an increase in the frequencies of homozygous genotypes and a decrease of heterozygous genotypes.

TABLE 14.1

SOME EXAMPLES OF INBREEDING DEPRESSION

The figures given show approximately the *decrease* of mean
phenotypic value per 10 per cent increase of the coefficient
of inbreeding: column (1) in absolute units; column (2) as
percentage of non-inbred mean; column (3) in terms of the
original phenotypic standard deviation (data not available
for all characters).

Character	Inbreeding depression per 10% increase of F		
	(1) units	(2) %	(3) $/\sigma_P$
Cattle (A. Robertson, 1954)			
Milk-yield	29·6 gal.	3·2	0·17
Pigs (Dickerson *et al.* 1954)			
Litter size at birth	0·38 young	4·6	0·15
Weight at 154 days	3·64 lb.	2·7	0·12
Sheep (Morley, 1954)			
Fleece weight	0·64 lb.	5·5	0·51
Length of wool	0·12 cm.	1·3	0·14
Body weight at 1 year	2·91 lb.	3·7	0·36
Poultry (Shoffner, 1948)			
Egg-production	9·26 eggs	6·2	—
Hatchability	4·36 %	6·4	—
Body weight	0·04 lb.	0·8	—
Mice (Original data)			
Litter size at birth	0·60 young	8·0	0·28
Weight at 6 weeks (♀♀)	0·58 gm.	2·6	0·26
Drosophila melanogaster			
(Tantawy and Reeve, 1956)			
Fertility (per pair per day)	2·2 offspring	6·7	—
Viability (egg to adult)	2·6 %	3·7	—
Wing length	2·8 ($\frac{1}{100}$) mm.	1·4	0·80
Drosophila subobscura			
(Hollingsworth and Smith, 1955)			
Fertility (per pair per day)	6·0 offspring	12·5	—
Egg hatchability	8·3 %	8·3	—

Therefore a change of population mean on inbreeding must be connected with a difference of genotypic value between homozygotes and heterozygotes. Let us now see more precisely how the population mean depends on the degree of inbreeding, which we may conveniently express as the inbreeding coefficient, F.

Consider a population, subdivided into a number of lines, with a coefficient of inbreeding, F. The expression for the population mean is derived by putting together the reasoning set out in Tables 3.1 and 7.1, in the following way. Table 14.2 shows the three genotypes of a two-allele locus with their genotypic frequencies in the whole population. These frequencies come from Table 3.1, $\bar{p}$ and $\bar{q}$ being the gene frequencies in the whole population. Then the third column gives the genotypic values assigned as in Fig. 7.1. The value and

TABLE 14.2

Genotype	Frequency	Value	Frequency × Value
A_1A_1	$\bar{p}^2 + \bar{p}\bar{q}F$	$+a$	$\bar{p}^2a + \bar{p}\bar{q}aF$
A_1A_2	$2\bar{p}\bar{q} - 2\bar{p}\bar{q}F$	d	$2\bar{p}\bar{q}d - 2\bar{p}\bar{q}dF$
A_2A_2	$\bar{q}^2 + \bar{p}\bar{q}F$	$-a$	$-\bar{q}^2a - \bar{p}\bar{q}aF$

$$\text{Sum} = a(\bar{p} - \bar{q}) + 2d\bar{p}\bar{q} - 2d\bar{p}\bar{q}F$$
$$= a(\bar{p} - \bar{q}) + 2d\bar{p}\bar{q}(1 - F)$$

frequency of each genotype are multiplied together in the right-hand column, the summation of which gives the contribution of this locus to the population mean. Thus, referring still to the effects of a single locus, we find that a population with inbreeding coefficient F has a mean genotypic value:

$$M_F = a(\bar{p} - \bar{q}) + 2d\bar{p}\bar{q}(1 - F) \qquad \dots\dots(14.1)$$
$$= M_0 - 2d\bar{p}\bar{q}F \qquad \dots\dots(14.2)$$

where M_0 is the population mean before inbreeding, from equation 7.2. The change of mean resulting from inbreeding is therefore $-2d\bar{p}\bar{q}F$. This shows that a locus will contribute to a change of mean value on inbreeding only if d is not zero; in other words if the value of the heterozygote differs from the average value of the homozygotes. This conclusion, though demonstrated in detail only for two alleles at a locus, is equally valid for loci with more than two alleles. The following general conclusions can therefore be drawn: that a change of mean value on inbreeding is a consequence of dominance at the loci concerned with the character, and that the direction of the change is toward the value of the more recessive alleles. The dominance may be partial or complete, or it may be overdominance; all that is necessary for a locus to contribute to a change of mean is that the heterozygote should not be exactly intermediate between the two homozygotes. Equation *14.2* shows also that the magnitude of the change of mean depends on the gene frequencies. It is greatest when $\bar{p}\bar{q}$ is maximal: that is, when $\bar{p} = \bar{q} = \frac{1}{2}$. Genes at intermediate frequencies therefore contribute more to a change of mean than genes at high or low frequencies, other things being equal.

Now let us consider the combined effect of all the loci that affect the character. In so far as the genotypic values of the loci combine additively, the population mean is given by summation of the contributions of the separate loci, thus:

$$M_F = \Sigma a(\bar{p} - \bar{q}) + 2(\Sigma d\bar{p}\bar{q})(1 - F) \qquad \ldots \ldots (14.3)$$
$$= M_0 - 2F\Sigma d\bar{p}\bar{q} \qquad \ldots \ldots (14.4)$$

and the change of mean on inbreeding is $-2F\Sigma d\bar{p}\bar{q}$.

These expressions show what are the circumstances under which a metric character will show a change of mean value on inbreeding. The chief one is if the dominance of the genes concerned is preponderantly in one direction; i.e. if there is directional dominance. If the genes that increase the value of the character are dominant over their alleles that reduce the value, then inbreeding will result in a reduction of the population mean, i.e. a change in the direction of the more recessive alleles. The contribution of each locus, however, depends also on its gene frequencies, those with intermediate frequencies having the greatest effect on the change of mean value.

We have now reached two conclusions about the effects of inbreeding, one from observation—that inbreeding reduces fitness; the other from theory—that the change is in the direction of the more recessive alleles. Putting these two conclusions together leads to the generalisation, already familiar from Mendelian genetics, that deleterious alleles tend to be recessive.

Another conclusion that can be drawn from equation *14.4* is that when loci combine additively the change of mean on inbreeding should be directly proportional to the coefficient of inbreeding. In other words the change of mean should be a straight line when plotted against F. Two examples of experimentally observed inbreeding depression are illustrated in Fig. 14.1.

On the whole the observed inbreeding depression does tend to be linear with respect to F, and this might be taken as evidence that epistatic interaction between loci is not of great importance. There are, however, several practical difficulties that stand in the way of drawing firm conclusions from observations of the rate of inbreeding depression. One is that as inbreeding proceeds and reproductive capacity deteriorates, it soon becomes impossible to avoid the loss of

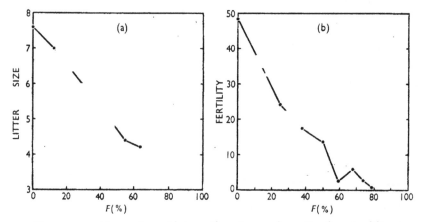

FIG. 14.1. Examples of inbreeding depression affecting fertility. (*a*) Litter-size in mice (original data). Mean number born alive in 1st litters, plotted against the coefficient of inbreeding of the litters. The first generation was by double-first-cousin mating; thereafter by full-sib mating. No selection was practised. (*b*) Fertility in *Drosophila subobscura*. Mean number of adult progeny per pair per day, plotted against the inbreeding coefficient of the parents. Consecutive full-sib matings. (Redrawn from Hollingsworth & Smith, 1955.)

some lines. The survivors are then a selected group to which the theoretical expectations no longer apply. Thus precise measurement of the rate of inbreeding depression can generally be made only over the early stages, before the inbreeding coefficient reaches high levels. Another difficulty, met with particularly in the study of mammals, arises from maternal effects. Maternal qualities are among the most sensitive characters to inbreeding depression. The effect of inbreeding on another character that is influenced by maternal effects is therefore two-fold: part being attributable to the inbreeding of the individuals measured and part to the inbreeding in the mothers. So the relationship between the character measured and the coefficient of inbreeding cannot be depicted in any simple manner. In consequence of these difficulties reliable conclusions cannot easily be drawn from the exact form of the inbreeding depression observed in experiments.

EXAMPLE 14.1. The complications arising from maternal effects may be illustrated by litter size in pigs and mice. Litter size is a composite character, which is partly an attribute of the mother and partly an attribute of the young in the litter. It is therefore influenced both by the inbreeding of the mother and by the inbreeding of the young, and these two influences are difficult to disentangle in practice. Studies on pigs (Dickerson *et al.*, 1954) have shown that the reduction of litter size due to inbreeding in the mother alone is about 0·20 young per 10 per cent of inbreeding; and the reduction due to inbreeding in the young alone is about 0·17 young per 10 per cent of inbreeding. Thus the effects of inbreeding in the mother and in the young are about equally important. A small experiment with mice (original data) gave much the same picture. A rough separation of the effects of inbreeding in the mother and in the young was made by means of crosses between lines after 2 or 3 generations of sib mating. (The justification for regarding this as a measure of the inbreeding depression will be explained in the next section.) The mean litter sizes, arranged according to the coefficient of inbreeding of the mothers and of the young, are given in the table.

		Inbreeding coefficient of mothers		
		0%	37·5%	50%
Inbreeding coefficient of young	0%	8·2	7·5	7·3
	50%	—	6·3	—
	59%	—	—	5·8

The three comparisons in the first row show the effect of inbreeding in the mothers, and give values of 0·19, 0·18 and 0·16 for the reduction of litter size per 10 per cent of inbreeding. The comparisons in the second and third column show the effect of inbreeding in the young, and give values of 0·24 and 0·25 for the reduction per 10 per cent of inbreeding. Thus inbreeding in the young had rather more effect than inbreeding in the mother. These results, however, should not be taken as being characteristic of mice in general.

The effect of selection. The neglect of selection during inbreeding is an unrealistic omission because natural selection cannot be wholly avoided even in laboratory experiments. Since inbreeding tends to reduce fitness, natural selection is likely to oppose the inbreeding process by favouring the least homozygous individuals.

The balance between selection and the dispersion of gene frequencies was discussed in Chapter 4, and the only further point that need be added here is that the operation of natural selection makes the inbreeding depression dependent on the rate of inbreeding. One must distinguish between the state of dispersion of gene frequencies and the coefficient of inbreeding as computed from the population size or the pedigree relationships. The state of dispersion is what determines the amount of inbreeding depression; the coefficient of inbreeding is a measure of the state of dispersion only in the absence of selection. When selection operates, the state of dispersion will be less than that indicated by the coefficient of inbreeding, and the discrepancy between the two will be greater when the rate of inbreeding is slower, because the selection will then be relatively more potent. Therefore one must expect the inbreeding depression caused by a given increase of the computed coefficient of inbreeding to be less when inbreeding is slow than when it is rapid.

HETEROSIS

Complementary to the phenomenon of inbreeding depression is its opposite, "hybrid vigour" or *heterosis*. When inbred lines are crossed, the progeny show an increase of those characters that previously suffered a reduction from inbreeding. Or, in general terms, the fitness lost on inbreeding tends to be restored on crossing. That the phenomenon of heterosis is simply inbreeding depression in reverse can be seen by consideration of how the population mean depends on the coefficient of inbreeding, as shown in equation *14.4*. Consider, as before, a population subdivided into a number of lines. If the lines are crossed at random, the average coefficient of inbreeding in the cross-bred progeny reverts to that of the base population. Thus, if a number of crosses are made at random between the lines, the mean value of any character in the cross-bred progeny is expected to be the same as the population mean of the base population. In other words, the heterosis on crossing is expected to be equal to the depression on inbreeding. Furthermore, if the population is continued after the crossing by random mating among the cross-bred and subsequent generations, the coefficient of inbreeding will remain unchanged, and the population mean is consequently expected to remain at the level of the base population. We may, thus, make the following generalisation on theoretical grounds: that, in the absence of selection, inbreeding followed by crossing of the lines in a large population is not expected to make any permanent change in the population mean.

EXAMPLE 14.2. An experiment with mice (R. C. Roberts, unpublished) was designed to test the theoretical expectation that in the absence of selection the heterosis on crossing should be equal to the depression on inbreeding. The character studied was litter size. Thirty lines taken from a random-bred population were inbred by 3 consecutive generations of full-sib mating, bringing the coefficient of inbreeding up to 50 per cent in the litters and 37·5 per cent in the mothers. No selection was practised during the inbreeding, and only 2 of the 30 lines were lost as a consequence of their inbreeding depression.

	Litter size
Before inbreeding	8·1
Inbred (litters: $F = 50\%$)	5·7
Cross-bred	8·5

After the third generation of inbreeding, crosses were made at random between the lines, and in the next generation crosses between the F_1's were made so as to give cross-bred mothers with non-inbred young. The mean litter sizes observed at the different stages are given in the table. The inbreeding depression was 2·4 and the heterosis 2·8; the two are equal within the limits of experimental error.

Single crosses. The foregoing theoretical conclusions refer to the average of a large number of crosses between lines derived from a single base population. In practice, however, one is often interested in a somewhat different problem, namely the heterosis shown by a particular cross between two lines, or between two populations which may have no known common origin. To refer the changes of mean value to changes of inbreeding coefficient would be inappropriate under these circumstances, and the theoretical basis of the heterosis is better expressed in terms of the gene frequencies in the two lines. We may recall from Chapter 3 that inbreeding leads to a dispersion of gene frequencies among the lines, the lines becoming differentiated in gene frequency as inbreeding proceeds; and the coefficient of inbreeding is a means of expressing the degree of differentiation (equation *3.14*). In turning from the inbreeding coefficient to the gene frequencies as a basis for discussion we are therefore turning from the general, or average, consequence of crossing, to the particular circumstances in two lines.

Let us, then, consider two populations, referred to as the "parent populations," both random-bred though not necessarily large. The parent populations are crossed to produce an F_1 or "first cross-bred generation," and the F_1 individuals are mated together at random to produce an F_2 or "second cross-bred generation." The amount of heterosis shown by the F_1 or the F_2 will be measured as the deviation from the mid-parent value, i.e. as the difference from the mean of the two parent populations. First consider the effects of a single locus with two alleles whose frequencies are p and q in one population, and p' and q' in the other. Let the difference of gene frequency between the two populations be y, so that $y = p - p' = q' - q$. The algebra is then simplified by writing the gene frequencies p' and q' in the second population as $(p - y)$ and $(q + y)$. Let the genotypic values be $a, d, -a$, as before. They are assumed to be the same in the two populations, epistatic interaction being disregarded. We have to find the mean of each parent population and the mid-parent value; then the mean of the F_1 and the mean of the F_2. The parental means, M_{P_1} and M_{P_2}, are found from equation 7.2. They are

$$M_{P_1} = a(p - q) + 2dpq$$
$$M_{P_2} = a(p - y - q - y) + 2d(p - y)(q + y)$$
$$= a(p - q - 2y) + 2d[pq + y(p - q) - y^2]$$

The mid-parent value is

$$M_{\bar{P}} = \tfrac{1}{2}(M_{P_1} + M_{P_2})$$
$$= a(p - q - y) + d[2pq + y(p - q) - y^2] \quad \ldots\ldots(14.5)$$

When the two populations are crossed to produce the F_1, individuals taken at random from one population are mated to individuals taken at random from the other population. This is equivalent to taking genes at random from the two populations, as shown in Table 14.3. The F_1 is therefore constituted as follows:

TABLE 14.3

FREQUENCIES OF ZYGOTES IN THE F_1

			Gametes from P_1	
			A_1	A_2
			p	q
Gametes	A_1	$p - y$	$p(p - y)$	$q(p - y)$
from P_2	A_2	$q + y$	$p(q + y)$	$q(q + y)$

duals taken at random from the other population. This is equivalent to taking genes at random from the two populations, as shown in Table 14.3. The F_1 is therefore constituted as follows:

Genotypes	A_1A_1	A_1A_2	A_2A_2
Frequencies	$p(p-y)$	$2pq+y(p-q)$	$q(q+y)$
Genotypic values	a	d	$-a$

The mean genotypic value of the F_1 is therefore:

$$M_{F_1} = a(p^2 - py - q^2 - qy) + d[2pq + y(p-q)]$$
$$= a(p - q - y) + d[2pq + y(p-q)] \qquad \cdots \cdots (14.6)$$

The amount of heterosis, expressed as the difference between the F_1 and the mid-parent values, is obtained by subtracting equation 14.5 from equation 14.6:

$$H_{F_1} = M_{F_1} - M_{\bar{P}}$$
$$= dy^2 \qquad \cdots \cdots (14.7)$$

Thus heterosis, just like inbreeding depression, depends for its occurrence on dominance. Loci without dominance (i.e. loci for which $d = 0$) cause neither inbreeding depression nor heterosis. The amount of heterosis following a cross between two particular lines or populations depends on the square of the difference of gene frequency (y) between the populations. If the populations crossed do not differ in gene frequency there will be no heterosis, and the heterosis will be greatest when one allele is fixed in one population and the other allele in the other population.

Now consider the joint effects of all loci at which the two parent populations differ. In so far as the genotypic values attributable to the separate loci combine additively, we may represent the heterosis produced by the joint effects of all the loci as the sum of their separate contributions. Thus the heterosis in the F_1 is

$$H_{F_1} = \Sigma dy^2 \qquad \cdots \cdots (14.8)$$

If some loci are dominant in one direction and some in the other their effects will tend to cancel out, and no heterosis may be observed, in spite of the dominance at the individual loci. The occurrence of heterosis on crossing is therefore, like inbreeding depression, dependent on directional dominance, and the absence of heterosis is not sufficient ground for concluding that the individual loci show no dominance.

Before we go on to consider the F_2 it is perhaps worth noting that the formulation of the heterosis in terms of the square of the difference of gene frequency, in equations 14.7 and 14.8, is quite in line with the previous formulation of the inbreeding depression in terms of the coefficient of inbreeding. If we envisage once more the whole population subdivided into lines, and we suppose pairs of lines to be taken at random, then the mean squared difference of gene frequency between the pairs of lines will be equal to twice the variance of gene frequency among the lines. That is: $(\overline{y^2}) = 2\sigma_q^2$. And, by equation 3.14, $2\sigma_q^2 = 2\bar{p}\bar{q}F$. Therefore the mean amount of heterosis shown by crosses between random pairs of lines is equal to the inbreeding depression as given in equation 14.2, though of opposite sign.

Now let us consider the F_2 of a particular cross of two parent populations, the F_2 being made by random mating among the individuals of the F_1. In consequence of the random mating, the genotype frequencies in the F_2 will be the Hardy-Weinberg frequencies corresponding to the gene frequency in the F_1. The mean genotypic value of the F_2 is then easily derived by application of equation 7.2. The gene frequency in the F_1, being the mean of the gene frequencies in the two parent populations, is $(p - \frac{1}{2}y)$ for one allele, and $(q + \frac{1}{2}y)$ for the other. Putting these gene frequencies in place of p and q respectively in equation 7.2 gives the mean genotypic value of the F_2 as:

$$M_{F_2} = a(p - \tfrac{1}{2}y - q - \tfrac{1}{2}y) + 2d(p - \tfrac{1}{2}y)(q + \tfrac{1}{2}y)$$
$$= a(p - q - y) + d[2pq + y(p - q) - \tfrac{1}{2}y^2] \quad \ldots\ldots(14.9)$$

The amount of heterosis shown by the F_2 is the difference between the F_2 and mid-parent values. So, from equations 14.5 and 14.9,

$$H_{F_2} = M_{F_2} - M_{\bar{P}}$$
$$= \tfrac{1}{2}dy^2$$
$$= \tfrac{1}{2}H_{F_1} \quad \ldots\ldots(14.10)$$

We find therefore that the heterosis shown by the F_2 is only half as great as that shown by the F_1. In other words, the F_2 is expected to drop back half-way from the F_1 value toward the mid-parent value. At first sight this conclusion may seem to contradict the one arrived at earlier, when we were considering crosses between many lines, the F_1 and F_2 means then being equal. The difference between the two situations is that an F_2 made by random mating among a large number of different crosses has the same inbreeding coefficient as the F_1. But an F_2 made from an F_1 derived from a single cross has inevitably an increased inbreeding coefficient. If the inbreeding coefficient is worked out in the manner described in Example 5.2, it will be found to be half the inbreeding coefficient of the parent lines. The change of mean from F_1 to F_2 may therefore be regarded as inbreeding depression. It cannot be overcome by having a large number of parents of the F_2 because the restriction of population size that causes the inbreeding has already been made in the single cross of only two lines, or parent populations. There need, however, be no further rise of the inbreeding coefficient in the F_3 and subsequent generations. Provided, therefore, that there is no other reason for the gene frequency to change, the population mean will be the same in the generations following as in the F_2.

That the heterosis expected in the F_2 is half that found in the F_1 is equally true when the joint effects of all loci are considered, provided that epistatic interaction is absent. The conclusion for a single locus was based on the principle that Hardy-Weinberg equilibrium is attained by a single generation of random mating. It will be remembered from Chapter 1 (p. 19), however, that this is not true with respect to genotypes at more than one locus considered jointly. Therefore if there is epistatic interaction, the population mean will not reach its equilibrium value in the F_2, but will approach it more or less rapidly according to the number of interacting loci and the closeness of the linkage between them. The existence of epistatic interaction is intimately connected with the scale of measurement, but this matter will not be discussed until Chapter 17. Here we need only note that for reasons connected with the scale of measurement the halving of the heterosis in the F_2 expected on theoretical grounds is not often found at all exactly in practice, though the F_2 usually falls somewhere between the F_1 and mid-parent values. Some examples from plants of the heterosis observed in the F_1 and F_2 generations are illustrated in Fig. 14.2. It will be noticed that with some of the characters shown, the F_1 and F_2 are lower in value than the mid-parent, and the heterosis is consequently negative in sign. This is in no way inconsistent with our definition of heterosis as the difference between the F_1 or F_2 and the mid-parent value. The sign of the difference depends simply on the nature of the measurement. For example, the character "days to first fruit," represented in the lower graphs, shows heterosis of negative sign: but if the character were called "speed of development" and expressed as a reciprocal of time the heterosis would be positive in sign.

The relative amount of heterosis observed in the F_1 and F_2

generations is complicated also by the existence of maternal effects, particularly in mammals. A character subject to a maternal effect,

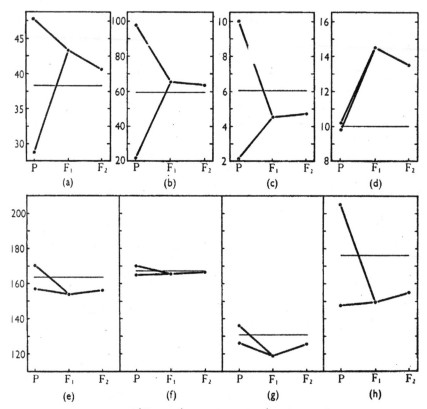

FIG. 14.2. Some illustrations of heterosis observed in crosses between pairs of highly inbred strains of plants. The points show the mean values of the two parent strains, the F_1 and the F_2 generations. The mid-parent values are shown by horizontal lines. Graph (a) refers to tobacco, *Nicotiana rustica* (data from Smith, 1952). All the other graphs refer to tomatoes, *Lycopersicon* (Data from Powers, 1952). The characters represented are:

    (a) Height of plant (in.)
    (b) Mean weight of one fruit (gm.)
    (c) Number of locules per fruit
    (d) Mean weight per locule (gm.)
    (e)–(h) Mean time in days between the planting of the seed and the ripening of the first fruit, in 4 different crosses.

such as litter size, is divided between two generations. The maternally determined component of the character may be expected to follow the same general pattern of heterosis in the F_1 and F_2 as we have just discussed, but it will be one generation out of phase with the non-maternal part of the character. Thus the heterosis observed in the F_1 is attributable to the non-maternal part, the maternal effect being still at the inbred level. In the F_2, however, the non-maternal part will lose half the heterosis as explained above, but the maternal effect will now show the full effect of its heterosis since the mothers are now in the F_1 stage. This rather complicated situation may perhaps be more

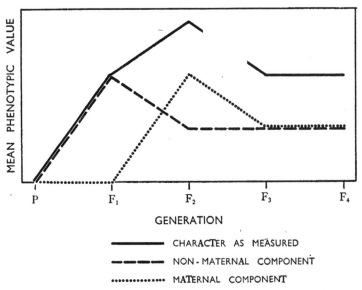

FIG. 14.3. Diagram of the heterosis expected in a character subject to a maternal effect, when two lines are crossed and the F_2 is made by random mating among the F_1. The maternal and non-maternal components of the character separately are here supposed to show equal amounts of heterosis, and to combine by simple addition to give the character as it is measured.

readily grasped from the diagrammatic representation in Fig. 14.3. As a result of maternal effects, therefore, the loss of heterosis in the F_2 and subsequent generations is usually less noticeable with animals than with plants, and experiments of great precision would be required to detect any regular pattern.

Wide crosses. We have seen that the amount of heterosis shown by a particular cross depends, among other things, on the differences of gene frequency between the two populations crossed. This would seem to indicate that the amount of heterosis would increase with the degree of genetic differentiation between the two populations and would be limited only by the barrier of interspecific sterility. This, however, is not true. Crosses between subspecies, or between local races, taken from the wild often fail to show heterosis, particularly in characters closely related to fitness which show heterosis in crosses between less differentiated laboratory populations. Indeed the F_1's of wide crosses are often less fit than the parent populations. Much of the evidence about such crosses comes from studies of wild populations of *Drosophila pseudoobscura* and other species, (see Dobzhansky, 1950; Wallace and Vetukhiv, 1955). Though wide crosses may not show heterosis in fitness, they do often show heterosis in certain characters, particularly growth rate in plants. Dobzhansky (1950, 1952), who drew attention to this, refers to heterosis in fitness as "euheterosis" and to heterosis in a character that does not confer greater fitness as "luxuriance."

The error in extending our earlier conclusion to wide crosses arises from the fact that we have assumed epistatic interaction between loci to be negligible, an assumption that is probably justified for crosses between breeds of domestic animals or between laboratory populations, but is obviously not justified in the case of crosses between differentiated wild populations. The existing genetic differentiation between wild populations has, for the most part, arisen by evolutionary adaptation to the local conditions. Adaptation to local conditions or to a particular way of life involves many different characters, both structural and functional, because the fitness of the organism depends on the harmonious interrelations of all its parts. If two populations adapted to different ways of life are crossed, the cross-bred individuals will be adapted to neither, and will consequently be less fit than either of the parent populations. The effect of this evolutionary adaptation on the genetic structure of the populations is as follows. The genes A_1 and B_1, say, are selected in one population because together they increase fitness, though either one separately may not; while, in another population living under different conditions, the genes A_2 and B_2 are selected for similar reasons. In respect of fitness, therefore, there is epistatic interaction between these two loci. But if these pairs of genes become fixed throughout the two populations, A_1 and B_1 in one and A_2 and B_2 in the other, and so become part of their constant genetic structure, the variation arising from this interaction will disappear. Within any one population, therefore, we may find very little epistatic variation, and the interaction will become apparent as a cause of variation between individuals only in a cross-bred population in which there is segregation at both interacting loci.

The idea that the genetic structure of a natural population evolves as a whole, so that the selection pressure on any one locus is dependent on the alleles present at many of the other loci, is expressed in the terms "coadaptation" and "integration," used to describe the genetic structure of natural populations. (For general discussions of these concepts, see Dobzhansky, 1951b; Lerner, 1954, 1958; Wright, 1956.) The important point for us to note is this. The property of coadaptation, or integration, assumes primary importance only when different populations are to be compared and when the results of crossing adaptively differentiated populations are to be studied; it is of less importance in the genetic study of a single population. In this book we are chiefly concerned with the genetic variation within a population: that is, the variation arising from the segregation of genes in the population. Some of this variation arises from epistatic ineraction between the genes segregating at different loci, which is the raw material, as it were, from which coadaptation could evolve if the population were to become subdivided. But the amount of this epistatic variation within a population is probably seldom very large, and moreover it is seldom necessary to distinguish it from other sources of non-additive genetic variance.

CHAPTER 15

INBREEDING AND CROSSBREEDING:
II. Changes of Variance

The effect of inbreeding on the genetic variance of a metric character is apparent, in its general nature, from the description of the changes of gene frequency given in Chapter 3. Again, we have to imagine the whole population, consisting of many lines. Under the dispersive effect of inbreeding, or random drift, the gene frequencies in the separate lines tend toward the extreme values of 0 or 1, and the lines become differentiated in gene frequency. Since the mean genotypic value of a metric character depends on the gene frequencies at the loci affecting it, the lines become differentiated, or drift apart, in mean genotypic value. And, since the genetic components of variance diminish as the gene frequencies tend toward extreme values (see Fig. 8.1), the genetic variance within the lines decreases. The general consequence of inbreeding, therefore, is a redistribution of the genetic variance; the component appearing between the means of lines increases, while the component appearing within the lines decreases. In other words, inbreeding leads to genetic differentiation between lines and genetic uniformity within lines. The differentiation is illustrated from experimental data in Fig. 15.1.

The subdivision of an inbred population into lines introduces an additional observational component of variance, the between-line component, and it is not surprising that this adds a considerable complication to the theoretical description of the components of genetic variance. Indeed, a full theoretical treatment of the redistribution of variance has not yet been achieved. Here we shall attempt no more than a brief description of the main outlines, and for this we shall have to make some simplifications. In particular we shall entirely neglect the interaction component of genetic variance arising from epistasis. For detailed treatment of various aspects of the problem, and for references, see Kempthorne (1957, Ch. 17). After this description of the redistribution of genetic variance we shall consider changes of environmental variance. The greater sensitivity of inbred individuals to environmental sources of variation was mentioned earlier, in Chapter 8. This phenomenon interferes with the experimental study of the changes of variance, and until it is better understood we cannot put much reliance on the theoretical

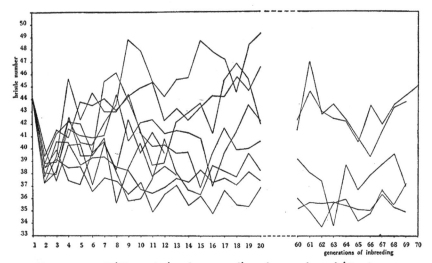

FIG. 15.1. Differentiation between lines by random drift, shown by abdominal bristle number in *Drosophila melanogaster*. The graphs show the mean bristle number in each of 10 lines during full-sib inbreeding without artificial selection. (From Rasmuson, 1952; reproduced by courtesy of the author and the editor of *Acta Zoologica*.)

expectations concerning variance being manifest in the observable phenotypic variance. Finally, in this chapter, we shall discuss the use of inbred animals for experimental purposes.

REDISTRIBUTION OF GENETIC VARIANCE

The redistribution of variance arising from additive genes (i.e. genes with no dominance) is easily deduced. This is because with additive genes the proportions in which the original variance is distributed within and between lines does not depend on the original gene frequencies. When there is dominance, however, we cannot deduce the changes of variance without a knowledge of the initial gene frequencies. This not only adds considerably to the mathematical complexity, but it renders a general solution impossible. We shall first consider the case of additive genes, and then very briefly indicate the conclusions arrived at for dominant genes. The effect of selection will not be specifically discussed. We need only note that natural selection will tend to render the actual state of dispersion of gene frequencies less than that indicated by the inbreeding coefficient computed from the population size or pedigree relationships. Therefore we must expect the redistribution of genetic variance to proceed at a slower rate than the theoretical expectation, and we must expect the discrepancy to be greater when inbreeding is slow than when it is rapid.

No dominance. What follows refers to the variance arising from additive genes: it does not apply to the additive variance arising from genes with dominance. The conclusions therefore apply, strictly speaking, only to characters which show no non-additive variance. They serve, however, to indicate the general effect of inbreeding on variance, and may be taken as a fair approximation to what is expected of characters such as bristle number in *Drosophila*, that show little non-additive genetic variance. The description to be given refers to slow inbreeding, and is not strictly true of rapid inbreeding by sib-mating or self-fertilisation. The redistribution of the variance under rapid inbreeding is, however, not very different except in the first few generations.

Consider first a single locus. When there is no dominance the genotypic variance in the base population, given in equation 8.7, becomes

$$V_G = 2p_0 q_0 a^2$$

The variance within any one line is

$$V_G = 2pq a^2$$

where p and q are the gene frequencies in that line. The mean variance within lines is

$$V_{Gw} = 2(\overline{pq}) a^2$$

where $(\overline{pq})$ is the mean value of pq over all lines. Now, $2(\overline{pq})$ is the overall frequency of heterozygotes in the whole population, which, by Table 3.1, is equal to $2p_0 q_0 (1 - F)$, where F is the coefficient of inbreeding. Therefore

$$V_{Gw} = 2p_0 q_0 a^2 (1 - F)$$
$$= V_G (1 - F)$$

and this remains true when summation of the variances is made over all loci. Thus the within-line variance is $(1-F)$ times the original variance, and as F approaches unity the within-line variance approaches zero.

Now let us consider the between-line variance. This is the variance of the true means of lines, and would be estimated from an analysis of variance as the between-line component. For a single locus, still with no dominance, the mean genotypic value of a line with gene frequency p and q is obtained from equation 7.2 as

$$M = a(p-q)$$
$$= a(1-2q)$$

Thus we want to find the variance of $(a-2aq)$. Now, in general, $\sigma^2_{(X-Y)} = \sigma^2_X + \sigma^2_Y$, if X and Y are uncorrelated. Since in this case a is constant from line to line (epistasis being assumed absent) it has no variance, and so

$$\sigma^2_M = \sigma^2_{(2aq)}$$

Again, in general, $\sigma^2_{(KX)} = K^2 \sigma^2_X$ when K is a constant. So

$$\sigma^2_M = 4a^2 \sigma^2_q$$
$$= 4a^2 p_0 q_0 F \quad \text{(from 3.14)}$$
$$= 2F V_G$$

and this also remains true when summation is made over all loci. Thus the between-line genetic variance is $2F$ times the genetic variance in the base population.

The partitioning of the genetic variance into components as explained above is summarised in Table 15.1. The total genetic

TABLE 15.1

Partitioning of the variance due to additive genes in a population with inbreeding coefficient F, when the variance due to additive genes in the base population is V_G.

Between lines	$2F V_G$
Within lines	$(1-F)V_G$
Total	$(1+F)V_G$

variance in the whole population is the sum of the within-line and between-line components, and is equal to $(1+F)$ times the original genetic variance. (This is true also of close inbreeding.) Thus when inbreeding is complete the genetic variance in the population as a whole is doubled, and all of it appears as the between-line component.

The genetic variance within lines, before inbreeding is complete, is partitioned within and between the families of which the lines are composed. Under slow inbreeding with random mating within the lines, it is partitioned equally within and between full-sib families. The covariance of relatives within the lines is just as described in Chapter 9, each line being a separate random-breeding population with a total genetic variance of $(1-F)V_G$, on the average. From this we can deduce what the heritability is expected to be within any one line. It will be $(1-F)V_G/[(1-F)V_G+V_E]$, and this reduces to

$$h_t^2 = \frac{h_0^2(1-F_t)}{1-h_0^2 F_t} \qquad \ldots\ldots(15.1)$$

where h_t^2 and F_t are the heritability within lines and the inbreeding coefficient at time t, and h_0^2 is the original heritability in the base population. This shows how the heritability is expected to decline with the inbreeding in a small population. The formula, however, is applicable only to characters with no non-additive variance, and in the absence of selection. The operation of natural selection renders the reduction of the heritability less than expected, especially under slow inbreeding. This point has been demonstrated experimentally with *Drosophila* (Tantawy and Reeve, 1956).

Dominance. The components of variance arising from additive genes will have been seen to be independent of the gene frequencies in the base population. When we consider genes with any degree of dominance, however, we find that the changes of variance on inbreeding depend on the initial gene frequencies, and this makes it impossible to give a general solution in terms of the genetic variance present in the base population. We shall therefore do no more than give the conclusions arrived at by A. Robertson (1952) for the case of fully dominant genes, when the recessive allele is at low frequency. This is the situation most likely to apply to variation in fitness arising from deleterious recessive genes, though the effects of selection are here disregarded. Fig. 15.2 shows the redistribution of variance arising from recessive genes at a frequency of $q=0.1$ in the base population. Fig. 15.2(*a*) refers to full-sib mating with only one family in each line, and Fig. 15.2(*b*) refers to slow inbreeding. A surprising feature of the conclusions is that the within-line variance at first increases, reaching a maximum when the coefficient of inbreeding is a little under 0.5, and it remains at a fairly high level until the coefficient of inbreeding approaches 1. The reason, in general terms, for the apparent anomaly that the variation within lines increases during the first stages of inbreeding, can be seen from a consideration of the relationship between the gene frequency and the variance arising from a dominant gene shown in Fig. 8.1(*b*). The gene frequency is taken to start at a value of 0·1, and on inbreeding it

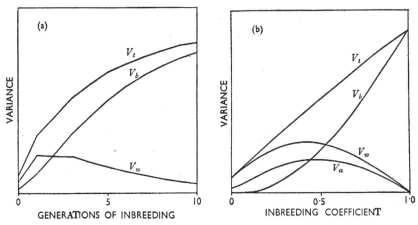

FIG. 15.2. Redistribution of variance arising from a single fully recessive gene with initial frequency $q_0 = 0·1$. (*a*) with full-sib mating, (*b*) with slow inbreeding. (From A. Robertson, 1952; reproduced by courtesy of the author and the editor of *Genetics*.)

V_t = total genetic variance.
V_b = between-line component.
V_w = within-line component.
V_a = additive genetic variance within lines.

will increase in some lines and decrease in others, the increase being on the average equal in amount to the decrease. But examination of the graph shows that an increase of gene frequency by a certain amount will increase the variance more than a decrease of the same amount will reduce it. Therefore, on the average, the variance within the lines will increase in the early stages of inbreeding. This increase of variance would be detectable in practice only if a substantial part of the genetic variance were due to recessive genes at low frequencies.

Practical considerations. The extent to which the theoretical changes of variance described in this chapter can be observed in practice depends on how much environmental variance is present. The precise estimation of variance requires a large number of observations and the estimates obtained in practice are usually subject to rather large deviations due to the chances of sampling. Consequently the changes of variance must usually be quite substantial before they are likely to be readily detected. The genotypic variance, moreover, seldom constitutes the major part of the phenotypic variance. Therefore, in relation to the original phenotypic variance, the expected changes due to inbreeding are usually rather small, and this renders their detection all the more difficult. Furthermore, the detection of the expected changes of phenotypic variance is entirely dependent on the constancy of the environmental variance, and this cannot be assumed without evidence, as we shall show in the next section. For these reasons, and also because of the simplifications we have had to make, we must bear in mind the uncertainties in the connexion between what is expected and what may be observed in the phenotypic variance.

CHANGES OF ENVIRONMENTAL VARIANCE

Several times in previous chapters we have referred to the fact that the environmental component of variance may differ according to the genotype; in particular that inbred individuals often show more environmental variation than non-inbred individuals. This fact has been revealed by many experiments in which the variances of inbreds and of hybrids have been compared. Any difference of phenotypic variance between highly inbred lines and the F_1 between them (i.e. the "hybrid") must be attributed to a difference of the environmental component, because the genetic variance is negligible in amount in the hybrids as well as in the inbred lines. The greater susceptibility of inbreds than of hybrids to environmental sources of variation has been observed in a wide variety of characters and organisms. Some examples are cited in Table 15.2; others will be found in the review by Lerner (1954).

The cause of the greater environmental variance of inbreds is not yet fully understood. It has been suggested that the possession of different alleles at specific loci endows the hybrids with greater "biochemical versatility" (Robertson and Reeve, 1952b), which enables them to adjust their development and physiological mechanisms to the circumstances of the environment: in other words that developmental and physiological homeostasis is improved by allelic diversity. On the other hand, it has been suggested (Mather, 1953a) that the reduced homeostatic power of inbreds is to be regarded as a manifestation of inbreeding depression: homeostatic power is likely to be an important aspect of fitness, and would therefore be expected, like other aspects of fitness, to decline on inbreeding. The underlying mechanism, we may presume, would be directional dominance, genes that increase homeostatic power tending on the average to be

TABLE 15.2

COMPARISONS OF PHENOTYPIC VARIANCE IN INBREDS AND HYBRIDS

The figures are the averages of the inbred lines, and of the F_1's where more than one cross was made. $(C.V.)^2 = $ Squared coefficient of variation.

			Inbreds	Hybrids
Drosophila melanogaster—wing length (Robertson and Reeve, 1952*b*) $(C.V.)^2$. 6 inbreds and 6 F_1's			2·35	1·24
Mice—duration of "Nembutal" anaesthesia (McLaren and Michie, 1956*b*). Log minutes. 2 inbreds and 1 F_1			0·0665	0·0165
Mice—age at opening of vagina (Yoon, 1955). Days. 3 inbreds and 2 F_1's			51·7	17·4
Mice—weight at ages given (Chai, 1957) $(C.V.)^2$. 2 inbreds and 1 F_1	Birth		119	59
	3 weeks		98	47
	60 days		24	19
Rats—weight at 90 days (Livesay, 1930.) $(C.V.)^2$. 3 inbreds and 2 F_1's			522	170

dominant over their alleles that decrease it. Lerner (1954) sees a causal connexion between variability and fitness. He believes greater stability to be a general property of heterozygotes and regards it as the cause of their greater fitness. Though the increase of environmental variance on inbreeding is a phenomenon of great theoretical interest and some practical importance, too little is known about it to justify a more detailed discussion of its causes here. Comprehensive discussions will be found in Lerner (1954) and Waddington (1957).

There are, however, two further points in connexion with the phenomenon that should be mentioned. The first is a technical matter. If the mean value of the character differs between inbreds and hybrids, as it frequently does, then it may be difficult to decide on a proper basis for the comparison of the variances. It is necessary to find a measure of the variance that does not merely reflect the difference of mean value, and for this purpose the coefficient of variation is often an appropriate measure. The problem is basically a matter of the choice of scale, and will be discussed again in Chapter 17.

The second point concerns the nature of the environmental variation that is being measured. There is a distinction to be made between the "developmental" variation arising from "accidents of development" on the one hand, and adaptive reponses to changed conditions on the other. The developmental variation is a manifestation of incomplete buffering, or canalisation, of development and is generally regarded as being harmful. Inbreds, in so far as they show a greater amount of developmental variation, are therefore less fit than hybrids; they are less well able to adjust their development to different conditions of the environment so as to achieve the optimal phenotype. An adaptive response, in contrast, is a modification of the phenotypic value that is beneficial to the individual, such as for example the thickening of the coat of mammals in response to low temperature. If the greater fitness of hybrids over inbreds extends to adaptive responses we should therefore expect hybrids to show more variation of this sort than inbreds. Thus the nature of the environmental variation has an important bearing on the interpretation of a difference of variability between inbreds and hybrids.

UNIFORMITY OF EXPERIMENTAL ANIMALS

Inbred strains of laboratory animals, particularly of mice, are widely used as experimental material in pharmacological, physiological, and nutritional laboratories, when uniformity of biological material is desired. In some kinds of work, work for example which demands the absence of immunological reactions, it is genetic uniformity that is required, and abundant experience has shown that the inbred strains of mice fully satisfy this requirement. In spite of doubts about how effective natural selection for heterozygotes may be in delaying the progress towards homozygosity, these strains have been proved in practice to be genetically uniform. In the course of their maintenance, however, strains inevitably become split up into sublines, and it is only within a subline that their genetic uniformity can be relied on.   Recent work, described in the two following

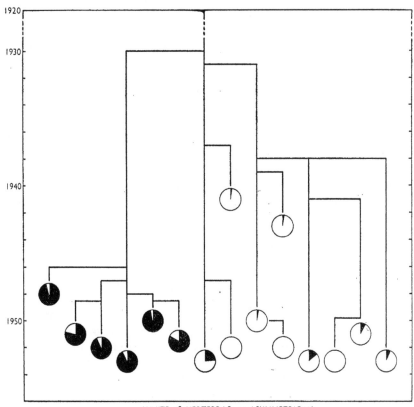

WHITE = 5 VERTEBRAE (+ ASYMMETRICAL)

BLACK = 6 VERTEBRAE

Fig. 15.3.   Differentiation between sublines of the C3H inbred strain of mice, in the number of lumbar vertebrae. Each circle represents a sample of individuals classified for the number of lumbar vertebrae. The proportions of black and white in the circles show the proportions of individuals with 6 and with 5 lumbar vertebrae respectively. (Small proportions of asymmetrical individuals are included with the 5-vertebra classes.) The circles are positioned according to the date of clasification, and arranged according to their pedigree relationships. (Data from McLaren and Michie, 1954.)

examples, has revealed genetic differentiation within two widely used strains of mice, and has shown that differences can sometimes be detected between sublines separated by only a few generations.

EXAMPLE 15.1. The inbred strain of mice known as C3H exhibits variability in the number of lumbar vertebrae, and the sublines differ markedly in this character. Some sublines consist entirely of mice with 5 vertebrae, others entirely of mice with 6, and others with different proportions. The strain originated in 1920 and was split into three main groups of sublines in about 1930, each group being later subdivided further. The number of lumbar vertebrae has been studied in 16 sublines maintained in America and Britain (McLaren and Michie, 1954). The pedigree relationships between these sublines, and the proportions of the two vertebral types in them, are shown in Fig. 15.3. One of the three main groups of sublines has predominantly 6 lumbar vertebrae, and the other two groups predominantly 5. This differentiation between the main groups may have been due to residual segregation in the strain at the time when the main groups became separated. The strain had, however, been full-sib mated for 10 years—probably between 20 and 30 generations—before the separation of the groups, and residual segregation therefore seems unlikely. The sublines within the main groups are differentiated in a manner that points to mutation rather than residual segregation as the cause. The mutational origin of differentiation is more clearly proved in the study described in the next example.

EXAMPLE 15.2. Another inbred strain of mice, known as C57BL, has been the subject of a thorough study by Grüneberg and co-workers (Deol, Grüneberg, Searle, and Truslove, 1957; Carpenter, Grüneberg, and Russell, 1957). Twenty-seven skeletal characters were examined in four main groups of sublines, three maintained in America and one in Britain, the British group being studied in greater detail. The nature and extent of the differentiation found cannot be easily summarised, and therefore we shall only state the conclusions reached about the cause of the differentiation. Each of the four main groups differed from the others in between 7 and 17 out of the 27 characters. The following conclusions were drawn: (1) The differentiation could not reasonably be attributed to residual segregation before the separation of the sublines; and segregation following an accidental outcross was conclusively disproved. (2) Sublines that had been separated for a longer time tended to differ by a greater number of characters than sublines more recently separated. But the magnitude of the difference in any one character was no greater between long-separated sublines than between sublines only recently separated. From this it was concluded that the differences in each character were caused by mutations at single loci. The average difference caused by one mutational step amounted to about 0·6 standard deviation of the character affected.

The study cited in the above example shows that the differences between sublines, though they may be readily detectable, are probably caused by rather few loci. The differentiation is quite small in comparison with the differences between strains or between individuals in a non-inbred population.

In much of the work for which inbred strains are used it is not the genetic uniformity alone that matters, but the phenotypic uniformity. The more variable the animals the larger the number that must be used to attain a given degree of precision in measuring their mean response to a treatment. The value of uniformity is therefore in reducing the number of animals that must be used in an experiment or a test. Inbred animals, however, are costly to produce because of their poor breeding qualities, and the advantage gained from genetic uniformity has to be weighed against the extra cost of the material. If the character to be measured is one of which the phenotypic variance is chiefly environmental in origin, then the absence of genetic variation in an inbred strain will reduce the phenotypic variance by only a small amount. The extra cost of the inbred animals may then outweigh the advantage of their being slightly more uniform than non-inbred animals. The phenotypic uniformity of inbred animals, however, has been taken on trust from the genetical theory of inbreeding, and it seems now that this trust has, to some extent at least, been misplaced. In some characters inbred animals are more phenotypically variable than non-inbred (see Table 15.4) on account of their greatly increased environmental variation. It seems now that for some, perhaps for many, characters the greatest phenotypic uniformity is found in hybrids (i.e. F_1's) produced by crossing two inbred strains. The value of hybrids for work requiring phenotypic uniformity has been discussed by Grüneberg (1954); and by Biggers and Claringbold (1954).

One final point about the use of inbred and hybrid animals may be noted. An inbred strain or the F_1 of two inbred strains has a unique genotype; and that of an inbred, moreover, is one that cannot occur in a natural population. Testing the response to any treatment on one inbred strain or one hybrid is therefore testing it on one genotype. If there are appreciable differences of response between different genotypes, the experimenter is then not justified in describing his results as referring, for example, to "the mouse."

CHAPTER 16

INBREEDING AND CROSSBREEDING:

III. The Utilisation of Heterosis

The crossing of inbred lines plays a major role in the present methods of plant improvement, though in animal improvement it plays a much less important part. In this chapter the genetic principles underlying the use of inbreeding and crossing will be explained, and the various methods described in outline. Technical details, however, will not be given: for these the reader should consult a textbook of plant breeding (e.g. Hayes, Immer, and Smith, 1955). We shall be concerned with outbreeding plants and with animals. But since at first sight the methods applicable to naturally self-fertilising plants are superficially rather like those applicable to outbreeding plants and animals, it will be advisable first to consider very briefly the improvement of self-fertilising plants.

Self-fertilising plants. Each variety of a naturally self-fertilising plant is a highly inbred line, and the only genetic variation within it is that arising from mutation. Genetic improvement can therefore be made only by choosing the best of the existing varieties or by crossing different varieties. The purpose of the crossing is to produce genetic variation on which selection can operate. After a cross has been made, the F_1 and subsequent generations are allowed to self-fertilise naturally. A new population, subdivided into lines, is thus made, and the lines become differentiated as the inbreeding proceeds. Selection is applied by choosing the best lines, which become new and improved varieties. The essential point to note is that what is sought is an improved inbred line, and not a superior crossbred generation: the purpose of the crossing is to provide genetic variation and not to produce heterosis. The process of crossing and selection among the subsequent lines may be repeated cyclically. If two good lines are selected out of the first cross, these may be crossed and a second cycle of selection applied to the derived lines. The genetic properties of a population derived from a cross of two highly inbred lines, such as two varieties of a self-fertilising plant, are peculiar in that all segregating genes have a frequency of 0.5 in the population as a whole. This greatly simplifies the theoretical description of the variances and covariances. Special methods of analysis applicable to such populations have been developed which lead to a separation of the additive, dominance, and epistatic effects, and so provide a guide to the possibilities of improvement in the population of lines derived from a particular cross. For a description of these methods, see Mather (1949), Hayman (1958), and Kempthorne (1957, Ch. 21) where other references are given.

Outbreeding plants, and animals. Applied to naturally outbreeding plants and to animals, the purpose of crossing inbred lines is to produce superior cross-bred, or F_1, individuals. The utilisation of heterosis in this way depends on selection as well as on the inbreeding and crossing. The selection is applied, in principle, to the crosses, with the aim of finding pairs of lines that cross well, so that the lines may be perpetuated and provide cross-bred individuals for commercial use. In practice, however, the performance of the lines themselves has to be taken into account, because the lines must be reasonably productive if they are to be maintained and used for crossing. This method has been very successful with plants, and has led to an improvement of 50 per cent in the yield of maize grown commercially in the United States, since hybrid seed started to be used in the early 1930's (Mangelsdorf, 1951). Its success with animals, however, has been much less notable. The reasons probably lie chiefly in the greater amount of space and labour required by animals and in their lower reproductive rate, both of which add greatly to the difficulty of producing and testing the inbred lines. During the inbreeding a large proportion of the lines die out from inbreeding depression before a reasonably high degree of inbreeding has been attained. Consequently the inbreeding programme must start with a very large number of lines if enough are to be left after the wastage to give some scope for the selection of good crosses. Another point is that with plants that can be self-fertilised, such as maize, the inbreeding proceeds much faster than with animals. To attain an inbreeding coefficient of, say, 90 per cent would require only 4 years for maize, but 11 years for pigs or chickens, and about 50 years for cattle with a 4- or 5-year generation interval.

Let us now consider the genetic principles on which the utilisation of heterosis depends. It was shown in Chapter 14 that crosses made at random between lines inbred without selection are expected to have a mean value equal to that of the base population. This is the reason why inbreeding and crossing alone cannot be expected to lead to an improvement, but must be supplemented by selection. In practice some improvement can be expected from the effects of natural selection. It eliminates lethal and severely deleterious genes during the inbreeding, and in so far as these genes affect the desired character an improvement of the cross-bred mean over that of the base population is to be expected. But this improvement will not be very great, because the deleterious genes eliminated will have been at low frequencies in the base population—and the more harmful, the lower the frequency—so that their effect on the population mean will be small. It has been calculated, on the basis of assumptions about the number of loci concerned and their mutation rates, that an improvement of 5 per cent in fitness is the most that could be expected from the elimination of deleterious recessive genes (Crow, 1948, 1952). The bulk of the improvement, therefore, must come from artificial selection applied to the economically desirable characters.

The crossing of inbred lines produces no genotypes that could not occur in the base population. But whereas the best genotypes occur only in certain individuals in the base population, they are replicated in every individual of certain crosses. It is in this replication of a desirable genotype that the chief merit of the method lies. Let us, for simplicity, consider crosses between fully inbred lines. The gametes produced by a highly inbred line are all identical, except for mutation. And the gene content of the gametes of any one line could in principle be found in a gamete from the base population. Therefore the genotype of the F_1 of two lines could in principle be found in an individual of the base population. Thus, provided there has been no selection during the inbreeding, a set of crosses made at random is genetically equivalent to a set of individuals taken at random from the base population; and the individuals of one cross are replicates of one individual in the base population. This replication of a genotype in the individuals of a cross allows the genotypic value to be measured with little error; whereas the genotypic value of an individual in the base population is only crudely measured by its phenotypic value. Further, it is the genotypic value that is measured in the cross and can be reproduced indefinitely, as long as the inbred lines are maintained; whereas only the breeding value can be reproduced by selection of individuals in a non-inbred population. Therefore the condition under which inbreeding and crossing are likely to be a better means of improvement than selection without inbreeding is when much of the genetic variance of the character is non-additive.

The amount of improvement that can be made by selection among a number of crosses depends on the amount of variation between the crosses. The same relationship holds between the intensity of selection, the standard deviation, and the selection differential as was described in Chapter 11 and illustrated in Fig. 11.3. In the following section the variance between crosses made at random between pairs of lines inbred without selection will be examined.

VARIANCE BETWEEN CROSSES

The variance between crosses to be considered is the variance of the true means of the crosses, or the between-cross component as estimated from an analysis of variance. The variance of the observed means will contain a fraction of the within-cross component for the reasons explained in connexion with family selection in Chapter 13. We shall assume that the experimental design has eliminated all non-genetic sources of variation from the between-cross component.

If the lines crossed are fully inbred there will be no genetic variance within the crosses, and the variance between crosses will be equal to the genotypic variance in the base population, since each cross is equivalent to an individual of the base population. When the lines are only partially inbred, however, some genetic variance will appear within the crosses, and the between-cross variance will be less than with fully inbred lines. It is therefore important to know in what manner the between-cross variance increases as inbreeding proceeds, since this will tell us how much is to be gained by proceeding to high levels of inbreeding.

We noted that crosses between fully inbred lines are genetically equivalent to single individuals of the base population. Crosses between partially inbred lines are analogous, not to individuals, but to families, with degrees of relationship dependent on the inbreeding coefficient of the lines. The variance between families can be formulated in terms of the degree of relationship in the families (Kempthorne, 1954), and this formulation may be extended to crosses by regarding the crosses as families with a relationship depending on the inbreeding coefficient of the lines. The following expression is then obtained for the component of variance between crosses:

T   F.Q.G.

Between-cross variance

$$=FV^A + F^2V_D + F^2V_{AA} + F^3V_{AD} + F^4V_{DD} + \ldots \quad\ldots\ldots(16.1)$$

In this expression V_A and V_D are the additive and dominance variances in the base population; V_{AA}, V_{AD} and V_{DD} are the interaction components as explained in Chapter 8; and F is the inbreeding coefficient of the lines as specified below. The interaction components are included because epistasis may have important effects. Only two-factor interactions, however, are shown: the higher interactions have coefficients in correspondingly higher powers of F. (For every A in the subscript there is a factor F, and for every D a factor F^2.) The formulation in equation 16.1 is conditional on the following specifications about how the crosses are made. 1. All lines have the same coefficient of inbreeding. 2. All lines have independent ancestry back to the base population; i.e. there is no relationship between the lines. 3. Each cross is made from many individuals of the parent lines; and these individuals are not related to each other within their lines. This means that the genetic variance within the lines is fully represented within the crosses. 4. The coefficient of inbreeding, F, refers not to the individuals used as parents of the crosses, but to their progeny if they were mated within their own lines; in other words, F is the inbreeding coefficient of the next generation of the lines.

Let us now examine the expression 16.1 and consider what it tells us about the variance between crosses. When the inbreeding coefficient is unity the between-cross variance is, as we have already stated, simply the sum of all the components of genetic variance in the base population. During the progress of the inbreeding the contribution of the additive variance increases linearly with F; those of the dominance variance and of $A \times A$ interactions increases with the square of F; and the other interaction components with the third or fourth power of F. This means that the dominance and interaction components contribute proportionately more at higher levels of inbreeding than at lower levels. If the character is one with predominantly non-additive variance, the crosses will differ little in merit during the early stages but will differentiate rapidly in the final stages. Since this is the sort of character for which inbreeding and crossing is likely to be the most effective means of improvement, it is clear that inbreeding must be taken to a fairly high level if anything approaching its full benefit is to be realised. Some idea of the level of inbreeding required can be obtained by noting that with $F = 0.5$ the between-cross variance is equal to the variance between full-sib families in the base population. At this level of inbreeding, therefore, the best cross would do no more than replicate the best full-sib family in a non-inbred population.

Combining ability. The components of genetic variance making up the between-cross variance that we have been discussing are causal components, in the sense explained in Chapter 9. The variance between crosses, however, can also be analysed into observational components in the following way. Suppose a set of lines are crossed at random, each line being simultaneously crossed with a number of others. We can then calculate for each line its mean performance, i.e. the mean value of the F_1's in crosses with other lines. This is known as the *general combining ability* of the line. The performance of a particular cross may deviate from the average general combining ability of the two lines, and this deviation is known as the *special* (or *specific*) *combining ability* of the cross. Or, if we measure the mean values as deviations from the general mean of all crosses, we can express the value of a certain cross as the sum of the general combining abilities of the two lines and the special combining ability of the pair of lines. Thus the mean value of the cross of line X with line Y is

$$M_{XY} = G.C._X + G.C._Y + S.C._{XY} \qquad \ldots\ldots(16.2)$$

where G.C. and S.C. stand for the general and special combining abilities. The variance between crosses can therefore be analysed into two components: variance of general combining abilities and variance of special combining abilities; the latter being, in statistical terms, the interaction component.

The observational components of variance attributable to general and special combining ability are made up of the causal components in the following way.

Variance of crosses attributable to:

General combining ability $= FV_A + F^2 V_{AA} + \ldots$

Special combining ability $= F^2 V_D + F^3 V_{AD} + F^4 V_{DD} + \ldots$ $\left.\right\}\ldots\ldots(16.3)$

So differences of general combining ability are due to the additive genetic variance in the base population, and to $A \times A$ interactions; and differences of special combining ability are attributable to the non-additive genetic variance. Consequently the variance of general combining ability increases linearly with F (apart from the interaction component), while the variance of special combining ability increases with higher powers of F. It is therefore the special, and not the general, combining ability that is expected to increase more rapidly as the inbreeding reaches high levels.

EXAMPLE 16.1. An analysis of egg-laying in crosses between highly inbred lines of *Drosophila melanogaster* is reported by Gowen (1952). Five lines were crossed in all ways, including reciprocals, and the numbers of eggs laid by females in the fifth to ninth days of adult life were recorded. The analysis of the crosses yielded the following percentage composition of the variance of egg number:

Variance component	*% of total*
General combining ability	11·3
Special combining ability	9·7
Differences between reciprocals	2·3
Within crosses	76·6

Thus about half the variance between crosses was due to general, and half to special, combining ability.

Some of the methods of improvement by crossing aim at utilising only the variance of general combining ability, and then the measurement of the general combining ability of the lines becomes an important procedure. In addition to the making of specific crosses between the lines, there are two other methods of measuring general combining ability. A method convenient for use with plants is known as the *polycross* method. A number of plants from all the lines to be tested are grown together and allowed to pollinate naturally, self-pollination being prevented by the natural mechanism for cross-pollination, or by the arrangement of the plants in the plot. The seed from the plants of one line are therefore a mixture of random crosses with other lines, and their performance when grown tests the general combining ability of that line. Another method, applicable also to animals, is known as *top-crossing*. Individuals from the line to be tested are crossed with individuals from the base population. The mean value of the progeny then measures the general combining ability of the line, because the gametes of individuals from the base population are genetically equivalent to the gametes of a random set of inbred lines derived without selection from the base population.

These methods are essentially methods for comparing the general combining abilities of different lines, and so leading to the choice of the lines most likely to yield the best cross, among all the crosses that might be made between the available lines. But if much of the variation between crosses is due to special combining ability, then the general combining ability of two lines will not provide a reliable guide to the performance of their cross.

METHODS OF SELECTION FOR COMBINING ABILITY

The methods of improvement by inbreeding and crossing fall into two groups, according to whether they are designed to utilise only the variation in general combining ability or to utilise also the variation in special combining ability.

Selection for general combining ability. When the improvement of general combining ability only is sought the procedure of selection is much simplified. The general combining abilities of all available lines can be measured, as already explained, without the necessity of making and testing all the possible crosses between them. Some selection can usefully be applied to the lines before they are tested in crosses. There is some degree of correlation between a line's performance as an inbred and its general combining ability, so a proportion of lines can be discarded on the basis of their own performance before the crosses are made. And, finally, there is less to be lost by making the crosses at a relatively low coefficient of inbreeding. Selection for general combining ability may be repeated in cycles, a procedure known in plant breeding as *recurrent selection*. (In animal breeding this term has come to have a different meaning, as will be explained below.) Lines are inbred by self-fertilisation for one or two generations and their general combining abilities tested. The lines with the best general combining abilities are then crossed and a second cycle of inbreeding and selection carried out. A review of the progress made by this method is given by Sprague (1952).

The seed for commercial use is usually not made by a single cross of two lines, but by a *3-way* or *4-way cross*. The object of this is to overcome the generally low production of an inbred used as seed parent. In a 3-way cross the F_1 of two lines is used as seed parent and crossed with a third inbred line. In a 4-way cross two F_1's of different pairs of lines are crossed. The performance of 3-way and 4-way crosses can be reliably predicted from the performance of the constituent single crosses.

Even though selection for general combining ability is widely used in plant breeding and has abundantly proved its success, it is not, perhaps, altogether clear why it is preferred to selection without inbreeding, made either by individual selection or by family selection. Since the variation in general combining ability is attributable to additive variance in the population from which the lines were derived, selection should be effective without inbreeding. Comparisons of the two methods by experiment have not been made on a scale sufficient to prove convincingly the superiority of selection with inbreeding (see Robinson and Comstock, 1955).

Selection for general and specific combining ability. The specific combining ability of a cross cannot be measured without making and testing that particular cross. Therefore to achieve a reasonably high intensity of selection for specific combining ability a large number of crosses must be made and tested. Is no short-cut possible? Could the superior combining ability not be, as it were, built into the lines by selection? From the causes of heterosis explained in Chapter 14 it is clear that what is wanted is two lines that differ widely in the gene frequencies at all loci that affect the character and that show dominance. It should therefore be possible to build up these differences of gene frequency in two lines by selection. Instead of the differences of gene frequency being produced by the random process of inbreeding, they would be produced by the directed process of selection, which would be both more effective and more economical. Two methods based on this idea have been devised. These methods, though originating from plant breeding, provide— in theory at least—the most hopeful means of utilising heterosis in animals. We shall first describe the method known as *reciprocal recurrent selection*, or simply as *reciprocal selection*. In outline, the procedure is as follows.

The start is made from two lines, say A and B. (We shall call them "lines" even though they will not be deliberately inbred.) Crosses are made reciprocally, a number of A ♂♂ being mated to B ♀♀, and a number of B ♂♂ to A ♀♀. The cross-bred progeny are then measured for the character to be improved and the parents are judged from the performance of their progeny. The best parents are selected and the rest discarded, together with all the cross-bred progeny, which are used only to test the combining ability of the parents. The selected individuals must then be remated, to members of their own line, to produce the next generation of parents to be tested. These are crossed again as before and the cycle repeated. It is seldom practicable to select among the female parents, and the selection is chiefly applied to the males. Each male is mated to several females of the other line so that the judgment of his combining ability may be based on a reasonably large number of progeny. Most of these females are needed to mate to the selected males of their own line for the continuation of the line. Deliberate inbreeding is avoided as far as possible, for the reason to be explained below. The use of all the females as parents in their own lines helps to reduce the rate of inbreeding and allows relatively few males to be used, which intensifies the selection.

An essential prerequisite is that there should be some difference of gene frequency between the two lines at the beginning, or else selection for combining ability will be unable to produce a differentiation of the lines. Any locus at which the gene frequencies are the same in the two lines will be in equilibrium, though an unstable equilibrium. Any shift in one direction or the other will give the selection something to act on and the difference will be increased. The initial difference between the lines may be obtained by starting from two different breeds or varieties, choosing two that already cross well; or by deliberate inbreeding, up to perhaps 25 per cent, and relying on random differentiation of gene frequencies.

Though the performance of the cross is expected to increase under this method of selection, the performance of the lines themselves in respect of the character selected is expected to decrease, for this reason. Characters to which selection would be applied in this way are those subject to inbreeding depression and heterosis; that is to say, those in which dominance is directional. The changes of gene frequency brought about by the selection are toward the extremes, and consequently the mean values of the lines will decline for the reasons explained in connexion with inbreeding in Chapter 14. This decline in the performance of the lines, however, should not be quite as deleterious as the effects of deliberate inbreeding. Inbreeding, as a random process, affects all loci, and the mean values of all characters showing directional dominance decline. But under reciprocal selection it is only the selected character that should decline, except in so far as linked loci are carried along. Nevertheless, reproductive fitness is nearly always a component of economic value, and it is doubtful how far the distinction will hold. This, however, is the reason why deliberate inbreeding of the lines is to be avoided.

The second method is simpler in procedure than reciprocal selection described above. It was devised as a modification of recurrent selection, intended to utilise special as well as general combining ability (Hull, 1945), and as yet it has no distinctive name. It is known variously as "Hull's modification of recurrent selection," "recurrent selection to inbred tester," "recurrent selection for special combining ability," and in animal breeding simply as *"recurrent selection."* It differs from reciprocal selection in the following way. Instead of starting with two lines and selecting both for combining ability with the other, one starts with only one line and selects it for combining ability with a "tester" line which has previously been inbred. This reduces the amount of effort spent on the testing, and is expected to yield more rapid progress at the beginning because the initial differences of gene frequency between the line and the tester are likely to be more marked. But the ultimate gain is expected to be less than under reciprocal selection, because the general combining ability of the tester line is predetermined, and only the general combining ability of the selected line and the special combining ability of the cross can be improved.

The two methods of selection for special combining ability described in this section are comparatively new methods of improvement and very little practical experience of them has yet been gained. The account of them given here is consequently based almost entirely on theory. Theoretical assessments of their merits in relation to other methods have been made by Comstock, Robinson, and Harvey (1949) and by Dickerson (1952). Though on theoretical grounds they seem promising, the results of the only experiments so far published (Bell, Moore, and Warren, 1955; Rasmuson, 1956) are not encouraging.

Before we leave the subject of inbreeding we must give some further consideration to the particular genetic property that makes selection with inbreeding and crossing preferable to selection without inbreeding. From the theoretical point of view, and leaving all practical considerations aside, the crucial genetic property is overdominance of the genes concerned. The following section is devoted to a consideration of overdominance and its significance.

Overdominance

Overdominance is the property shown by two alleles when the heterozygote lies outside the range of the two homozygotes in genotypic value with respect to the character under discussion. Its meaning was illustrated in Fig. 2.3 with respect to fitness as the character, and it has been mentioned from time to time in other chapters. We saw in Chapter 2 how selection favouring heterozygotes leads to a stable gene frequency at an intermediate value, and how this overdominance with respect to fitness probably accounts for much of the stable polymorphism found in natural populations. And in Chapter 12 we saw how overdominance may be a source of non-additive genetic variance in populations that have reached their limit under artificial selection. It is, however, in connexion with the utilisation of heterosis by inbreeding and crossing, or by reciprocal selection, that overdominance has its most important practical consequences. In earlier chapters two basic methods of improvement were distinguished, one being selection without inbreeding, and the other inbreeding followed by crossing. In this chapter we have seen that selection is an integral part of the second method also. The essential distinction therefore lies in the crossing, rather than in the selection. Now, crossing two lines in which different alleles are fixed gives an F_1 in which all individuals are heterozygotes; and this is the only way of producing a group of individuals that are all heterozygotes. In a non-inbred population no more than 50 per cent of the individuals can be heterozygotes for a particular pair of alleles. Consequently, if heterozygotes of a particular pair of alleles are superior in merit to homozygotes, inbreeding and crossing will be a better means of improvement than selection without inbreeding. Furthermore, it is only when there is overdominance with respect to the desired character, or combination of characters, that inbreeding and crossing can achieve what selection without inbreeding cannot. Under any other conditions of dominance the best genotype is one of the homozygotes, and all individuals can be made homozygous by selection, without the disadvantages attendant on inbreeding and much more simply than by methods dependent on crossing. It was stated earlier in this chapter that the potentialities of inbreeding and crossing are greatest when there is much non-additive genetic variance and little additive. Now we see that this is only part of the truth:

in principle inbreeding and crossing can surpass selection without in-breeding only when a substantial part of the non-additive variance is due to overdominance. It is therefore of great practical importance to know whether overdominance with respect to economically desirable characters is a major source of variation. It is also of great theoretical interest to know whether overdominance with respect to natural fitness is a common phenomenon affecting many loci, because natural selection favouring heterozygotes would be a potent factor tending to maintain genetic variation in populations. This point will be discussed further in Chapter 20.

The contribution of overdominance to the variance, and the pro-portion of loci that show overdominance, are really two different questions. Genes that are overdominant with respect to fitness will be at intermediate frequencies and will therefore contribute much more variation than genes at low frequencies. So overdominance may be a major source of variation and yet be a property of only a few loci.

The evidence concerning overdominance has been compre-hensively reviewed by Lerner (1954), who reaches the conclusion that overdominance with respect to fitness and characters closely con-nected with it is widespread and very important. A contrary view is expressed by Mather (1955b) on the grounds that much of what appears to be overdominance with respect to certain characters in plants can be attributed to epistatic interaction. These two conflicting opinions will be enough to show that the problem of overdominance remains still an open question. The aim here is not to discuss the opinions, but to indicate briefly the nature of the evidence.

The evidence concerning overdominance is broadly speaking of two sorts, direct and indirect. The direct evidence comes from the comparison of heterozygotes and homozygotes in identifiable geno-types. The indirect evidence comes from the study of the expected consequences of overdominance as they affect the genetic properties of a population, or the outcome of certain breeding methods. Both sorts of evidence are complicated by linkage. We have to distinguish between overdominance as a property of a single locus, and over-dominance as a property of a segment of chromosome, which we shall refer to as apparent overdominance. Unequivocal evidence of over-dominance arising from a single locus is scarce because it can only be obtained from a locus that has mutated in a highly inbred line, or from a population in which coupling and repulsion linkages are in equilibrium. The segregation that can be observed in practice, and that gives rise to the genetic variation in a population, is usually not a segregation of single loci but of segments of chromosome, longer or shorter according to the amount of crossing-over. These segments of chromosome, or units of segregation, can show overdominance even though the separate loci do not. All that is needed to produce some degree of apparent overdominance is two genes, linked in repulsion, and both partially recessive. Its most extreme form is produced by two lethal genes linked in repulsion—a "balanced lethal" system—when the heterozygote of the segment spanned by the two loci is the only viable genotype.

In considering the direct evidence it is necessary to recognise that overdominance may be manifested at different "levels" according to the complexity of the character under discussion. A pair of alleles with pleiotropic effects may be found not to exhibit overdominance when any of the characters they affect is examined separately; yet if natural fitness or economic merit is founded on a combination of these characters, the alleles may show overdominance with respect to fitness or merit. Thus there may be no overdominance at the lower level of the simpler characters, but overdominance at the higher level of the more complex character.

EXAMPLE 16.2. An example of overdominance due to pleiotropy is provided by the pygmy gene in mice, already referred to in several examples in earlier chapters. The gene reduces body size and in the homozygote it causes sterility (King, 1955). In respect of body size it is nearly, but not quite, recessive. In respect of sterility it is probably also nearly recessive, though this was not proved. In neither body size nor sterility separately is there overdominance. But if small size were desirable (as it was in the experiment in which the gene was discovered), then under these conditions the genotype with the highest merit is the heterozygote, since the sterile homozygotes cannot reproduce. With respect to merit, or fitness under these conditions, the gene therefore shows overdominance. The lethal gene in the line of *Drosophila* selected for high bristle number, mentioned in Chapter 12, is another case of the same sort of overdominance; and so also is the sickle-cell anaemia described in Example 2.4.

The observations that provide direct evidence concerning overdominance may be briefly summarised as follows. The experience of Mendelian genetics shows that mutant genes are not commonly overdominant with respect to their main effects. Nor is overdominance with respect to natural fitness at all obvious. Indeed, if there were more than a mild degree of overdominance with respect to fitness a gene would not be rare enough to be classed as a "mutant." Though the evidence of Mendelian genetics suggests that overdominance is not a very common property of genes, many cases are nevertheless known. Overdominance due to pleiotropy, such as the cases mentioned in the above example, are not infrequent. And, overdominance with respect to certain components of natural fitness has been proved for some of the blood group genes in poultry (see Briles, Allen, and Millen, 1957; Gilmour, 1958).

The nature of the indirect evidence concerning overdominance is, in brief summary, as follows.

1. Experiments on the rate of loss of genetic variance during inbreeding point to the operation of natural selection in favour of heterozygotes (Tantawy and Reeve, 1956; Briles, Allen, and Millen, 1957; Gilmour, 1958). This indicates apparent overdominance, but it does not prove overdominance at the individual loci.

2. Crow (1948, 1952) has given reasons for thinking that the yield of grain obtained from the best crosses between inbred lines of maize is too high to be accounted for without overdominance at some loci. The reasoning depends on assumptions about the number of loci affecting yield and the mutation rates, and the conclusion is therefore tentative. Robinson et al. (1956) point out that the reasoning cannot justifiably be applied to maize crosses because the lines crossed generally come from different varieties and not from the same base population as required by Crow's hypothesis.

3. Comstock and Robinson (1952) have devised methods for measuring the average degree of dominance from measurements made on non-inbred populations. Preliminary results from maize (Robinson and Comstock, 1955) suggest that there cannot be overdominance (as distinct from apparent overdominance) at more than a small proportion of the loci that influence the yield of grain.

4. The existence of polymorphism in natural populations, as described in Chapter 2, cannot readily be explained except by supposing that the genes concerned are overdominant with respect to fitness.

From the foregoing outline of the evidence it is clear that the problem of how important overdominance is remains unsolved. Some of the differences of opinion about it may arise from different views of what phenomena are to be included under the term— whether apparent overdominance due to linkage, or overdominance

to be regarded as overdominance or not.
of how important overdominance is means
ng to whether we are concerned with its
of genes, or with the amount of variation it

CHAPTER 17

SCALE

The choice of a suitable scale for the measurement of a metric character has been mentioned several times in the foregoing chapters. The explanation of what is involved in the choice of a scale and a discussion of the criteria of suitability have, however, been deferred till this point because these are matters that cannot be properly appreciated until the nature of the deductions to be made from the data are understood. In other words the choice of a scale has to be made in relation to the object for which the data are to be used. The data from any experimental or practical study are obtained in the form most convenient for the measurement of the character. That is to say the phenotypic values are recorded in grams, pounds, centimetres, days, numbers, or whatever unit of measurement is most convenient. The point at issue is whether these raw data should be transformed to another scale before they are subjected to analysis or interpretation. A transformation of scale means the conversion of the original units to logarithms, reciprocals, or some other function, according to what is most appropriate for the purpose for which the data are to be used.

It is tempting to suppose that each character has its "natural" scale, the scale on which the biological process expressed in the character works. Thus, growth is a geometrical rather than an arithmetical process, and a geometric scale would appear to be the most "natural." For example, an increase of 1 gm. in a mouse weighing 20 gm. has not the same biological significance as an increase of 1 gm. in a mouse weighing 2 gm.: but an increase of 10 per cent has approximately the same significance in both. For this reason a transformation to logarithms would seem appropriate for measurements of weight. This, however, is largely a subjective judgment, and some objective criterion for the choice of a scale is needed. There are several recognised criteria (see Wright, 1952b); but, as Wright points out, the different criteria are often inconsistent in the scale they indicate. And, moreover, the same criterion applied to the same character may indicate different scales in different populations. Therefore the idea that every character must have its "natural" and correct scale is largely illusory.

In the first chapter on metric characters, Chapter 6, it was stated that we should assume throughout that any metric character under discussion would be measured on an "appropriate" scale, the criterion being that the distribution of phenotypic values should approximate to a normal curve. This is, in principle, the chief criterion, and a markedly asymmetrical, or skewed, distribution is a certain indication that the data may have to be transformed if they are to be used in certain ways. But a transformation may still be required even if the distribution is not markedly asymmetrical: we shall see below that the most important criterion then is that the variance should be independent of the mean. We shall treat the choice of scale in this chapter by showing what will arise if the transformation required is not made. We shall find that certain phenomena arise, called *scale effects*, which disappear when the appropriate transformation is made. For the sake of clarity we shall discuss in particular the logarithmic transformation which converts an arithmetic to a geometric scale. This is probably the commonest and most useful transformation. The general principles, outlined by reference to the log transformation, will, however, apply equally to other transformations. Let us first consider the distribution of phenotypic values.

Fig. 17.1 shows three distributions plotted as if from the original data on an arithmetic scale. They would all three be symmetrical and normal if the data were first transformed to logarithms, or plotted on logarithmic paper. There are two points of importance to notice. First, the degree of departure from normality depends on the amount of variation in relation to the mean. This may be seen from a comparison of the two upper graphs, (a) and (b), which are not very noticeably asymmetrical, with the lower graph, (c), which is. The relationship between the amount of variation and the mean, which determines the degree of departure from normality, is best expressed as the coefficient of variation; i.e. the ratio of standard deviation to mean, often multiplied by 100 to bring it to a percentage. The coefficient of variation of the two upper graphs is 20 per cent, while that of the lower graph is 50 per cent. Thus, a transformation to logarithms does not make an appreciable difference to the shape of the distribution unless the coefficient of variation is fairly high—that is, above about 20 per cent or so. Consequently, statistical procedures which do not rely on a strictly normal distribution, such as the analysis of variance, can be carried out on the untransformed data when the coefficient of variation is not above about 20 per cent. Transformations to other scales are also less necessary when the coefficient of variation is low than when it is high.

The second point to notice in Fig. 17.1 is that the variance, when computed in arithmetic units, increases when the mean increases. This may be seen in the two upper graphs, (a) and (b). These have

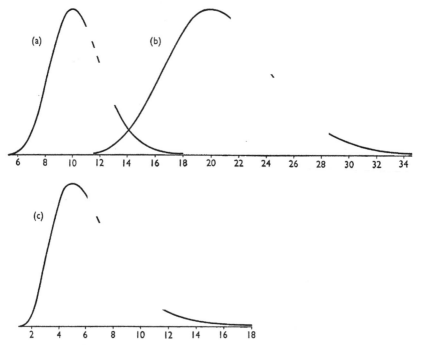

FIG. 17.1. Distributions that are symmetrical and normal on a logarithmic scale shown plotted on an arithmetic scale. Explanation in text.

both the same variance in logarithmic units, but different means. The mean—or strictly speaking the mode—of (b) is double that of (a) and the standard deviation in arithmetic units is correspondingly doubled. Though the distributions are not very noticeably skewed and a transformation does not seem to be very strongly indicated, yet in consequence of the difference of mean the variances differ very greatly. Here, then, is one of the commonest scale effects, namely a change of variance following a change of the population mean. The two graphs (a) and (b) in Fig. 17.1 might well represent two popula-

ions which have diverged by some generations of two-way selection,
f the character were something like body weight measured in grams
r pounds. Such characters are commonly found to increase in
ariance when the mean increases and to decrease in variance when
he mean decreases. Fig. 17.2 shows an example from an experiment
ith mice (MacArthur, 1949), the character being weight at 60 days.

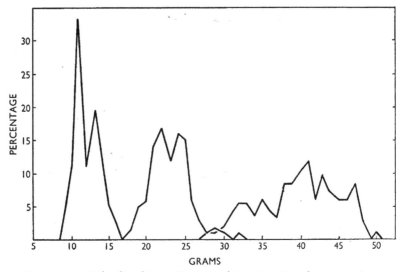

FIG. 17.2. Distributions of body weight of male mice at 60 days.
Centre: base population before selection. Left and right: small
and large strains after 21 generations of two-way selection. (Redrawn from MacArthur, 1949.)

	Small	Unselected	Large
Standard deviation	1·71	2·56	5·10
Coeff. of variation, %	14·3	11·1	12·8

Phenomena such as the change of variance discussed above are
called scale effects if they disappear when the measurements are
appropriately transformed: in other words, if their cause can be
attributed to the scale of measurement. But they are none the less
real, though labelled as a scale effect or removed by transformation.
The large mice, for example, are really more variable than the small
when their weights are measured in grams. What is gained by recog-
nising this as a scale effect is that there is no need to look deeper into
the genetic properties of the character for an explanation.

A convenient test for the appropriateness of a logarithmic transformation is provided by the proportionality of standard deviation

u   F.Q.G.

and mean, which we noted in connexion with graphs (a) and (b) in Fig. 17.1. If two distributions have the same variance on a logarithmic scale then the coefficients of variation in arithmetic units will be the same. Thus, constancy of the coefficient of variation indicates constancy of variance on a logarithmic scale. And, if variances are to be compared, we may simply compare the coefficients of variation instead of expressing the variances in logarithmic units. The standard deviations and coefficients of variation of the distributions shown in Fig. 17.2 are given in the legend to the figure. The coefficients of variation, though not identical, are much more alike than the standard deviations, and this shows that the changes of variance that have resulted from the selection can be attributed, in large part at least, to the scale of measurement.

The effect of scale on the connexion between variance and mean complicates the comparison of the variances of two populations that differ also in mean, as for example the comparison of the variances of inbreds and hybrids discussed in Chapter 15. If a difference of variance is to be unambiguously attributed to a difference of homeostatic power, for example, there must be independent grounds for believing that a similar difference would not be expected as a scale effect connected with the difference of mean.

Let us return to the consequences of selection and pursue them a little further. If the variance changes with the change of mean as a result of selection, so also will the selection differential and the response. The response per generation of a character such as we have been considering would therefore be expected to increase with the progress of selection in the upward direction, and to decrease correspondingly in the downward direction. The response to two-way selection would then be asymmetrical. An example of an asymmetrical response which can most probably be attributed to a scale effect in this way is shown in Fig. 17.3. Plotted in arithmetic units, as in (a), the response is much greater in the upward than in the downward direction. A transformation to logarithms, shown in (b), renders the response much more nearly symmetrical. This does not do away with the fact that the character as measured increased much more than it decreased under selection. But it accounts for the asymmetry without the need for more elaborate hypotheses. A convenient way of eliminating scale effects from the graphical presentation of a response to selection is to plot the response in the form of the realised heritability, as explained in Chapter 11 and illustrated in Fig. 11.5. The

realised heritability, which is the ratio of response to selection differential, is very little influenced by scale effects (Falconer, 1954*a*).

When means or variances are to be compared, for example in a
comparison of two populations or in following the changes resulting
from selection, and a transformation to logarithms is indicated, it is
not necessary to convert each individual measurement. On the other

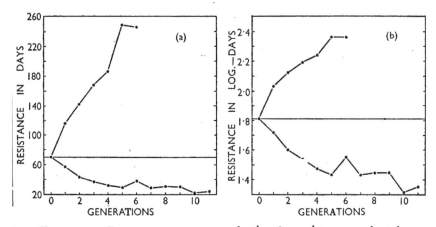

FIG. 17.3.   Response to two-way selection for resistance to dental
caries in rats. Resistance is measured in days and plotted on an arithmetic scale in (*a*), and on a logarithmic scale in (*b*). The arithmetic
means were converted to logarithmic means by formula 17.1. The
coefficient of variation was high—about 50%—and was approximately constant. The reason why the upward selection has not
covered so many generations as the downward is simply that the
increased resistance lengthened the generation interval. (Data
from Hunt, Hoppert, and Erwin, 1944.)

hand it is not sufficient to convert the arithmetic mean or variance to
logarithms, unless the coefficient of variation is very low. The con-
versions may be conveniently made by the two following formulae,
given by Wright (1952*b*). The first converts the mean of arithmetic
values to the mean of logarithmic values, and the second converts the
variance as computed from the arithmetic values to the variance as it

$$\overline{(\log x)} = \log \bar{x} - \tfrac{1}{2} \log (1 + C^2) \qquad \dots\dots(17.1)$$
$$\sigma^2_{(\log x)} = 0.4343 \log (1 + C^2) \qquad \dots\dots\dots(17.2)$$

would be computed from logarithmic values.   In these formulae C is
the coefficient of variation in the form $\sigma/\bar{x}$ computed from arithmetic
values, and the logarithms are to the base 10.

We turn now to what is perhaps a more fundamental effect of a scale transformation—its effect on the apparent nature of the genetic variance. To understand this we must go back to a single locus and consider the effect, or mode of action, of the genes. Let us imagine a locus with two alleles whose mode of action is geometric, the genotypic value of A_2A_2 being 50 per cent greater than A_1A_2 and that of A_1A_2 being also 50 per cent greater than A_1A_1. Thus on the logarithmic scale there is no dominance, the heterozygote being exactly midway between the two homozygotes. Now suppose the genotypic values are measured in arithmetic units, such as grams, and that A_1A_1 has a value of 10 units. Then A_1A_2 will be 15 units and A_2A_2 22·5 units. On the arithmetic scale, therefore, A_1 is partially dominant to A_2, the heterozygote no longer falling mid-way between the homozygotes. Thus the degree of dominance is influenced by the scale of measurement, and so also is the proportionate amount of dominance variance. This effect of a scale transformation, however, is normally rather small. A gene that causes a 50 per cent difference between the genotypic values, such as we have considered, would be a major gene, easily recognisable individually. But even so the degree of dominance on the arithmetic scale is not very great. Minor genes with effects of perhaps 1 per cent or 10 per cent would be scarcely influenced in their dominance.

In the same way that the dominance is affected by the scale, so also is the epistatic interaction between different loci. Loci with geometric effects would combine without interaction if the genotypic values were measured in logarithmic units. But when measured in arithmetic units there would be interaction deviations due to epistasis. Thus the amount of interaction variance is also influenced by the scale of measurement. The following example illustrates the dependence of interaction on scale.

EXAMPLE 17.1. The pygmy gene in mice is a major gene affecting body size, homozygotes being much reduced in size. The effect of this gene was studied in different genetic backgrounds (King, 1955). The gene was transferred from the strain selected for small size where it arose, to a strain selected for large size, by repeated backcrosses. The mean difference between pygmy homozygotes and normals (i.e. heterozygotes and normal homozygotes together) was measured in the two strains and during the transference, the comparisons being made between pygmies and normals in the same litters. The results are shown in Fig. 17.4. The difference between pygmies and normals increases with the weight of the normals.

In the background of the small strain the pygmies were about 7 gm. smaller than normals, but in the background of the large strain they were about 12 gm. smaller. Thus the pygmy gene shows epistatic interaction with the other genes that affect body size. But if the effect of the gene is expressed as a proportion, it is constant and independent of the other genes present.

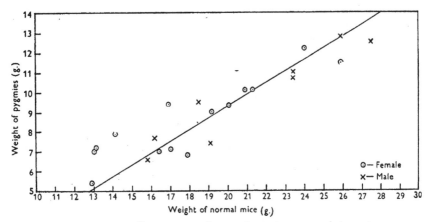

Fig. 17.4. Intra-litter comparisons of the 6-week weights of pygmies and normals. Mean of pygmies plotted against mean of normals in the same litter. (From King, 1955; reproduced by courtesy of the author and the editor of the *Journal of Genetics*.)

Pygmies are about half the weight of their normal litter-mates, no matter what the actual weights are. Thus if the comparisons are made in logarithmic units there is no epistatic interaction.

In general, therefore, a scale transformation may remove or reduce the variance attributable to epistatic interaction, and this variance might then be labelled as a scale effect. A transformation which removes or reduces interaction variance may be useful if conclusions are to be drawn from an analysis that depends for its validity on the absence of interaction. A detailed treatment of the relationship between scale and epistatic interaction is given by Horner, Comstock, and Robinson (1955).

In this chapter we have outlined some of the scale effects most commonly met with, and have indicated the circumstances under which a transformation of scale may be helpful to the interpretation of results and the drawing of conclusions. Transformations of scale, however, should not be made without good reason. The first purpose of experimental observations is the description of the genetic properties of the population, and a scale transformation obscures rather than illuminates the description. If epistasis, for example, is found, this is an essential part of the description, and it is better labelled as epistasis than as a scale effect. The transformation of scale is essentially a statistical device to be employed for the purpose of simplifying the analysis of the data, or to make possible the drawing of valid conclusions from the analysis. It is sometimes helpful also in the interpretation of results. If epistasis, for example, were found to disappear on transformation to a logarithmic scale we could conclude that the effects of different loci combined by multiplication rather than by addition. Or, if there were good reasons for attributing a difference of variance to a scale effect we should not need to invoke more complicated genetic explanations. The choice of scale, however, raises troublesome problems in connexion with the interpretation of results. Logical justification of a scale transformation can only come from some criterion other than the property about which the conclusions are to be drawn. If there is no independent criterion the argument becomes circular, and the distinction between a scale effect and some other interpretation becomes meaningless. There is also a more fundamental difficulty: the scale appropriate for one population may not be appropriate for another, and the scale appropriate to the genetic and environmental components of the variation may be different. This difficulty is strikingly illustrated by an analysis of the character "weight per locule" in a number of crosses between varieties of tomato (Powers, 1950). By the same criterion—normality of the distribution—this character was found to require an arithmetic scale in some crosses and a geometric scale in others; and, moreover, in the F_2 generations of some crosses the genetic variation required one scale while the environmental variation required another.

CHAPTER 18

THRESHOLD CHARACTERS

There are many characters of biological interest or economic importance whose inheritance is multifactorial but whose distribution is discontinuous. For example: resistance to disease, a character expressed either in survival or in death with no intermediate; "litter" size in the larger mammals that bear usually one young at a time but sometimes two or three; or the presence or absence of any organ or structure. Characters of this sort appear at first sight to be outside the realm of quantitative genetics because they do not exhibit continuous variation; yet when subjected to genetic analysis they are found to be under the influence of many genes just as any metric character. For this reason they have been called "quasi-continuous variations" (Grüneberg, 1952): the phenotypic values are discontinuous but the mode of inheritance is like that of a continuously varying character.

The clue to the understanding of the inheritance of such characters lies in the idea that the character has an underlying continuity with a "threshold" which imposes a discontinuity on the visible expression of the character, as depicted in Fig. 18.1. The underlying continuous variation is both genetic and environmental in origin, and may be thought of as the concentration of some substance or the speed of some developmental process—of something, that is to say, that could in principle be measured and studied as a metric character in the ordinary way. The hypothetical measurement of this variation is supposed to be made on a scale that renders its distribution normal, and the unit of measurement is the standard deviation of the distribution. This provides what may be called the *underlying scale*. We now have two scales for the description of the phenotypic values: the underlying scale which is continuous, and the visible scale which is discontinuous. The two are connected by the threshold, or point of discontinuity. This is a point on the continuous scale which corresponds with the discontinuity in the visible scale. The idea will be clearer from an inspection of Fig. 18.1, which depicts a character whose visible expression can take only two forms, such as alive versus dead, or present versus absent. Individuals whose phenotypic values on the underlying scale exceed the threshold will appear in one visible class, while individuals below the threshold will appear in the other.

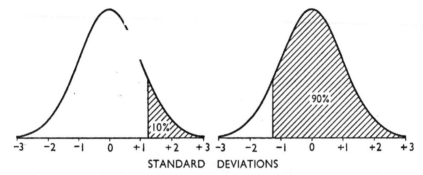

STANDARD DEVIATIONS

FIG. 18.1. Illustrations of a threshold character with two visible classes. The vertical line marks the theshold between the two phenotypic classes, one of which is cross-hatched. The population depicted on the left has an incidence of 10 %; that on the right, an incidence of 90 %.

On the visible scale individuals can have only two values, 0 or 1. Groups of individuals, however, such as families or the population as a whole can have any value, in the form of the proportion or percentage of individuals in one or other class. This may be referred to as the *incidence* of the character. Susceptibility to disease, for example, can be expressed as the percentage mortality in the population or in a family. The incidence is quite adequate as a description of the population or group, but the percentage scale in which the incidence is expressed is inappropriate for some purposes because on a percentage scale variances differ according to the mean. The interpretation of genetic analyses of threshold characters is therefore facilitated by the transformation of incidences to values on the underlying scale. The transformation is easily made by reference to a table of probabilities of the normal curve. The threshold is a point of truncation whose deviation from the population mean can be found from the proportion of the population falling beyond it. A table of "probits" (Fisher and Yates, 1943, Table IX) is convenient to use because it refers to a single tail of the distribution and obviates confusion over the sign of the deviation. The transformation from the visible to the underlying scale enables us to state the mean phenotypic value of a population or family in terms of its standard deviation, and to compare the means of different populations or families provided they have the same standard deviation. It is convenient to take the position of the threshold as the origin, or zero-point, on the underlying scale and to express the mean as a deviation from the threshold. Thus if the incidence of the character is, for example, 10 per cent, a table of the normal curve shows that the threshold exceeds the mean by 1·28 standard deviations. The population mean, referred to the threshold as origin, is therefore $-1·28\sigma$. Or, if the incidence were 90 per cent then the population mean would be $+1·28\sigma$, as shown in Fig. 18.1. For any comparison of means, however, it is necessary to assume that the populations compared have the same variance on the underlying scale. If reasons are known for the variances not being equal—in comparisons, for example, between inbreds, F_1's and F_2's—then the means cannot be expressed on a common scale that allows a valid comparison to be made.

This is as far as we can go with a character that is visibly expressed in only two classes. The mean of a population or group can be stated, but not the variance, because the mean has to be stated in terms of the standard deviation. We can, however, subject the observed means of families to analysis and compute the heritability of the character. The heritability of threshold characters is treated by A. Robertson and Lerner (1949) and by Dempster and Lerner (1950), and will not be further discussed here.

If a character has three classes in its visible scale then comparisons can be made between the variances of populations as well as between the means. The number of lumbar vertebrae in mice is a character of this sort that has been extensively studied (Green, 1951; McLaren and Michie, 1955). The number is usually either 5 or 6, but some individuals have 5 on one side and 6 on the other. This comes about through the last vertebra being sacralised on one side and not on the other. The asymmetrical mice have $5\frac{1}{2}$ lumbar vertebrae and are regarded as being intermediate between the 5-class and the 6-class.

When the visible scale has three classes there are two thresholds, as shown in Fig. 18.2. If the assumption is made that the difference between the two thresholds represents a constant difference on the underlying scale, then we have not only a fixed origin of the scale but also a fixed unit, and this provides a basis for the comparison of variances as well as of means. The underlying scale then has one of the thresholds as origin and the threshold difference as the unit of measurement. The idea is most easily explained by a numerical example. Consider the two populations illustrated in Fig. 18.2. Let their standard deviations on a common underlying scale be σ_1 and σ_2 respectively, and let them have the following incidences in the three visible classes, X, I, and Z, of which I is the intermediate class:

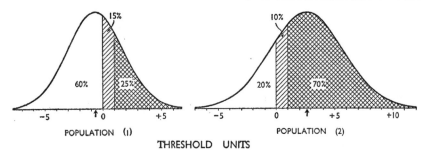

POPULATION (1)   POPULATION (2)

THRESHOLD   UNITS

FIG. 18.2.   Illustrations of a threshold character with three visible classes, in two populations with incidences as shown. The axes are marked in threshold units, and the population means are indicated by arrows. Further explanation in text.

	Class		
	X	I	Z
Incidence, %. Population (1)	60	15	25
Population (2)	20	10	70

The deviations of the thresholds from the population means, found from a table of the normal curve, are as follows:

	X/I	I/Z	*Threshold interval*
Population (1)	$+0.25\sigma_1$	$+0.67\sigma_1$	$0.42\sigma_1$
Population (2)	$-0.84\sigma_2$	$-0.52\sigma_2$	$0.32\sigma_2$

The intervals between the two thresholds, given above on the right, are found by subtraction of the deviations of the two thresholds in each population. These threshold intervals are supposed by hypothesis to be equal on the common underlying scale. By assigning the threshold interval the value of one "threshold unit" we can therefore express the standard deviations of the two populations on a common basis in terms of threshold units. The standard deviations then become

$$\sigma_1 = 2.38 \text{ threshold units}$$
$$\sigma_2 = 3.12 \text{ threshold units.}$$

The means of the populations can also be expressed in threshold units. Reckoned from the X/I threshold as origin they are

$$M_1 = -0.25 \quad \sigma_1 = -0.60 \text{ threshold units}$$
$$M_2 = +0.84 \quad \sigma_2 = +2.62 \text{ threshold units.}$$

The standard deviation and population mean of a character with three visible classes may be put in general form in the following way. Let X be the incidence in one visible class, and Y the incidence in this class together with the intermediate class. Let the threshold between these two classes be the origin of the underlying scale. Let x and y be the deviations of the two thresholds corresponding to the incidences X and Y respectively. Then the standard deviation is

$$\sigma = \frac{1}{x-y} \text{ threshold units} \qquad \ldots\ldots(18.1)$$

and the mean is

$$M = -x\sigma$$
$$= \frac{-x}{x-y} \text{ threshold units} \qquad \ldots\ldots(18.2)$$

The comparison of variances in this way depends entirely, as we have pointed out, on the assumption that the interval between the two thresholds is constant from one population to another. If we think again of the hypothetical substance or process whose concentration or rate determines the value on the underlying scale, the assumption is that the intermediate class spans the same difference of concentration or of rate in the two populations compared. Whether this assumption is a reasonable one or not is hard to judge. It may, nevertheless, lead to reasonable results, as the following example shows.

EXAMPLE 18.1. The number of lumbar vertebrae was studied in two inbred lines of mice and their cross (Green and Russell, 1951). The inbred lines were a branch of the C3H strain with predominantly 5 lumbar vertebrae, and the C57BL strain with predominantly 6 lumbar vertebrae. Crosses were made reciprocally, and F_2 generations were made from each F_1. The incidences of the 5-vertebra class and of the intermediate class of asymmetrical mice with $5\frac{1}{2}$ are given in the table. The reciprocal F_1's were found to differ and are listed separately. The F_2's did not differ and their results are pooled. The table gives also the positions of the two thresholds in standard deviations; and the mean and standard deviation computed in threshold units, the mean being reckoned from the threshold between the 5-class and the asymmetrical class as origin. The distribu-

Population	Incidence, %		Deviation of thresholds from mean, in σ		Mean and standard deviation in threshold units	
	5	$5\frac{1}{2}$	$5/5\frac{1}{2}$	$5\frac{1}{2}/6$	M	σ
Inbreds C3H	96·9	2·3	+1·87	+2·41	−3·44	1·84
C57	1·3	2·0	−2·23	−1·84	+5·74	2·58
F_1's						
C3H♀ × C57♂	57·4	15·5	+0·19	+0·61	−0·44	2·36
C57♀ × C3H♂	29·0	25·0	−0·55	+0·10	+0·85	1·53
F_2 (pooled)	46·7	12·2	−0·08	+0·23	+0·27	3·25

tions of the populations, based on the computed means and standard deviations, are shown graphically in Fig. 18.3. It should be noted that the means and standard deviations of the inbreds are not very precisely estimated because the incidences are low. The computed properties of the populations follow the expected pattern. The F_1 generation is intermediate in mean between the two parental populations, though there is a maternal effect causing a difference between the reciprocal F_1's. This maternal effect has been further studied and confirmed by McLaren and Michie (1956a). The variance of the F_1 is somewhat lower than that of the parental inbreds, as might be expected from a reduction of environmental variance in the hybrids. This was further studied and confirmed by McLaren and Michie (1955). The F_2 is equal in mean to the F_1, but shows an increased variance as would be expected from the segregation of genes. If we take 2·00 as the mean standard deviation of the F_1, representing purely environmental variation, then the environmental variance is 4·00, and the total phenotypic variance given by the F_2 is 10·56; therefore the genotypic variance works out at 6·56, or 62 per cent of the total. Thus the analysis of the threshold character studied in this cross leads to very reasonable results, and the assumptions on which it rests do not seem to be very seriously wrong.

The meaning of the threshold unit in which values on the underlying scale are expressed may conveniently be discussed by reference to the number of lumbar vertebrae in mice, described in the above example. From the graduation of the scale at the foot of Fig. 18.3 it appears that the threshold interval corresponds to one vertebra. It is therefore tempting to regard the scale as indicating "potential" vertebrae, ranging from 5 at the origin to 15 at the upper extreme

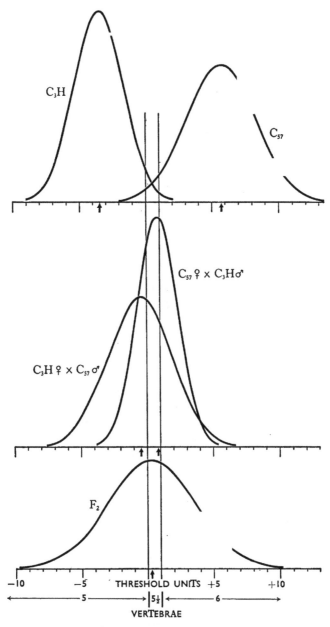

FIG. 18.3. Distributions of number of lumbar vertebrae in mice transformed to the underlying scale of threshold units. The upper distributions are two inbred lines, the two middle ones are the two reciprocal F_1's, and the lower distribution is the F_2. (Data from Green & Russell, 1951.) See example 18.1 for further explanation.

and to − 5 at the lower extreme. We should then regard the developing vertebral column as being protected by canalisation against this wide range of potential variation, so that the vertebrae actually formed are restricted to the narrow range between 5 and 6. This interpretation, however, assumes that individuals with a potential number anywhere between 5 and 6 will be asymmetrical with $5\frac{1}{2}$ vertebrae; and for this there is no justification. The asymmetrical individuals may equally well, or more probably, be those with almost exactly $5\frac{1}{2}$ potential vertebrae. Suppose, for example, that the range of potential vertebrae that gave rise to an asymmetrical individual were between 5·4 and 5·6. Then 1 threshold unit would correspond to 0·2 potential vertebrae; the origin of the underlying scale would be at 5·4 and the variation would range from 7·4 potential vertebrae at one extreme to 3·4 at the other. Or, if the asymmetrical individuals covered a range of only 0·1 potential vertebrae, the whole distribution would lie within the potential numbers of 5 and 6, just as the actual range does. Thus the threshold unit is purely arbitrary in nature; though useful for the comparison of populations, it cannot be given any concrete interpretation.

From what has been said so far in this chapter it will be clear that threshold characters do not provide ideal material for the study of quantitative genetics, because the genetic analyses to which they can be subjected are limited in scope and subject to assumptions that one would be unwilling to make except under the force of necessity. We turn now to a consideration of some aspects of selection for threshold characters, which has more practical importance than the genetic analyses that we have been considering, and does not involve the same theoretical difficulties.

SELECTION FOR THRESHOLD CHARACTERS

Selection for threshold characters has some practical importance in connexion with the improvement of viability and with changing the response of experimental animals to treatments, such as, for example, increasing or decreasing drug resistance. We shall consider only characters with two visible classes; and we shall assume that there is no means of measuring some aspect of the character that varies continuously, such as measuring the time of survival instead of classifying simply dead versus alive.

The response to selection depends in the usual way on the selection differential. But the selection differential does not depend primarily on the proportion selected, as with a continuously varying character, but on the incidence, for the following reason., We may breed exclusively from those individuals in the desired phenotypic class, but we cannot discriminate between those with high and those with low values on the underlying scale. The selected individuals are therefore a random sample from the desired class, and the mean of the selected individuals is the mean of the desired class, irrespective of whether we select all of the desired class or only a portion of it. The point will be made clearer by reference to Fig. 18.1, letting the cross-hatching represent the desired class. Let us suppose that the replacement rate allows us to select 10 per cent of the population. If we select out of the population on the right, with an incidence of 90 per cent, the mean of the selected individuals will be the same as if we had selected 90 per cent. But if we select out of the population on the left, with an incidence of 10 per cent, we shall use all of the individuals in the desired class and none of the others. The selection differential will then be the same as if we had selected on the basis of a continuously varying character. Thus the selection differential is greatest when the incidence is exactly equal to the proportion selected. If it is less we shall be forced to use some individuals of the undesired class; and if it is greater we shall do no better than we should by selecting the whole of the desired class.

With some characters, however, the incidence can be altered and this provides a means of improving the response to selection. If the character is, for example, a reaction to some treatment, the treatment can be increased or reduced in intensity, so that the incidence is altered. This is an alteration of the mean level of the environment, and its effect is in principle to shift the distribution of phenotypic values with respect to the fixed threshold. But it is more convenient to regard it as changing the nature of the character and shifting the threshold with respect to a fixed mean phenotypic level. When the level of the threshold can be controlled in this way, the maximum speed of progress under selection will be attained by adjusting the threshold so that the incidence is kept as nearly as possible equal to the minimum proportion that must be selected for breeding. The progress made can be assessed by subjecting the population, or part of it, to the original treatment under which the threshold is at its original level.

Genetic assimilation. A very interesting result of the application of this principle of changing the threshold by environmental
means is the phenomenon known as "genetic assimilation" (Waddington, 1953). If a threshold character appears as a result of an
environmental stimulus, and selection is applied for this character, it
may eventually be made to appear spontaneously, without the necessity of the environmental stimulus. In this way what was originally
an "acquired character" becomes by perfectly orthodox principles of
selection an "inherited character" (Waddington, 1942). In such a
situation there are two thresholds, one spontaneous and the other

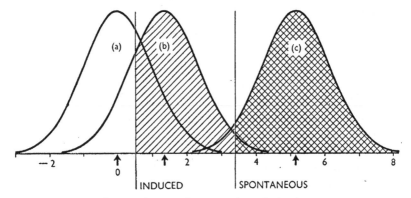

Fig. 18.4.   Diagram illustrating genetic assimilation of a threshold
character. Distributions on the underlying scale, which is marked
in standard deviations. The vertical lines show the positions of the
induced and spontaneous thresholds, and the arrows mark the
population means at three stages of selection.
(a) before selection: incidence—induced = 30 %, spontaneous = 0 %
(b) after some selection: incidence—induced = 80 %, spontaneous = 2 %
(c) after further selection: incidence—induced = 100 %, spontaneous = 95 %

induced, as shown in Fig. 18.4. The spontaneous threshold is at first
outside the range of variation of the population, so that there is no
variation of phenotype and no selection can be applied, (Fig. 18.4, a).
The induced threshold, however, is within the range of the underlying scale covered by the population, and it allows individuals toward
one end of the distribution to be picked out by selection. In this way
the mean genotypic value of the population is changed. If this change
goes far enough some individuals will eventually cross the spontaneous threshold and appear as spontaneous variants, (Fig. 18.4, b).
When the spontaneous incidence becomes high enough selection may

be continued without the aid of the environmental stimulus, and the spontaneous incidence may be further increased, (Fig. 18.4, *c*).

EXAMPLE 18.2. An experimental demonstration of genetic assimilation in *Drosophila melanogaster* is described by Waddington (1953). The character was the absence of the posterior cross-vein of the wing. In the base population no flies with this abnormality were present, but treatment of the puparium by heat shock caused about 30 per cent of cross-veinless individuals to appear. Selection in both directions was applied to the treated flies, and after 14 generations the incidence of the induced character had risen to 80 per cent and fallen to 8 per cent. At this time cross-veinless flies began to appear in small numbers among untreated flies of the upward-selected line, and by generation 16 the spontaneous incidence was between 1 and 2 per cent. Selection was then continued without treatment, the population being subdivided into a number of lines. The best four of the lines, selected without further treatment, reached spontaneous incidences ranging from 67 per cent to 95 per cent. The distributions in Fig. 18.4 illustrate the progress of the upward selection. Graph (*b*) shows a spontaneous incidence of 2 per cent and an induced incidence of 80 per cent and thus corresponds approximately with generation 16. On the assumption of constant variance, the change of mean at this stage amounted to 1·36 standard deviations. Graph (*c*) shows a spontaneous incidence of 95 per cent and represents the line that finally showed the greatest progress. Its mean on the underlying scale is 5·15 standard deviations above that of the initial population.

The idea of genetic assimilation is not confined to threshold characters; but for its wider significance the reader must be referred to Waddington (1957).

CORRELATED CHARACTERS

This chapter deals with the relationships between two metric characters, in particular with characters whose values are correlated—either positively or negatively—in the individuals of a population. Correlated characters are of interest for three chief reasons. Firstly in connexion with the genetic causes of correlation through the pleiotropic action of genes: pleiotropy is a common property of major genes, but we have as yet had little occasion to consider its effects in quantitative genetics. Secondly in connexion with the changes brought about by selection: it is important to know how the improvement of one character will cause simultaneous changes in other characters. And thirdly in connexion with natural selection: the relationship between a metric character and fitness is the primary agent that determines the genetic properties of that character in a natural population. This last point, however, will be discussed in the next chapter.

GENETIC AND ENVIRONMENTAL CORRELATIONS

In genetic studies it is necessary to distinguish two causes of correlation between characters, genetic and environmental. The genetic cause of correlation is chiefly pleiotropy, though linkage is a cause of transient correlation particularly in populations derived from crosses between divergent strains. Pleiotropy is simply the property of a gene whereby it affects two or more characters, so that if the gene is segregating it causes simultaneous variation in the characters it affects. For example, genes that increase growth rate increase both stature and weight, so that they tend to cause correlation between these two characters. Genes that increase fatness, however, influence weight without affecting stature, and are therefore not a cause of correlation. The degree of correlation arising from pleiotropy expresses the extent to which two characters are influenced by the same genes. But the correlation resulting from pleiotropy is the overall, or net, effect of all the segregating genes that affect both characters. Some genes may increase both characters, while others increase one and reduce the other; the former tend to cause a positive correlation, the latter a negative one. So pleiotropy does not necessarily cause a detectable correlation. The environment is a cause of correlation in so far as two characters are influenced by the same differences of environmental conditions. Again, the correlation resulting from environmental causes is the overall effect of all the environmental factors that vary; some may tend to cause a positive correlation, others a negative one.

The association between two characters that can be directly observed is the correlation of phenotypic values, or the *phenotypic correlation*. This is determined from measurements of the two characters in a number of individuals of the population. Suppose, however, that we knew not only the phenotypic values of the individuals measured, but also their genotypic values and their environmental deviations for both characters. We could then compute the correlation between the genotypic values of the two characters and the correlation between the environmental deviations, and so assess independently the genetic and environmental causes of correlation. And if, in addition, we knew the breeding values of the individuals, we could determine also the correlation of breeding values. In principle there are also correlations between dominance deviations, and between the various interaction deviations. To deal with all these correlations, even in theory, would be unmanageably complex, and fortunately is not necessary, since the practical problems can be quite adequately dealt with in terms of two correlations. These are the *genetic correlation*, which is the correlation of breeding values, and the *environmental correlation*, which is not strictly speaking the correlation of environmental deviations, but the correlation of environmental deviations together with non-additive genetic deviations. In other words, just as the partitioning of the variance of one character into the two components, additive genetic versus all the rest, was adequate for many purposes, so now the covariance of two characters need only be partitioned into these same two components. The "genetic" and "environmental" correlations thus correspond to the partitioning of the covariance into the additive genetic component versus all the rest. The methods of estimating these two correlations will be explained later. Let us consider first how they combine together to give the directly observable phenotypic correlation.

The following symbols will be used throughout this chapter:

X and Y: the two characters under consideration.

r_P the phenotypic correlation between the two characters, X and Y.

r_A the genetic correlation between X and Y (i.e. the correlation of breeding values).

r_E the environmental correlation between X and Y (including non-additive genetic effects).

cov the covariance of the two characters X and Y, with subscripts P, A, or E, having the same meaning as for the correlations.

σ^2 and σ variance and standard deviation, with subscripts P, A, or E, as above, and X or Y according to the character referred to. E.g. σ^2_{PX} = phenotypic variance of character X.

h^2 the heritability, with subscript X or Y, according to the character.

e^2 $= 1 - h^2$.

(The customary symbol for the genetic correlation is r_G, but since the genetic correlation is almost always the correlation of breeding values we shall use the symbol r_A for the sake of consistency with previous chapters.)

A correlation, whatever its nature, is the ratio of the appropriate covariance to the product of the two standard deviations. For example, the phenotypic correlation is

$$r_P = \frac{cov_P}{\sigma_{PX}\sigma_{PY}}$$

The phenotypic covariance is the sum of the genetic and environmental covariances, so we can write the phenotypic correlation as

$$r_P = \frac{cov_A + cov_E}{\sigma_{PX}\sigma_{PY}}$$

The denominator can be differently expressed by the following device: $\sigma^2_A = h^2\sigma^2_P$, and $\sigma^2_E = e^2\sigma^2_P$. So $\sigma_P = \sigma_A/h = \sigma_E/e$. The phenotypic correlation then becomes

$$r_P = h_X h_Y \frac{cov_A}{\sigma_{AX}\sigma_{AY}} + e_X e_Y \frac{cov_E}{\sigma_{EX}\sigma_{EY}}$$

Therefore

$$r_P = h_X h_Y r_A + e_X e_Y r_E \qquad \qquad \dots\dots(19.1)$$

This shows how the genetic and environmental causes of correlation combine together to give the phenotypic correlation. If both characters have low heritabilities then the phenotypic correlation is determined chiefly by the environmental correlation: if they have high heritabilities then the genetic correlation is the more important.

The genetic and environmental correlations are often very different in magnitude and sometimes different even in sign, as may be seen from the examples given in Table 19.1. A difference in sign between the two correlations shows that genetic and environmental sources of variation affect the characters through different physiological mechanisms. The correlations between body-weight and egg-laying characters in poultry provide striking examples. Pullets that are larger at 18 weeks from genetic causes reach sexual maturity later and lay fewer eggs, but the eggs are larger. Pullets that are larger from environmental causes reach sexual maturity earlier and lay more eggs, which however are very little different in size.

The dual nature of the phenotypic correlation makes it clear that the magnitude and even the sign of the genetic correlation cannot be determined from the phenotypic correlation alone. Let us therefore consider the methods by which the genetic correlation can be estimated.

Estimation of the genetic correlation. The estimation of genetic correlations rests on the resemblance between relatives in a manner analogous to the estimation of heritabilities described in Chapter 10. Therefore only the principle and not the details of the procedure need be described here. Instead of computing the components of variance of one character from an analysis of variance, we compute the components of covariance of the two characters from an analysis of covariance which takes exactly the same form as the analysis of variance. Instead of starting from the squares of the individual values and partitioning the sums of squares according to the source of variation, we start from the product of the values of the two characters in each individual and partition the sums of products according to the source of variation. This leads to estimates of the observational components of covariance, whose interpretation in

TABLE 19.1

SOME EXAMPLES OF PHENOTYPIC, GENETIC, AND
ENVIRONMENTAL CORRELATIONS

The environmental correlations (except those marked*) were calculated for this table from the genetic correlations and heritabilities given in the papers cited, by equation *19.1*. They are not purely environmental in causation but include correlation due to non-additive genetic causes, as explained in the text. Those marked* are true environmental correlations, estimated directly from the phenotypic correlation in inbred lines and crosses.

	r_P	r_A
Cattle (Johansson, 1950)		
Milk-yield : butterfat-yield.	·93	·85
Milk-yield : butterfat %.	− ·14	− ·20
Butterfat-yield : butterfat %.	·23	·26
Pigs (Fredeen and Jonsson, 1957)		
Body length : backfat thickness.	− ·24	− ·47
Growth rate : feed efficiency.	− ·84	− ·96
Backfat thickness : feed efficiency.	·31	·28
Sheep (Morley, 1955)		
Fleece weight : length of wool.	·30	− ·02
Fleece weight : crimps per inch.	− ·21	− ·56
Fleece weight : body weight.	·36	− ·11
Poultry (Dickerson, 1957)		
Body weight : egg-production.	·09	− ·16
(at 18 weeks) (to 72 weeks of age)		
Body weight : egg weight.	·16	·50
(at 18 weeks)		
Body weight : age at first egg.	− ·30	·29
(at 18 weeks)		
Mice (Falconer, 1954*b*)		
Body weight : tail length (within litters).	·44	·59
Drosophila melanogaster		
Bristle number, abdominal : sternopleural.	·06	·08
(Clayton, Knight, Morris, and Robertson, 1957)		
Number of bristles on different abdominal segments. (Reeve and Robertson, 1954)	—	·96
Thorax length : wing length. (Reeve and Robertson, 1953)	—	·75

terms of causal components of covariance is exactly the same as that of the components of variance given in Table 10.4. Thus, in an analysis of half-sib families the component of covariance between sires estimates $\frac{1}{4}cov_A$, i.e. one quarter of the covariance of breeding values of the two characters. For the estimation of the correlation the components of variance of each character are also needed. Thus the between-sire components of variance estimate $\frac{1}{4}\sigma^2_{AX}$ and $\frac{1}{4}\sigma^2_{AY}$. Therefore the genetic correlation is obtained as

$$r_A = \frac{cov_{XY}}{\sqrt{var_X \, var_Y}} \qquad \ldots\ldots(19.2)$$

where *var* and *cov* refer to the components of variance and covariance.

The offspring-parent relationship can also be used for estimating the genetic correlation. To estimate the heritability of one character from the resemblance between offspring and parents we compute the covariance of offspring and parent for the one character by taking the product of the parent or mid-parent value and the mean value of the offspring. To estimate the genetic correlation between two characters we compute what might be called the "cross-covariance," obtained from the product of the value of X in parents and the value of Y in offspring. This "cross-covariance" is half the genetic covariance of the two characters, i.e. $\frac{1}{2}cov_A$. The covariances of offspring and parents for each of the characters separately are also needed, and then the genetic correlation is given by

$$r_A = \frac{cov_{XY}}{\sqrt{cov_{XX} \, cov_{YY}}} \qquad \ldots\ldots(19.3)$$

where cov_{XY} is the "cross-covariance," and cov_{XX} and cov_{YY} are the offspring-parent covariances of each character separately.

The genetic correlation can also be estimated from responses to selection in a manner analogous to the estimation of realised heritability. This will be explained in the next section.

Data that provide estimates of genetic correlations provide also estimates of the heritabilities of the correlated characters, and of the phenotypic correlations. The environmental correlation can then be found from equation *19.1*. If highly inbred lines are available the environmental correlations can be estimated directly from the phenotypic correlation within the lines, or preferably within the F_1's of crosses between the lines.

Estimates of genetic correlations are usually subject to rather large sampling errors and are therefore seldom very precise. The sampling variance of genetic correlations is treated by Reeve (1955c) and by A. Robertson (1959b). The standard error of an estimate is given approximately by the following formula :

$$\sigma_{(r_A)} = \frac{1 - r_A^2}{\sqrt{2}} \cdot \sqrt{\frac{\sigma_{(h_X^2)} \, \sigma_{(h_Y^2)}}{h_X^2 h_Y^2}}$$

where σ denotes standard error. Since the standard errors of the two heritabilities appear in the numerator, an experiment designed to minimise the sampling variance of an estimate of heritability, in the manner described in Chapter 10, will also have the optimal design for the estimation of a genetic correlation.

Correlated Response to Selection

The next problem for consideration concerns the response to selection: if we select for character X, what will be the change of the correlated character Y? The expected response of a character, Y, when selection is applied to another character, X, may be deduced in the following way. The response of character X—i.e. the character directly selected—is equivalent to the mean breeding value of the selected individuals. This was explained in Chapter 11. The consequent change of character Y is therefore given by the regression of the breeding value of Y on the breeding value of X. This regression is

$$b_{(A)YX} = \frac{cov_A}{\sigma_{AX}^2} = r_A \frac{\sigma_{AY}}{\sigma_{AX}}$$

The response of character X, directly selected, by equation *11.4*, is

$$R_X = i h_X \sigma_{AX}$$

Therefore the correlated response of character Y is

$$CR_Y = b_{(A)YX} R_X$$

$$= i h_X \sigma_{AX} r_A \frac{\sigma_{AY}}{\sigma_{AX}}$$

$$= i h_X r_A \sigma_{AY} \qquad \qquad \dots(19.4)$$

Or, by putting $\sigma_{AY} = h_Y \sigma_{PY}$, the correlated response becomes

$$CR_Y = i h_X h_Y r_A \sigma_{PY} \qquad \qquad \dots(19.5)$$

Thus the response of a correlated character can be predicted if the genetic correlation and the heritabilities of the two characters are known. And, conversely, if the correlated response is measured by experiment, and the two heritabilities are known, the genetic correlation can be estimated. If the heritability of character Y is to be estimated as the realised heritability from the response to selection, then it is necessary to do a double selection experiment. Character X is selected in one line and character Y in another. Then both the direct and the correlated responses of each character can be measured. This type of experiment provides two estimates of the genetic correlation (by equation *19.5*), one from the correlated response of each character; and the two estimates should agree if the theory of correlated responses expressed in equation *19.5* adequately describes the observed responses (Falconer, 1954*b*). A joint estimate of the genetic correlation can be obtained from such double selection experiments, without the need for estimates of the heritabilities, from the following formula which may be easily derived from equations *11.4* and *19.4*:

$$r_A^2 = \frac{CR_X}{R_X} \frac{CR_Y}{R_Y} \qquad \ldots\ldots(19.6)$$

EXAMPLE 19.1. In a study of wing length and thorax length in *Drosophila melanogaster*, Reeve and Robertson (1953) estimated the genetic correlation between these two measures of body size from the responses to selection. There were two pairs of selection lines; one pair was selected for increased and for decreased thorax length, and the other pair for increased and for decreased wing length. In each line the correlated response of the character not directly selected was measured, as well as the response of the character directly selected. Two estimates of the genetic correlation were obtained by equation *19.6*, one from the responses to upward selection and the other from the responses to downward selection. In addition, estimates of the genetic correlation in the unselected population were obtained from the offspring-parent covariance and also from the full-sib covariance. The four estimates were as follows:

Method	*Genetic correlation*
Offspring-parent	0·74
Full sib	0·75
Selection, upward	0·71
Selection, downward	0·73

The agreement between the estimates from selection and the estimates from the unselected population shows that the correlated responses were very close to what would have been predicted from the genetic analysis of the unselected population.

Close agreement between observed and predicted correlated responses, such as was shown in the above example, cannot always be expected, particularly if the genetic correlation is low. With a low genetic correlation the expected response is small and is liable to be obscured by random drift (see Clayton, Knight, Morris and Robertson, 1957). Also, if the genetic correlation is to any great extent caused by linkage, it is likely to diminish in magnitude through recombination, with a consequent diminution of the correlated response. There has not yet been enough experimental study of correlated responses to allow us to draw any conclusions about the number of generations over which they continue, nor about the total response when the limit is reached.

Indirect selection. Consideration of correlated responses suggests that it might sometimes be possible to achieve more rapid progress under selection for a correlated response than from selection for the desired character itself. In other words, if we want to improve character X, we might select for another character, Y, and achieve progress through the correlated response of character X. We shall refer to this as "indirect" selection; that is to say, selection applied to some character other than the one it is desired to improve. And we shall refer to the character to which selection is applied as the "secondary" character. The conditions under which indirect selection would be advantageous are readily deduced. Let R_X be the direct response of the desired character, if selection were applied directly to it. And let CR_X be the correlated response of character X resulting from selection applied to the secondary character, Y. The merit of indirect selection relative to that of direct selection may then be expressed as the ratio of the expected responses, CR_X/R_X. Taking the expected correlated response from equation *19.4* and the expected direct response from equation *11.4*, we find

$$\frac{CR_X}{R_X} = \frac{i_Y h_Y r_A \sigma_{AX}}{i_X h_X \sigma_{AX}}$$

$$= r_A \cdot \frac{i_Y}{i_X} \cdot \frac{h_Y}{h_X} \qquad \qquad .....(19.7)$$

If the same intensity of selection can be achieved when selecting for character Y as when selecting for character X, then the correlated response will be greater than the direct response if $r_A h_Y$ is greater than h_X. Therefore indirect selection cannot be expected to be superior to direct selection unless the secondary character has a substantially higher heritability than the desired character, and the genetic correlation between the two is high; or, unless a substantially higher intensity of selection can be applied to the secondary than to the desired character. The circumstances most likely to render indirect selection superior to direct selection are chiefly concerned with technical difficulties in applying selection directly to the desired character. Two such technical difficulties may be mentioned briefly.

1. If the desired character is difficult to measure with precision, the errors of measurement may so reduce the heritability that indirect selection becomes advantageous. Threshold characters in general are likely for this reason to repay a search for a suitable correlated character, unless the position of the threshold can be adjusted in the manner described in the last chapter. An interesting experimental result which may well prove to be an example of indirect selection being superior to direct selection concerns sex ratio in mice. The sex ratio among the progeny may be regarded as a metric character of the parents. Selection applied directly to sex ratio was ineffective in changing it (Falconer, 1954c), but selection for blood-pH produced a correlated change of sex ratio (Weir and Clark, 1955; Weir, 1955). The reason for the ineffectiveness of direct selection is probably that the true sex ratio of a family is subject to a large error of estimation resulting from the sampling variation, and the heritability is consequently very low.

2. If the desired character is measurable in one sex only, but the secondary character is measurable in both, then a higher intensity of selection will be possible by indirect selection. Other things being equal, the intensity of selection would be twice as great by indirect as by direct selection; but a better plan would be to select one sex directly for the desired character and the other indirectly for the secondary character.

Though indirect selection has been presented above as an alternative to direct selection, the most effective method in theory is neither one nor the other but a combination of the two. The most effective use that can be made of a correlated character is in combination with the desired character, as an additional source of information about the breeding values of individuals. This, however, is a special case of a more general problem which will be dealt with in the final section of this chapter. First we shall show how the idea of indirect selection can be extended to cover selection in different environments.

Genotype-Environment Interaction

The concept of genetic correlation can be applied to the solution of some problems connected with the interaction of genotype with environment. The meaning of interaction between genotype and environment was explained in Chapter 8, where it was discussed as a source of variation of phenotypic values, which in most analyses is inseparable from the environmental variance. The chief problem which it raises and which we are now in a position to discuss concerns adaptation to local conditons. The existence of genotype-environment interaction may mean that the best genotype in one environment is not the best in another environment. It is obvious, for example, that the breed of cattle with the highest milk-yield in temperate climates is unlikely also to have the highest yield in tropical climates. But it is not so obvious whether smaller differences of environmental conditions also require locally adapted breeds; nor is it intuitively obvious how much of the improvement made in one environment will be carried over if the breed is then transferred to another environment. These matters have an important bearing on breeding policy. If selection is made under good conditions of feeding and management on the best farms and experimental stations, will the improvement achieved be carried over when the later generations are transferred to poorer conditions? Or would the selection be better done in the poorer conditions under which the majority of animals are required to live? The idea of genetic correlation provides the basis for a solution of these problems in the following way.

A character measured in two different environments is to be regarded not as one character but as two. The physiological mechanisms are to some extent different, and consequently the genes required for high performance are to some extent also different. For example, growth rate on a low plane of nutrition may be principally a matter of efficiency of food-utilisation, whereas on a high plane of nutrition it may be principally a matter of appetite. By regarding performance in different environments as different characters with genetic correlation between them we can in principle solve the problems outlined above from a knowledge of the heritabilities of the different characters and the genetic correlations between them (Falconer, 1952). If the genetic correlation is high, then performance in two different environments represents very nearly the same character, determined by very nearly the same set of genes. If it is low, then the characters are to a great extent different, and high performance requires a different set of genes. Here we shall consider only two environments, but the idea can be extended to an indefinite number of different environments (A. Robertson, 1959*b*).

Let us consider the problem of the "carry-over" of the improvement from one environment to another. Let us suppose that we select for character X—say growth rate on a high plane of nutrition—and we look for improvement in character Y—say growth rate on a low plane of nutrition. The improvement of character Y is simply a correlated response and the expected rate of improvement was given in equation *19.5* as

$$CR_Y = i h_X h_Y r_A \sigma_{PY}$$

The improvement of performance in an environment different from the one in which selection was carried out can therefore be predicted from a knowledge of the heritability of performance in each environment and the genetic correlation between the two performances. We can also compare the improvement expected by this means with that expected if we had selected directly for character Y, i.e. for performance in the environment for which improvement is wanted. This is simply a comparison of indirect with direct selection, which was explained in the previous section. The comparison is made from the ratio of the two expected responses given in equation *19.7*, i.e.

$$\frac{CR_Y}{R_Y} = r_A \frac{i_X h_X}{i_Y h_Y}$$

This shows how much we may expect to gain or lose by carrying out the selection in some environment other than the one in which the improved population is required to live. If we assume that the intensity of selection is not affected by the environment in which the selection is carried out, then the indirect method will be better if $r_A h_X$ is greater than h_Y, where h_X is the square root of the heritability in the environment in which selection is made, and h_Y is the square root of the heritability in the environment in which the population is required subsequently to live. If the genetic correlation is high, then the two characters can be regarded as being substantially the same; and if there are no special circumstances affecting the heritability or the intensity of selection it will make little difference in which environment the selection is carried out. But if the genetic correlation is low, then it will be advantageous to carry out the selection in the environment in which the population is destined to live, unless the heritability or the intensity of selection in the other environment is very considerably higher.

This is the theoretical basis for dealing with selection in different environments. So far, however, there has been little experimental work to substantiate the theory. The results of the experiments that have been carried out do not appear to be fully in agreement with theoretical expectations, and this suggests that other factors not yet understood are probably operating. (See Falconer and Latyszewski, 1952; Falconer, 1952.)

SIMULTANEOUS SELECTION FOR MORE THAN ONE CHARACTER

When selection is applied to the improvement of the economic value of animals or plants it is generally applied to several characters simultaneously and not just to one, because economic value depends on more than one character. For example, the profit made from a herd of pigs depends on their fertility, mothering ability, growth rate, efficiency of food-utilisation, and carcass qualities. How, then, should selection be applied to the component characters in order to achieve the maximum improvement of economic value? There are several possible procedures. One might select in turn for each character singly ("tandem" selection); or one might select for all the characters at the same time but independently, rejecting all individuals that fail to come up to a certain standard for each character regardless of their values for any other of the characters ("independent culling levels"). It has been shown, however, that the most rapid improvement of economic value is expected from selection applied simultaneously to all the component characters together, appropriate weight being given to each character according to its relative economic importance, its heritability and the genetic and phenotypic correlations between the different characters, (Hazel and Lush, 1942; Hazel, 1943). The practice of selection for economic value is thus a matter of some complexity. The component characters have to be combined together into a score, or *index*, in such a way that selection applied to the index, as if the index were a single character, will yield the most rapid possible improvement of economic value. If the characters are uncorrelated there is no great problem: each character is weighted by the product of its relative economic value and its heritability. This is the best that can be done in the absence of information about the genetic correlations, but if the genetic correlations are known the efficiency of the index can be improved. The following account gives an outline of the principles on which the construction of a selection index is based. For a fuller account the reader should consult Lerner (1950) and the original papers of Fairfield Smith (1936) and Hazel (1943).

For the sake of simplicity we shall consider only two component characters of economic value, but the conclusions can readily be extended to any number of characters. Let the economic value be determined by two characters X and Y, and let w be the additional profit expected from one unit increase of Y relative to that from one unit increase of X. The aim of selection therefore is to pick out individuals with the highest values of $(A_X + wA_Y)$, where A_X and A_Y are the breeding values of the two characters X and Y. Let us call this compound breeding value "merit," with the symbol H, so that

$$H = A_X + wA_Y \qquad \qquad \dots \dots (19.8)$$

The problem is to find out how the phenotypic values, P_X and P_Y, of the two component characters are to be combined into an index that gives the best estimate of an individual's merit, H. In Chapter 10 we saw how the best estimate of the breeding value of an individual for one character is the regression equation $A = b_{AP}P$, where b_{AP} is the regression of breeding value on phenotypic value, and is equal to the heritability (see p. 166). The present problem is essentially the same, only now we have to use partial regression coefficients. The multiple regression equation giving the best estimate of merit is

$$H = b_{HX.Y}P_X + b_{HY.X}P_Y \qquad \qquad \dots \dots (19.9)$$

where P_X and P_Y are phenotypic values measured as deviations from the population mean. (In this formula, and in those that follow, the symbol X has the same meaning as P_X, i.e. the phenotypic value of character X; and similarly Y and P_Y both mean the phenotypic value of character Y. Thus, $b_{HX.Y}$ is the regression of merit on the phenotypic value of X when the phenotypic value of Y is held constant, and $b_{HY.X}$ has a similar meaning with X and Y interchanged.) In practice it is convenient to have the index in a form that requires the manipulation of only one of the phenotypic values, i.e. in the form

$$I = P_X + WP_Y \qquad \qquad \dots\dots(19.10)$$

where I is the index by means of which individuals are to be chosen, and W is a factor by which the phenotypic value of character Y is to be multiplied. Since the absolute magnitude of the index is of no importance, but only its relative magnitude in different individuals, we can work with the phenotypic values as they stand instead of with deviations from the population mean. And we can put equation 19.9 into the form of equation 19.10 simply by dividing through by $b_{HX.Y}$. Then W in equation 19.10 is the ratio of the two partial regression coefficients,

$$W = \frac{b_{HY.X}}{b_{HX.Y}}$$

and our task now is to find a way of expressing W in terms of the genetic properties of the two characters.

First let us put the partial regression coefficients in terms of the total regression coefficients. For example,

$$b_{HY.X} = \frac{b_{HY} - b_{HX}b_{XY}}{1 - r_{XY}^2}$$

Therefore

$$W = \frac{b_{HY.X}}{b_{HX.Y}} = \frac{b_{HY} - b_{HX}b_{XY}}{b_{HX} - b_{HY}b_{YX}}$$

Now let us express these total regressions in terms of covariances and variances. For example, $b_{HY} = cov_{HY}/\sigma_Y^2$. After some simplification the expression reduces to

$$W = \frac{\sigma_X^2 cov_{HY} - cov_{HX}cov_{XY}}{\sigma_Y^2 cov_{HX} - cov_{HY}cov_{XY}} \qquad \dots\dots(19.11)$$

The variances σ_X^2 and σ_Y^2 here, and in what follows, are the phenotypic variances of characters X and Y. The covariances in the above expression can be expressed in terms of the phenotypic variance and the heritability of each character and of the phenotypic and genetic correlations between the two characters, all of which quantities can be estimated. Take, for example, the covariance of H with X. This may be written as follows:

$$cov_{HX} = \text{covariance of } (A_X + wA_Y) \text{ with } P_X$$
$$= cov_{(A_X.P_X)} + cov_{(wA_Y.P_X)}$$
$$= h_X^2\sigma_X^2 + wr_A h_X\sigma_X h_Y\sigma_Y$$

In this way the covariances in equation 19.11 can be expressed as follows, σ and σ^2 being phenotypic standard deviations and variances throughout:

$$\left.\begin{aligned}
cov_{HX} &= h_X^2\sigma_X^2 + wr_A h_X h_Y\sigma_X\sigma_Y \\
cov_{HY} &= wh_Y^2\sigma_Y^2 + r_A h_X h_Y\sigma_X\sigma_Y \\
cov_{XY} &= r_P\sigma_X\sigma_Y
\end{aligned}\right\} \dots\dots(19.12)$$

The procedure for selection is thus to compute the covariances given in 19.12, substitute them in 19.11 and use the value of W so obtained to compute the index of selection I given in 19.10. The value of the index for each individual then forms the basis of selection.

The index as formulated above is applicable only to individual selection. If family selection is applied then the heritabilities and correlations that go into the index must be those appropriate to the family means. Family selection, however, is not greatly improved by the use of an index, because the family heritabilities of the component characters are generally fairly high and the mean economic value of a family in terms of phenotypic values is not very different from its merit in terms of breeding values. Therefore family selection for economic value can be applied with little loss of efficiency if the phenotypic values are weighted only by w, the relative economic importance of each component character.

The complexity of selection by means of an index need hardly be emphasised, especially when the index is extended to cover many component characters. Even with two characters, estimates of no fewer than seven quantities are required for the construction of the index. Since some of these, particularly the genetic correlation, cannot usually be estimated with any great precision, the index cannot be regarded as much more than a rough guide to procedure. But since selection has to be applied to economic value by some means, it seems better to use a selection index, however imprecise, than to base selection on a purely arbitrary combination of component characters.

Use of a secondary character by means of an index. The

selection index described above can readily be adapted to meet the case where improvement of only one character is sought, the other character being used merely as an aid to more efficient selection. The use of a secondary character in this way was mentioned earlier, in connexion with indirect selection. Let X be the character it is desired to improve, and Y the secondary character. Then the relative "economic" value of character Y is zero, and we can substitute $w=0$ in the formulae of *19.12*. Substitution of the covariances in equation *19.11* then yields a formula which on simplification reduces to

$$W = \frac{\sigma_X(r_A h_Y - r_P h_X)}{\sigma_Y(h_X - r_A h_Y)} \qquad \dots\dots(19.13)$$

The selection index of equation *19.10* is then used with this value of W. The value of W in the index may be negative. This will arise if the phenotypic correlation between the two characters is chiefly environmental in origin. The secondary character then acts as an indicator of the environmental deviation rather than of the breeding value of the desired character (see Rendel, 1954; and Osborne, 1957*b*).

Genetic correlation and the selection limit. There is one important consequence of simultaneous selection for several characters to be discussed before we leave the subject. Just as the heritabilities are expected to change after selection has been applied for some time, so also are the genetic correlations. If selection has been applied to two characters simultaneously the genetic correlation between them is expected eventually to become negative, for the following reason. Those pleiotropic genes that affect both characters in the desired direction will be strongly acted on by selection and brought rapidly toward fixation. They will then contribute little to the variances or to the covariance of the two characters. The pleiotropic genes that affect one character favourably and the other adversely will, however, be much less strongly influenced by selection and will remain for longer at intermediate frequencies. Most of the remaining covariance of the two characters will therefore be due to these genes, and the resulting genetic correlation will be negative. The consequence of a negative genetic correlation, whether produced by selection in this way or present from the beginning, is that the two characters may each show a heritability that is far from zero, and yet when selection is applied to them simultaneously neither responds.

We have already discussed, in Chapter 12, what is essentially the same situation resulting from the combined effects of artificial and natural selection: a selection limit is reached even though the character to which artificial selection is applied still shows a substantial amount of additive genetic variance.

EXAMPLE 19.2. A practical example from a commercial flock of poultry is described by Dickerson (1955). Selection for economic value had been applied for many years, but recent progress in the component characters was much less than was to be expected from their heritabilities, which were found to be moderately high. Estimations of the genetic correlations between the component characters showed that many of these were negative. To take just one example, the relationships between egg-production and egg weight were as follows:

X	Y	h_X^2	h_Y^2	r_P	r_A	r_E
Production	Weight	0·32	0·59	−0·04	−0·39	+0·25

In spite of the high heritabilities neither character had shown any improvement over the last 10–15 years. The high negative genetic correlation would account for this failure to respond, if selection was applied to both characters simultaneously. It is interesting to note that environmental variation, unlike genetic variation, affects both characters in the same way and leads to a positive environmental correlation. The phenotypic correlation, which is almost zero, gives no clue to the genetic relationship between the two characters, and the failure to respond to selection could not have been predicted from it alone.

A population which has been subjected over a long period to selection for economic value throws light on the genetic properties to be expected in natural populations subject to natural selection for fitness. Fitness is a compound character with many components—far more than would appear in the most elaborate assessment of economic value—and so we should expect negative genetic correlations between its major components, a conclusion to be developed further in the next chapter. It is interesting to note, however, that natural selection takes no account of heritabilities or genetic correlations, and is therefore, in theory, less efficient in improving fitness than artificial selection by means of an index is in improving economic value.

CHAPTER 20

METRIC CHARACTERS UNDER NATURAL SELECTION

Throughout the discussion of the genetic properties of metric characters, which has occupied the major part of the book, very little attention has been given to the effects of natural selection, and something must now be done to remedy this omission. The absence of differential viability and fertility was specified as a condition in the theoretical development of the subject: that is to say, natural selection was assumed to be absent. Though for many purposes this assumption may lead to no serious error, a complete understanding of metric characters will not be reached until the effects of natural selection can be brought into the picture. The operation of natural selection on metric characters has, however, a much wider interest than just as a complication that may disturb the simple theoretical picture and the predictions based on it. It is to natural selection that we must look for an explanation of the genetic properties of metric characters which hitherto we have accepted with little comment. The genetic properties of a population are the product of natural selection in the past, together with mutation and random drift. It is by these processes that we must account for the existence of genetic variability; and it is chiefly by natural selection that we must account for the fact that characters differ in their genetic properties, some having proportionately more additive variance than others, some showing inbreeding depression while others do not. These, however, are very wide problems which are still far from solution, and in this concluding chapter we can do little more than indicate their nature. Any discussion of them, moreover, cannot but be controversial; the reader should therefore understand that the contents of this chapter are to a large extent matters of personal opinion, and that any conclusions to which the discussion may lead are open to dispute.

We shall refer throughout to a population that is in genetic equilibrium. Being in genetic equilibrium means that the gene frequencies are not changing, and therefore that the mean values of all metric characters are constant. (Changes of environmental conditions are assumed to be so slow as to be negligible.) The population is constantly subject to natural selection tending to increase fitness, but despite the selection the gene frequencies do not change and fitness does not improve. There can therefore be no additive genetic variance of fitness: in other words, if we could measure fitness itself as a character we should find that its genetic variance was entirely non-additive. For the purposes of discussion we may regard any natural population as being in genetic equilibrium, at least approximately, and also any population that has been subject to artificial selection consistently over a long period of time, provided that fitness is defined in terms of both the artificial and the natural selection. Fitness, crudely defined, is the "character" selected for, whether by natural selection alone or by artificial and natural selection combined.

If a population is in genetic equilibrium it follows that a reduction of fitness must in principle result from any change in the array of gene frequencies, apart from any genes that may have no effect on fitness. Natural selection must therefore be expected to resist any tendency to change of the gene frequencies, such as must result from artificial selection applied to any metric character other than fitness itself. This principle has been called "genetic homeostasis," and its consequences have been discussed, by Lerner (1954). Thus if we change any metric character by artificial selection we must expect a reduction of fitness as a correlated response. And if we then suspend the artificial selection before any of the variation has been lost by fixation, we must expect the population mean to revert to its original value. On the whole, the experience of artificial selection is in general agreement with this expectation, though under laboratory conditions the reduction of fitness may not be apparent in the early stages, and some characters appear to revert very slowly, if at all, toward the original value. Our domesticated animals and plants are perhaps the best demonstration of the effects of the principle. The improvements that have been made by selection in these have clearly been accompanied by a reduction of fitness for life under natural conditions, and only the fact that domestic animals and plants do not have to live under natural conditions has allowed these improvements to be made. The problems for discussion in this chapter must be seen against the background of this principle: that the existing array of gene frequencies, and consequently the existing genetic properties of the population, represent the best total adjustment to existing conditions that is possible with the available genetic variation.

The problem of how natural selection operates on metric characters has two aspects: the relation between any particular metric character and fitness, and the way in which natural selection operates on the individual loci concerned with a metric character. This latter aspect is part of a wider problem which concerns the reasons for the existence of genetic variation. We shall discuss these two aspects separately, because any conclusions that may be drawn about the second will depend on what can be discovered about the first.

Relation of Metric Characters to Fitness

The fitness of an individual is the final outcome of all its developmental and physiological processes. The differences between individuals in these processes are seen in variation of the measurable attributes which can be studied as metric characters. Thus the variation of each metric character reflects to a greater or lesser degree the variation of fitness; and the variation of fitness can theoretically be broken down into variation of metric characters. Let us consider for example a mammal such as the mouse, because this matter is more easily discussed in concrete terms. Fitness itself might be broken down into two or three major components, which could be measured and studied as metric characters. These might be the total number of young reared, and some measure of the quality of the young, such as their weaning weight. The variation of the major components would account for all the variation of fitness. Each of the major components might be broken down into other metric characters which would account for all their variation. Thus the total number of young weaned depends on the viability of the parent up to breeding age, its mating ability, average litter size, frequency of litters, and longevity. These characters in turn might be further broken down. For example, litter size depends on the number of eggs shed and the proportion that are brought to term. The number of eggs shed depends, again, on body size and endocrine activity, among other things. Thus each metric character has its place in one of a series of chains of causation converging toward fitness. And these chains of causation interconnect one with another: body size, for example, influences not only litter size, but also lactation, longevity, and probably many other characters. The relationship between any particular metric character and fitness is thus a very complicated matter. The following discussion of the problem is based largely on the ideas put forward by A. Robertson (1955*b*).

The way in which natural selection operates on a character depends on the part played by the character in the causation of differences of fitness: that is to say, on the manner and degree by which differences of value of the metric character cause differences of fitness. This we shall refer to as the "functional relationship" between the character and fitness. The functional relationship expresses the mode of operation of natural selection on the metric character; but it is not necessarily also the relationship that would be revealed if we could measure the fitness of individuals and compare their fitness with their values for the metric character. This point, however, will be more easily explained by an example to be given in a moment. Different characters must be expected to have different functional relationships with fitness, according to the nature of the character. In explanation of the kinds of relationship that may be envisaged let us take some examples of different sorts of character at different positions in the chain of causation.

1. **Neutral characters.** There may be some characters that have no functional relationship at all with fitness. This does not mean that, like vestigial organs, they have no function or use. It means that the variation in the character is not a cause of variation of fitness. Abdominal bristle number in *Drosophila* may be taken as an example of a character which is probably not far from this state, and two reasons can be given for regarding it thus. First, it is difficult to conceive of any biological reason why it should be important to have 18 bristles, or thereabouts, on each segment rather than more or fewer. And second, if we change the bristle number by artificial selection and then suspend the selection, the mean bristle number does not return to its original value—or returns only very slowly— under the influence of natural selection, even though it could be brought back rapidly by artificial selection (Clayton, Morris, and Robertson, 1957). In other words, genetic homeostasis in respect of bristle number is weak or non-existent. Such a metric character may be termed "neutral" with respect to fitness. The mean value of a neutral character in the population has little or nothing to do with the character itself, but is the outcome of the pleiotropic effects of the genes whose frequencies are controlled by their effects on other

characters. Though a neutral character has no functional relationship with fitness, we may nevertheless find that individuals with different values do in fact differ in fitness in a regular way. If the genetic variance of the character is predominantly additive then individuals with intermediate values will tend on the whole to be heterozygous at more loci than individuals with extreme values. Then if heterozygotes were superior in fitness for some other reason, unconnected with the character in question, this would result in intermediates being superior in fitness. At the level of observation there would be a relationship between values of the character and fitness, but this would not be a functional relationship because the values of the character are not the cause of the differences of fitness. The differences of fitness are the result of the functional relationships of other characters affected by the pleiotropic action of the genes.

2. Characters with intermediate optima. There are some characters for which an intermediate value is optimal for functional reasons. One might distinguish three sorts of intermediate optimum according to the reasons for intermediates being superior in fitness.

(i) *Optima determined by the character itself.* As an example we might take any character that measures the thermal insulation of a mammalian coat. Too dense a coat would be disadvantageous and so would too sparse a coat. An intermediate density would confer the highest fitness as a consequence of the function of the coat in thermoregulation. For such a character the mean value in the population is the optimal value, provided there are no complications of the sort to be considered later. Though irrefutable biological reasons might be given for supposing that a character such as the density of fur has an intermediate optimal value, we might nevertheless find that over the range of variation covered by the population there was very little variation in fitness. In practice therefore one could not expect always to draw a clear line between this sort of character and a neutral character such as we have taken bristle number to be.

(ii) *Optima imposed by the environment.* As an example we may take the clutch size of birds. It has been shown, particularly for the European robin and swift, that a larger number of young are reared from nests containing the average number of eggs than from nests with larger or smaller clutches (Lack, 1954). Thus individuals with intermediate values appear to be the fittest. If a character such as this has an optimal value that is intermediate there must obviously be some other factor interacting with it to determine fitness; for, otherwise, the individuals that lay more eggs must inevitably be the fitter. The other factor in this case is the supply of insects for feeding the young and the length of daylight available for their capture. With characters of this sort natural selection tends to eliminate individuals with extreme values and favours individuals with intermediate values. The mean value in the population is the optimal value under the environmental conditions to which the population is subjected. If the environment were to change, the population mean would change too in adjustment to the new optimum. In the case of clutch size it is noteworthy that the mean value varies with the latitude, being larger in the north than in the south.

(iii) *Optima imposed by a correlated character.* Body size in mice may be taken as an example. Larger mice have larger litters and, under laboratory conditions, they rear more young. Therefore if there were no other factor involved, larger mice would be fitter. Since body size can, as we have seen, be readily increased by artificial selection, there must be some other factor that prevents its being increased by natural selection in the wild. The other factor in this case is probably not environmental, but another character negatively correlated with size, namely wildness. A change of body size under artificial selection is always accompanied by a correlated change of wildness. Large mice are phlegmatic and unreactive to disturbance, whereas small mice are alert and react energetically to disturbance (MacArthur, 1949; Falconer, 1953). Therefore under natural conditions larger mice would more readily fall prey to cats and owls than small mice, and the advantage of greater fertility would be offset by the disadvantage of being less well fitted to escape predators. The body size of wild mice, it may be suggested, represents the best compromise between these two correlated characters. If we could measure the relationship between size and fitness in wild mice we should find that those of intermediate size were fittest. With characters of this sort also, the population mean represents the optimal value. But this value is optimal not because of this character itself but because of its genetic correlations with other characters. Large mice are selected against not because they are large but because, being large, they are inevitably also less wild. This example brings us to the point mentioned at the end of the last chapter: that we must expect to find negative genetic correlations between characters under simultaneous selection. In this case we find a negative genetic correlation between large size and wildness, both of which may reasonably be supposed to be favoured by natural selection. These two characters are "components" of fitness in the same way that characters of economic importance are components of total economic merit. What natural selection "aims at" is to increase both characters indefinitely, but the physiological connexions between them, which we see as a negative genetic correlation, limit the increase that is possible with the existing genetic variability.

3. Major components of fitness. If we could measure fitness itself—which is technically very difficult—we should obviously find no "optimal" value; the individuals most favoured by natural selection would not be those nearest to the population mean, but the most extreme. In spite of the selection toward higher values the mean fails to change under natural selection because there is no additive variance of fitness. If we measure as a metric character something that is a major component of fitness, in the sense that it accounts for a large part of the variation of fitness, we should probably find the same sort of relationship. Fitness would increase as the value of the character increased. At the very highest values, however, fitness would probably decline again slightly. Egg-laying in *Drosophila* might well be such a character, even if measured only over a few days, since the daily egg production is highly correlated with the total production (Gowen, 1952). We should almost certainly find that the fittest individuals were not those that laid an intermediate number of eggs, but those that laid almost the most. The most extreme individuals would probably be slightly less fit because of some environmental limitation or some correlated character, perhaps longevity. There must be many characters whose relationships with fitness fall between this and the previous type, characters with an optimal value above the population mean but yet below that of the most extreme individuals.

The foregoing discussion will be enough to explain the nature of the problem of the relationship between a metric character and fitness and to indicate the sort of solution that may be sought. Let us turn now to the connexion between the relationship with fitness and the nature of the genetic variation of a metric character. When we first discussed the heritability as a property of a character in Chapter 10, we noted a tendency toward lower heritabilities among characters more closely connected with fitness. But the precise meaning of a "close connexion" with fitness was not explained. It may now be suggested that the meaning of a close connexion with fitness may perhaps be seen in the functional relationships discussed above. Characters with the closest connexion are of the third type where the population mean is not at an optimal value; characters with a less close connexion are nearer to the second type; while characters with the least connexion are the neutral or nearly neutral characters. On the whole it does seem that characters with high heritabilities are to be found among the first type and characters with low heritabilities among the third. Differences of heritability are, however, not really relevant here. It is the genetic variance with which we are concerned; and the differences in the proportion of the genotypic variance that is additive, that we want to account for. But so little is known about how the genotypic variance is partitioned into additive and non-additive components that we can scarcely begin to tackle the problem. Four characters of *Drosophila*, however, seem to fit the picture fairly well, (see Table 8.2). For bristle number, which we have taken as a neutral character of the first type, 85 per cent of the genotypic variance is additive. Thorax length, which might perhaps be of the second type, has about the same proportion. For ovary size, however, only 43 per cent of the genotypic variance is additive, and this character might well be between the second and third types. For egg laying, which we have taken to be of the third type, the proportion is 29 per cent. These comparisons, of course, cannot be given much weight because in fact we know almost nothing of the functional relations of the characters with fitness. But they do suggest that the solution of the problem of why characters differ in their genetic properties may lie along these lines. The reaction of a character to inbreeding seems also to be connected with the proportion of non-additive genetic variance, those with most non-additive variance being those that suffer the greatest inbreeding depression. Some, perhaps most, of the non-additive variance must be attributed to dominance. Reasons for expecting the effects of genes on characters closely connected with fitness to show dominance, while the effects on characters not closely connected with fitness do not, have been put forward by A. Robertson (1955*b*); but it would take too much space here to summarise the argument. There we must leave the problem of the nature of the genetic variance and pass on to the second aspect of the operation of natural selection on metric characters.

Maintenance of Genetic Variation

The second aspect of the operation of natural selection on metric characters—its effects on the individual loci—is part of a wider problem, which concerns the mechanisms by which genetic variation is maintained. Almost every metric character, of the many that have been studied both in natural populations and in domesticated animals and plants, exhibits genetic variation. What are the reasons for the existence of this genetic variation? The coexistence in a population of different alleles at a locus is governed by the three processes of mutation, random drift, and selection. Allelic differences originate by mutation and are extinguished by random drift, since no natural population is infinite in size. Natural selection may tend to eliminate the differences by favouring one allele over all others at a locus; or it may tend to perpetuate the differences by favouring heterozygotes. Let us discuss the role of natural selection first and the roles of mutation and random drift later.

Effects of selection on individual loci. The way in which selection operates on any locus depends on the effects that the different alleles have on fitness itself, and not simply on their effects on one particular metric character. Therefore the functional relations between characters and fitness, which were discussed above, can indicate the action of selection only on those loci which affect fitness through the character in question and not through any pleiotropic effects on other characters. Let us consider the three types of character in turn.

1. *Neutral characters.* If there are genes whose only effects are on a neutral character, then selection plays no part in the existence of allelic differences at these loci. The gene frequencies at these loci must be controlled solely by mutation and random drift.

2. *Characters with intermediate optima.* The consequences of selection favouring individuals of intermediate value have been examined from different aspects by Wright (1935a, b), Haldane (1954a), and by A. Robertson (1956) who reaches the following conclusions. If the intermediate optimum is the result of the functional relations of the character to fitness, and the optimum is determined by the character itself or by the environment, then selection will tend toward fixation at all the loci whose only influence on fitness is through the character in question. This would apply to characters

of type 2 (i) and (ii) described above and exemplified by the density of mammalian fur and by clutch size in birds. Selection will thus tend to eliminate rather than to conserve variability arising from loci which affect fitness only through such characters. The rate at which the gene frequencies are expected to change toward fixation is very slow, and so the rate at which variation would be eliminated is also very slow; but on an evolutionary time-scale it would not be negligible. Characters of type 2 (iii), where an intermediate optimum is determined by a correlated character, have not yet been investigated in this connexion, and the mode of operation of selection on loci that affect them is not known.

3. *Major components of fitness.* The essential feature of a major component of fitness is that the population mean is not at the optimum. But we cannot deduce, from this fact alone, how selection operates on the individual loci. If the genes that affect these characters are at intermediate frequencies, it seems most probable that they are held there by selection favouring heterozygotes, because it seems hardly possible that the coefficients of selection are small enough to allow mutation alone to maintain intermediate frequencies. We do not know, however, whether these genes are at intermediate frequencies. It seems quite possible that a considerable portion of the genetic variation of these characters is due to genes at very low frequencies, where they are maintained by the balance between mutation and selection against the recessive homozygotes. Much evidence, however, has been presented by Lerner (1954) in support of the view that heterozygotes in general are superior in fitness; and Haldane (1954*b*) has pointed out that a general superiority of heterozygotes is a very reasonable expectation from biochemical considerations of gene action. Though the matter is not yet settled, the weight of evidence at present seems to point to superior fitness of heterozygotes, and consequently to natural selection favouring heterozygotes at most of the loci that affect fitness through its major components.

There are three other ways in which selection may influence genetic variability, to be discussed before we leave the subject. They are all subsidiary to the main effects on gene frequencies which we have been discussing; they may modify these main effects, but they do not in themselves provide a sufficient description of the operation of natural selection.

Variable selection. If characters have optimal values these optima are likely to vary from time to time and from season to season according to the environmental conditions. The selection pressures on the individual genes are therefore likely to change from generation to generation. The consequence of variable selection coefficients has been shown (Kimura, 1954) to be a tendency toward fixation—or more strictly, near-fixation—the favoured allele being the one that gives the highest average fitness. In this aspect selection would therefore tend to eliminate variability. The optimal values are likely to vary also from place to place within each generation, especially if different genotypes choose different environments in which to live, as Waddington (1957) suggests. This form of variable selection has been shown to be capable under certain conditions of maintaining stable polymorphism, as was mentioned in Chapter 2. Its effect on the variation of metric characters, however, has not been examined. It does not seem likely to be very great.

Balanced linkage. Mather's theory of "polygenic balance" is based on the idea of selection favouring intermediate values of metric characters and the effect this is likely to have on linkage (see for example, Mather, 1949, 1953*b*). In considering linkage between the loci affecting a metric character we have to take account of the linkage phase. We may say that two genes on the same chromosome are in coupling if they affect the character in the same direction, and in repulsion if they affect it in opposite directions. The two phases will be represented in equal frequencies in a random-breeding population subject to no selection, as was shown in Chapter 1. Now, chromosomes carrying genes in coupling will contribute more to the variation than chromosomes carrying genes in repulsion. And individuals with intermediate values will tend on the whole to carry repulsion chromosomes rather than coupling chromosomes. Therefore, if intermediates are favoured for functional reasons, selection will favour repulsion chromosomes and thus tend to build up "balanced" combinations of genes: that is, combinations in predominantly repulsion linkage, which contribute the minimal amount of variance. In this way, according to Mather, "potential" genetic variability is stored in latent form, and a compromise is reached between the conflicting needs of uniformity in adaptation to present circumstances and flexibility in adaptation to changing circumstances.

If, however, this supposed tendency of selection to build up balanced combinations is to have any significant effect on genetic variability it is necessary that the selection should be strong enough to maintain the balanced combinations in the face of recombination which must tend continuously to reduce them to a random arrangement. The selective forces required have been examined by Wright (1952*b*). It is clear, without going into the details, that coefficients of selection of the same order of magnitude as the recombination frequencies would be required. The balancing of linkage by natural selection therefore seems from Wright's reasoning to be relevant only to very short segments of chromosome. Loci with more than about 1 per cent recombination between them would not be expected to depart significantly from a random arrangement, unless they carried major genes with large effects on the character. Furthermore, if we consider a number of loci on the same chromosome, it is not clear how much difference of variance would be expected between fully balanced and fully random arrangements; it might well be very little. Experimental evidence on the matter is scanty. In two experiments, one with mice and the other with *Drosophila*, where artificial selection was applied for and against intermediates, no changes of variance were detected (Falconer and Robertson, 1956; Falconer, 1957*b*). Intensification of the selection against extremes therefore does not seem to have any effect on the variance within the time-span of a laboratory experiment.

Canalisation. Waddington's theory of "canalisation" is concerned with the developmental pathways through which the phenotypic values come to their expression (see Waddington, 1957). If intermediates are favoured because of their values of the metric character in question, then deviation from the optimal value is disadvantageous. Selection will therefore operate against the causes of deviation, and will tend to produce a greater stability so that development is canalised along the path that leads to the optimal phenotypic expression. The role ascribed to selection is its discrimination against alleles that increase variability. These may be at loci that affect the character in question or at other loci. Variation both of environmental and of genetic origin may be reduced in this way. The genetic variation is reduced not by eliminating the segregation, but by rendering the organism less sensitive to the effects of the segregation. A change in the proportion of genetic to environmental variation is therefore not necessarily to be expected. As a consequence of canalisation we should expect to find some characters less variable than others, the less variable being those for which deviation from the optimum has the more serious effect on fitness. This expected consequence of canalisation, however, cannot easily be tested experimentally, because, as Waddington (1957) points out, it is difficult to find a logical basis for comparing the variability of different characters.

Origin of variation by mutation. Before the reasons for the existence of genetic variability can be fully understood it will be necessary to know what part mutation plays in restoring what is lost by random drift or by selection. If there were no selection of any kind then the amount of genetic variation would come to equilibrium when its rate of origin was equal to its rate of extinction by random drift. The rate of extinction presents no very serious problem because we need know the population size only approximately. If, therefore, we knew the rate of origin by mutation we could decide whether a significant amount of the existing variation can be ascribed to mutation. Very little, however, is known about the rate of origin by mutation. The only evidence comes from two studies of *Drosophila* by Clayton and Robertson (1955) and Paxman (1957), which yielded very similar results. The following discussion is based on the experiment of Clayton and Robertson. Selection for abdominal bristle number was applied to an inbred line derived from the same base population on which the other studies of this character were made. From the rate of response to selection it was concluded that the average amount of variation arising by spontaneous mutation in one generation amounted to one thousandth part of the genetic variation present in the base population. In other words it would take about 1000 generations for mutation to restore the genetic variation to its original level. (We may note in passing that this proves mutation to have a negligible influence on the response of non-inbred populations to artificial selection, apart from the rare occurrence of mutants with major effects.) Now consider the loss of variance due to random drift in a population of effective size N_e, subject to no selection. If all the genetic variance is additive, as it very nearly is in the case of bristle number, then the rate of loss per generation is equal to the rate of inbreeding, which is $1/2N_e$. (This follows from the reasoning given in Chapter 15, where the variance within a line was shown to be $(1 - F)$ times the original variance.) Therefore the new variation arising by mutation at the rate found in this experiment would be lost at the same rate, if the rate of inbreeding were $1/1,000$: that is, in a population of effective size 500. The base population was roughly ten times this size and therefore the expected rate of extinction by random drift is less than the observed rate of origin by mutation. In other words, mutation alone seems to be capable of accounting for more variation of bristle number than was actually present in the base population. Therefore selection favouring heterozygotes does not seem to have been an important cause of the genetic variability of bristle number. This suggests that little of the variation of bristle number is due to the pleiotropic effects of genes that affect the major components of fitness. It suggests, in other words, that much of the variation of bristle number is due to genes that are not far from being neutral with respect to fitness. This conclusion, though only tentative, is in line with the fact, mentioned earlier, that bristle number shows little tendency to revert to the original mean value when artificial selection is relaxed. The conclusions to which the results of this experiment point cannot yet be extended to other characters. Characters more closely connected with fitness, when they have been studied from this point of view, may present a very different picture.

Evolutionary significance of variability. There can be little doubt that the existence of genetic variation is advantageous to the evolutionary survival of a species, the advantage it confers being the ability to evolve rapidly and so to meet the needs of a changing environment, both through the course of time and in the colonisation of new localities. Sexual reproduction and outbreeding are necessary conditions for the continued existence of genetic variation and it is noteworthy that the naturally inbreeding species among the higher plants are of comparatively recent origin. This suggests that the possession of genetic variability is necessary for the continued existence of a species over a long period of time; or in other words, that the prevalence of genetic variability among existing species is because those without it have not survived. The inbreeding plants, however, as we see them at present, compete successfully with the outbreeding species, and this proves that the possession of genetic variability does not confer much immediate advantage. The evolutionary significance of genetic variability, however, throws no light on the mechanisms that maintain it. It is these mechanisms, which have been discussed in this chapter, that are the concern of quantitative genetics.

THE GENES CONCERNED WITH QUANTITATIVE VARIATION

The genetic variation of metric characters appears from the results of experimental selection to be the product of segregation at some hundreds of loci, or more probably some thousands if the

variation of all characters is included. So natural populations probably carry a variety of alleles at a considerable proportion of loci, even
perhaps at virtually every locus. It seems unreasonable, therefore, to
think of genes having the control of a metric character as their
specific function: we cannot reasonably suppose that there are genes
whose only functions are the adjustment of, say, body size to an
optimal value. How, then, are we to think of the genes with which we
are concerned in quantitative genetics? Our knowledge of these
genes may be briefly summarised as follows.

The distinction between "major" and "minor" genes marks the
difference between those which we can study individually, and whose
properties are therefore fairly easily discovered, and those which we
cannot study individually and whose properties can only be deduced
by indirect means. Both are concerned with quantitative variation.
Among the major genes two sorts may be distinguished. There are
genes with more or less severely deleterious effects on fitness, and
these include nearly all the "mutants" of Mendelian genetics, as well
as lethals. Each may have pleiotropic effects on a variety of metric
characters. They are recessive, or nearly so, in their effects on fitness,
but not necessarily also in their effects on metric characters. They are
kept in equilibrium at low frequencies by natural selection balanced
against mutation. Being at low frequencies they contribute, individually, little to the genetic variance of any character; their total contribution, however, is unknown. They are probably an important cause
of inbreeding depression. Major genes of the second sort are those
responsible for the antigenic differences. The alleles at these loci are
at intermediate frequencies where they are probably maintained by
selection favouring heterozygotes. Their effects on fitness, however,
are probably fairly small—certainly small enough for all to be
regarded as "wild-type" alleles. Their effects on metric characters
are almost unexplored, and their importance as sources of variation is
consequently unknown. They presumably contribute to inbreeding
depression if heterozygotes are superior in fitness, but again their
relative importance in this respect is not known with certainty.
About the minor genes little is known. They do not necessarily
occupy loci different from those occupied by major genes. It seems
more likely, on the contrary, that they are isoalleles, capable of
mutating to major deleterious genes. They are performing their
primary functions perfectly adequately and may differ only in the rate
at which their primary product is synthesised. The variation of

metric characters which they produce may be quite incidental to their main biochemical functions. There is no reason at present to think that these minor genes differ in any essential way from the genes that determine antigenic differences. The fact that their effects are not individually recognisable, whereas the antigenic differences are, may be due only to the inadequacy of the techniques available for detecting biochemical differences among essentially normal individuals.

The problems that have been raised but left unanswered in this chapter will be sufficient indication of the directions which the future development of quantitative genetics may take. It does not seem to the present writer that much progress toward their solution is likely to be made by deductive reasoning, because most of the outstanding problems are not essentially theoretical in nature: the theoretical structure of the subject is now fairly clear, at least in its main outlines. Some of the outstanding problems are beyond the reach of the experimental techniques now at our command. New techniques, both more penetrating and more discriminating, will therefore be needed. Other problems arise from the paucity of experimental data and the consequent difficulty of deciding what phenomena are general and what are due to special circumstances. These problems will be solved not so much by deliberately designed experiments, but rather from the accumulated experience of experiments extended to a wider variety of characters and of organisms.

ined
owev

This list gives the meanings of most of the symbols used in the book. Many of the symbols listed are used also with other meanings in certain places, but these meanings, as well as the symbols not listed, do not appear more than a page or two removed from their definition. The more important differences from current usage are indicated where the equivalent symbols used by Lerner (1950)—denoted by (L)—and by Mather (1949) —denoted by (M)—are given.

A_1, A_2 Allelomorphic genes.

A Breeding value. $= G$ (L).

a Genotypic value of the homozygote A_1A_1, as deviation from the mid-homozygote value. $= d$ (M).

α Average effect of a gene-substitution.

α_1, α_2 Average effects of the alleles A_1 and A_2 respectively.

b Regression coefficient; e.g. b_{OP} = regression of offspring on parent.

CR Correlated response to selection.

D Dominance deviation.

d Genotypic value of the heterozygote A_1A_2, as deviation from the mid-homozygote value. $= h$ (M).

$\varDelta$ Change of $-$, as $\varDelta q$ = change of gene frequency, $\varDelta F$ = rate of inbreeding.

E Environmental deviation.

Ec Common environment; i.e. environmental deviation of family mean from population mean. $= C$ (L).

Ew Within-family environment; i.e. environmental deviation of individual from family mean. $= E'$ (L).

F Coefficient of inbreeding.

F_1 First generation of cross between lines or populations.

F_2 Second generation of cross, by random mating among F_1.

FS Full sibs.

f Coancestry; i.e. inbreeding coefficient of the progeny of the individuals concerned.

f (Chap. 13): Subscript referring to selection between families.

G Genotypic value. $= Ge$ (L).

Frequency of heterozygous genotype (A_1A_2).

Amount of heterosis; i.e. deviation of cross mean from mid-parent value.

Half sibs.

Heritability.

Interaction deviation, due to epistasis.

(Chap. 13 & 19): Index for selection.

Intensity of selection; i.e. selection differential in units of the phenotypic standard deviation. $=\bar{\imath}$ (L).

Population mean.

Immigration rate.

Population size; i.e. number of breeding individuals in a population or line.

(Chap. 10 & 13): Number of families.

Effective population size.

Number in various contexts. In Chapters 10 and 13, specifically number of offspring per family.

Offspring

Parent. $\bar{P} = $ Mid-parent.

Frequency of homozygous genotype (A_1A_1).

Panmictic index, ($=1 - F$).

Phenotypic value.

Gene frequency (of A_1). $=u$ (M).

(Chap. 11, part): proportion selected as parents from a normally distributed population. $=v$ (L).

Frequency of homozygous genotype (A_2A_2).

Gene frequency (of A_2). $=v$ (M).

Response to selection—specifically to individual selection. $=\Delta G$ (L).

(Chap. 8): Repeatability; i.e. correlation between repeated measurements of the same individual.

(Chap. 13): Coefficient of relationship; i.e. correlation of breeding values between related individuals. $=r^G$ (L).

(Chap. 19): Correlations between two characters:

r_A additive genetic correlation. $=r_G$ (L).

r_E environmental correlation.

r_P phenotypic correlation. $=r$ (L).

Selection differential in actual units of measurement. $=i$ (L).

Coefficient of selection against a particular genotype.

(Chap. 13): subscript referring to sib-selection.

Σ Summation of the quantity following the sign.

σ Standard deviation (σ^2 = variance) of the quantity indicated by subscript. Components of variance, from an analysis of variance are indicated by subscripts as follows:

σ_B^2 between groups, or families.

σ_D^2 between dams, within sires.

σ_S^2 between sires.

σ_T^2 total; i.e. the sum of all components.

σ_W^2 within groups, or families.

t Time in number of generations. As a subscript it means "at generation t".

t Phenotypic correlation between members of families.

u Mutation rate (from A_1 to A_2).

V Variance (causal component) of the value or deviation indicated by subscript. The most important are:

V_P Phenotypic variance. $= \sigma_P^2$ (L), $= V$ (M).

V_G Genotypic variance. $= \sigma_{G_e}^2$ (L).

V_A Additive genetic variance. $= \sigma_G^2$ (L), $= \frac{1}{2}D$ (M).

V_D Dominance variance.$\Big\}$ $= \sigma_{G'}^2$ (L)$\begin{cases} = \frac{1}{4}H \text{ (M)}. \\ = I \text{ (M)}. \end{cases}$

V_I Interaction variance.

V_E Environmental variance. $= \sigma_E^2$ (L), $= E$ (M).

v Mutation rate (from A_2 to A_1).

w (Chap. 13): subscript referring to selection within families.

X   (Chap. 19): One of two correlated characters.

Y   (Chap. 19): The other of two correlated characters.

y (Chap. 14): Difference of gene frequency between two lines.

z (Chap. 11): Height of the ordinate of a normal distribution, in units of the standard deviation.

INDEXED LIST OF REFERENCES

*The numbers in square brackets refer to pages in
the text where the work is mentioned*

ALLISON, A. C. 1954. Notes on sickle-cell polymorphism. *Ann. hum. Genet.*
[*Lond.*], **19**: 39–57.   [45]
     1955. Aspects of polymorphism in man. *Cold Spr. Harb. Symp. quant.
Biol.*, **20**: 239–252.   [44]
BARTLETT, M. S., and HALDANE, J. B. S. 1935. The theory of inbreeding
with forced heterozygosis. *J. Genet.*, **31**: 327–340.   [97]
BELL, A. E., MOORE, C. H., and WARREN, D. C. 1955. The evaluation of
new methods for the improvement of quantitative characteristics.
Cold Spr. Harb. Symp. quant. Biol., **20**: 197–211.   [286]
BIGGERS, J. D., and CLARINGBOLD, P. J. 1954. Why use inbred lines?
Nature [*Lond.*], **174**: 596.   [275]
BRILES, W. E., ALLEN, C. P., and MILLEN, T. W. 1957. The *B* blood group
system of chickens. I. Heterozygosity in closed populations.
Genetics, **42**: 631–648.   [290]
BRIQUET, R., and LUSH, J. L. 1947. Heritability of amount of spotting in
Holstein-Friesian cattle. *J. Hered.*, **38**: 99–105.   [167]
BRUMBY, P. J. 1958. Monozygotic twins and dairy cattle improvement.
Anim. Breed. Abstr., **26**: 1–12.   [183]
BRUMBY, P. J., and HANCOCK, J. 1956. A preliminary report of growth and
milk production in identical- and fraternal-twin dairy cattle. *N.Z.
J. Sci. Tech., Agric.*, **38**: 184–193.   [185]
BURI, P. 1956. Gene frequency in small populations of mutant *Drosophila.
Evolution*, **10**: 367–402.   [52, 53, 56, 59, 74]
BUTLER, L. 1952. A study of size inheritance in the house mouse. II. Analysis
of five preliminary crosses. *Canad. J. Zool.*, **30**: 154–171.   [216]
CAIN, A. J., and SHEPPARD, P. M. 1954a. Natural selection in Cepaea.
Genetics, **39**: 89–116.   [43, 83]
     1954b. The theory of adaptive polymorphism. *Amer. Nat.*, **88**: 321–
326.   [44]
CARPENTER, J. R., GRÜNEBERG, H., and RUSSELL, E. S. 1957. Genetical
differentiation involving morphological characters in an inbred
strain of mice. II. The American branches of the C57BL and
C57BR strains. *J. Morph.*, **100**: 377–388.   [274]
CASTLE, W. E., and WRIGHT, S. 1916. *Studies of inheritance in guinea-pigs
and rats. Publ. Carneg. Instn. Wash.*, No. 241: iv + 192 pp.   [168]
CEPPELLINI, R., SINISCALCO, M., and SMITH, C. A. B. 1955. The estimation
of gene frequencies in a random-mating population. *Ann. hum.
Genet.* [*Lond.*], **20**: 97–115.   [16]

CHAI, C. K. 1957. Developmental homeostasis of body growth in mice. *Amer. Nat.*, **91**: 49–55. [271]

CHAPMAN, A. B. 1946. Genetic and nongenetic sources of variation in the weight response of the immature rat ovary to a gonadotrophic hormone. *Genetics*, **31**: 494–507. [168]

CLAYTON, G. A., KNIGHT, G. R., MORRIS, J. A., and ROBERTSON, A. 1957. An experimental check on quantitative genetical theory. III. Correlated responses. *J. Genet.*, **55**: 171–180. [316, 320]

CLAYTON, G. A., MORRIS, J. A., and ROBERTSON, A. 1957. An experimental check on quantitative genetical theory. I. Short-term responses to selection. *J. Genet.*, **55**: 131–151.
[140, 168, 169, 177, 190, 195, 209, 210, 221, 245, 333]

CLAYTON, G. A., and ROBERTSON, A. 1955. Mutation and quantitative variation. *Amer. Nat.*, **89**: 151–158. [342]

1957. An experimental check on quantitative genetical theory. II. The long-term effects of selection. *J. Genet.*, **55**: 152–170.
[216, 223]

COCKERHAM, C. C. 1954. An extension of the concept of partitioning hereditary variance for analysis of covariances among relatives when epistasis is present. *Genetics*, **39**: 859–882. [138]

1956*a*. Effects of linkage on the covariances between relatives. *Genetics*, **41**: 138–141. [159]

1956*b*. Analysis of quantitative gene action. *Genetics in Plant Breeding. Brookhaven Symp. Biol.*, No. 9: 53–68. [140]

COMSTOCK, R. E., and ROBINSON, H. F. 1952. Estimation of average dominance of genes. *Heterosis*, ed. J. W. Gowen. Ames: Iowa State College Press. Pp. 494–516. [290]

COMSTOCK, R. E., ROBINSON, H. F., and HARVEY, P. H. 1949. A breeding procedure designed to make maximum use of both general and specific combining ability. *J. Amer. Soc. Agron.*, **41**: 360–367.
[286]

CROW, J. F. 1948. Alternative hypotheses of hybrid vigor. *Genetics*, **33**: 477–487. [278, 290]

1952. Dominance and overdominance. *Heterosis*, ed. J. W. Gowen. Ames: Iowa State College Press. Pp. 282–297. [278, 290]

1954. Breeding structure of populations. II. Effective population number. *Statistics and Mathematics in Biology*, ed. O. Kempthorne, T. A. Bancroft, J. W. Gowen, and J. L. Lush. Ames: Iowa State College Press. Pp. 543–556. [53, 60, 61, 64, 71]

1956. The estimation of spontaneous and radiation-induced mutation rates in man. *Eugen. Quart.*, **3**: 201–208. [38]

1957. Possible consequences of an increased mutation rate. *Eugen. Quart.*, **4**: 67–80. [39]

CROW, J. F., and MORTON, N. E. 1955. Measurement of gene frequency drift in small populations. *Evolution*, **9**: 202–214. [73, 74]

CRUDEN, D. 1949. The computation of inbreeding coefficients in closed populations. *J. Hered.*, **40**: 248–251. [88, 89]

DEMPSTER, E. R., and LERNER, I. M. 1950. Heritability of threshold characters. *Genetics*, **35**: 212–236. [303]

DEOL, M. S., GRÜNEBERG, H., SEARLE, A. G., and TRUSLOVE, G. M. 1957. Genetical differentiation involving morphological characters in an inbred strain of mice. I. A British branch of the C57BL strain. *J. Morph.*, **100**: 345–376. [274]

DICKERSON, G. E. 1952. Inbred lines for heterosis tests? *Heterosis*, ed. J. W. Gowen. Ames: Iowa State College Press. Pp. 330–351. [286]

—— 1955. Genetic slippage in response to selection for multiple objectives. *Cold Spr. Harb. Symp. quant. Biol.*, **20**: 213–224. [329]

—— 1957. (Two abstracts.) *Poult. Sci.*, **36**: 1112–1113. [316]

DICKERSON, G. E., et al. 1954. *Evaluation of selection in developing inbred lines of swine. Res. Bull. Mo. agric. Exp. Sta.*, No. 551: 60 pp. [249, 253]

DOBZHANSKY, TH. 1950. Genetics of natural populations. XIX. Origin of heterosis through natural selection in populations of *Drosophila pseudoobscura*. *Genetics*, **35**: 288–302. [262]

—— 1951a. *Genetics and the Origin of Species*. New York: Columbia University Press. 3rd edn. xi + 364 pp. [44]

—— 1951b. Mendelian populations and their evolution. *Genetics in the 20th Century*, ed. L. C. Dunn. New York: Macmillan Co. Pp. 573–589. [44, 263]

—— 1952. Nature and origin of heterosis. *Heterosis*, ed. J. W. Gowen. Ames: Iowa State College Press. Pp. 218–223. [262]

DOBZHANSKY, TH., and PAVLOVSKY, O. 1955. An extreme case of heterosis in a Central American population of *Drosophila tropicalis*. *Proc. nat. Acad. Sci. U.S.A.*, **41**: 289–295. [39]

DONALD, H. P., DEAS, D. W., and WILSON, A. L. 1952. Genetical analysis of the incidence of dropsical calves in herds of Ayrshire cattle. *Brit. vet. J.*, **108**: 227–245. [13]

EMIK, L. O., and TERRILL, C. E. 1949. Systematic procedures for calculating inbreeding coefficients. *J. Hered.*, **40**: 51–55. [88, 89]

FALCONER, D. S. 1952. The problem of environment and selection. *Amer. Nat.*, **86**: 293–298. [323, 324]

—— 1953. Selection for large and small size in mice. *J. Genet.*, **51**: 470–501. [96, 168, 199, 335]

—— 1954a. Asymmetrical responses in selection experiments. *Symposium on Genetics of Population Structure, Istituto di Genetica, Università di Pavia, Italy, August 20–23, 1953*. Un. int. Sci. biol., No. 15: 16–41. [31, 33, 203, 213, 297]

—— 1954b. Validity of the theory of genetic correlation. An experimental test with mice. *J. Hered.*, **45**: 42–44. [168, 316, 319]

—— 1954c. Selection for sex ratio in mice and *Drosophila*. *Amer. Nat.*, **88**: 385–397. [321]

—— 1955. Patterns of response in selection experiments with mice. *Cold Spr. Harb. Symp. quant. Biol.*, **20**: 178–196. [168, 201, 214, 216, 220]

1957*a*. Breeding methods—I. Genetic considerations. *The UFAW Handbook on the Care and Management of Laboratory Animals*, 2nd edn., edd. A. N. Worden and W. Lane-Petter. London: Universities Federation for Animal Welfare. Pp. 85–107.   [228]

1957*b*. Selection for phenotypic intermediates in *Drosophila*. *J. Genet.*, **55**: 551–561.   [341]

FALCONER, D. S., and LATYSZEWSKI, M. 1952. The environment in relation to selection for size in mice. *J. Genet.*, **51**: 67–80.   [324]

FALCONER, D. S., and ROBERTSON, A. 1956. Selection for environmental variability of body size in mice. *Z. indukt. Abstamm.-u. Vererblehre*, **87**: 385–391.   [341]

FISHER, R. A. 1918. The correlation between relatives on the supposition of Mendelian inheritance. *Trans. roy. Soc. Edinb.*, **52**: 399–433.   [2, 124]

1930. *The Genetical Theory of Natural Selection*. Oxford University Press. xiv + 272 pp.   [4]

1941. Average excess and average effect of a gene substitution. *Ann. Eugen. [Lond.]*, **11**: 53–63.   [124]

1949. *The Theory of Inbreeding*. Edinburgh: Oliver & Boyd. viii + 120 pp.   [90, 97, 99, 100]

FISHER, R. A., and YATES, F. 1943. *Statistical Tables*. Edinburgh: Oliver & Boyd. 2nd edn. viii + 98 pp.   [194, 302]

FORD, E. B. 1953. The genetics of polymorphism in the Lepidoptera. *Advanc. Genet.*, **5**: 43–87.   [44]

FREDEEN, H. T., and JONSSON, P. 1957. Genic variance and covariance in Danish Landrace swine as evaluated under a system of individual feeding of progeny test groups. *Z. Tierz. Züchtbiol.*, **70**: 348–363.   [167, 174, 175, 316]

GILMOUR, D. G. 1958. Maintenance of segregation of blood group genes during inbreeding in chickens. (Abstr.) *Heredity*, **12**: 141–142.   [290]

GOWE, R. S., ROBERTSON, A., and LATTER, B. D. H. 1959. Environment and poultry breeding problems. 5. The design of poultry control strains. *Poult. Sci.*, **38**: 462–471.   [72, 73]

GOWEN, J. W. 1952. Hybrid vigor in *Drosophila*. *Heterosis*, ed. J. W. Gowen. Ames: Iowa State College Press. Pp. 474–493.   [282, 336]

GREEN, E. L. 1951. The genetics of a difference in skeletal type between two inbred strains of mice (BalbC and C57blk). *Genetics*, **36**: 391–409.   [303]

GREEN, E. L., and RUSSELL, W. L. 1951. A difference in skeletal type between reciprocal hybrids of two inbred strains of mice (C57BLK and C3H). *Genetics*, **36**: 641–651.   [305, 307]

GRÜNEBERG, H. 1952. Genetical studies on the skeleton of the mouse. IV. Quasi-continuous variations. *J. Genet.*, **51**: 95–114.   [301]

1954. Variation within inbred strains of mice. *Nature [Lond.]*, **173**: 674.   [275]

HALDANE, J. B. S. 1924–32. A mathematical theory of natural and artificial selection. *Proc. Camb. phil. Soc.*, **23**: 19–41; 158–163; 363–372; 607–615; 838–844; **26**: 220–230; **27**: 131–142; **28**: 244–248. [2]

—— 1932. *The Causes of Evolution.* London: Longmans, Green & Co., Ltd. vii + 235 pp. [2, 4]

HALDANE, J. B. S. 1936. The amount of heterozygosis to be expected in an approximately pure line. *J. Genet.*, **32**: 375–391. [100]

—— 1937. Some theoretical results of continued brother-sister mating. *J. Genet.*, **34**: 265–274. [90, 97]

—— 1939. The spread of harmful autosomal recessive genes in human populations. *Ann. Eugen. [Lond.]*, **9**: 232–237. [41]

—— 1946. The interaction of nature and nurture. *Ann. Eugen. [Lond.]*, **13**: 197–205. [133]

—— 1949. The rate of mutation of human genes. *Proc. 8th int. Congr. Genet.* 1948 [*Stockh.*]. Lund: Issued as a supplementary volume of *Hereditas*, 1949. Pp. 267–273. [38]

—— 1954a. The measurement of natural selection. *Proc. 9th int. Congr. Genet.* [*Bellagio (Como)*], 1953, Pt. I (Suppl. to *Caryologia*, 6): 480–487. [338]

—— 1954b. *The Biochemistry of Genetics.* London: George Allen & Unwin Ltd. 144 pp. [339]

—— 1955. The complete matrices for brother-sister and alternate parent-offspring mating involving one locus. *J. Genet.*, **53**: 315–324. [90, 97]

HANCOCK, J. 1954. Monozygotic twins in cattle. *Advanc. Genet.*, **6**: 141–181. [183]

HARDY, G. H. 1908. Mendelian proportions in a mixed population. *Science*, **28**: 49–50. [9]

HAYES, H. K., IMMER, F. R., and SMITH, D. C. 1955. *Methods of Plant Breeding.* New York: McGraw-Hill Book Co., Inc. 2nd edn. xi + 551 pp. [276]

HAYMAN, B. I. 1955. The description and analysis of gene action and interaction. *Cold Spr. Harb. Symp. quant. Biol.*, **20**: 79–84. [140]

—— 1958. The theory and analysis of diallel crosses. II. *Genetics*, **43**: 63–85. [140, 277]

HAYMAN, B. I., and MATHER, K. 1953. The progress of inbreeding when homozygotes are at a disadvantage. *Heredity*, **7**: 165–183. [102]

HAZEL, L. N. 1943. The genetic basis for constructing selection indexes. *Genetics*, **28**: 476–490. [324, 325]

HAZEL, L. N., and LUSH, J. L. 1942. The efficiency of three methods of selection. *J. Hered.*, **33**: 393–399. [324]

HOLLINGSWORTH, M. J., and SMITH, J. M. 1955. The effects of inbreeding on rate of development and on fertility in *Drosophila subobscura*. *J. Genet.*, **53**: 295–314. [249, 252]

HORNER, T. W., COMSTOCK, R. E., and ROBINSON, H. F. 1955. *Non-allelic gene interactions and the interpretation of quantitative genetic data.* *Tech. Bull. N.C. agric. Exp. Sta.*, No. 118: v + 117 pp. [299]

HULL, F. H. 1945. Recurrent selection for specific combining ability in corn. *J. Amer. Soc. Agron.*, **37**: 134–145. [286]

HUNT, H. R., HOPPERT, C. A., and ERWIN, W. G. 1944. Inheritance of susceptibility to caries in albino rats (*Mus norvegicus*). *J. dent. Res.*, **23**: 385–401. [297]

JOHANSSON, I. 1950. The heritability of milk and butterfat yield. *Anim. Breed. Abstr.*, **18**: 1–12. [144, 167, 316]

KEMPTHORNE, O. 1954. The correlation between relatives in a random mating population. *Proc. roy. Soc.*, B, **143**: 103–113. [138, 279]

—— 1955a. The theoretical values of correlations between relatives in random mating populations. *Genetics*, **40**: 153–167.

[138, 152, 158, 174]

—— 1955b. The correlations between relatives in random mating populations. *Cold Spr. Harb. Symp. quant. Biol.*, **20**: 60–75. [138, 158]

—— 1957. *An Introduction to Genetic Statistics*. New York: John Wiley & Sons, Inc.; London: Chapman & Hall, Ltd. xvii + 545 pp. [4, 264, 277]

KEMPTHORNE, O., and TANDON, O. B. 1953. The estimation of heritability by regression of offspring on parent. *Biometrics*, **9**: 90–100. [171]

KERR, W. E., and WRIGHT, S. 1954a. Experimental studies of the distribution of gene frequencies in very small populations of *Drosophila melanogaster*: I. Forked. *Evolution*, **8**: 172–177. [74]

—— 1954b. Experimental studies of the distribution of gene frequencies in very small populations of *Drosophila melanogaster*. III. Aristapedia and spineless. *Evolution*, **8**: 293–302. [74]

KIMURA, M. 1954. Process leading to quasi-fixation of genes in natural populations due to random fluctuation of selection intensities. *Genetics*, **39**: 280–295. [57, 340]

—— 1955. Solution of a process of random genetic drift with a continuous model. *Proc. nat. Acad. Sci. U.S.A.*, **41**: 144–150. [54, 55, 57]

—— 1956. Rules for testing stability of a selective polymorphism. *Proc. nat. Acad. Sci. U.S.A.*, **42**: 336–340. [42]

KING, J. W. B. 1950. Pygmy, a dwarfing gene in the house mouse. *J. Hered.*, **41**: 249–252. [113]

—— 1955. Observations on the mutant "pygmy" in the house mouse. *J. Genet.*, **53**: 487–497. [113, 289, 298, 299]

KING, S. C., and HENDERSON, C. R. 1954a. Variance components analysis in heritability studies. *Poult. Sci.*, **33**: 147–154. [173]

—— 1954b. Heritability studies of egg production in the domestic fowl. *Poult. Sci.*, **33**: 155–169. [168]

KYLE, W. H., and CHAPMAN, A. B. 1953. Experimental check of the effectiveness of selection for a quantitative character. *Genetics*, **38**: 421–443. [225]

LACK, D. 1954. *The Natural Regulation of Animal Numbers*. Oxford: Clarendon Press. viii + 343 pp. [334]

LAMOTTE, M. 1951. Recherches sur la structure génétique des populations naturelles de *Cepaea nemoralis* (L.). *Bull. biol.*, Suppl. **35**: 238 pp.

[78, 83, 84]

LERNER, I. M. 1950. *Population Genetics and Animal Improvement*. Cambridge University Press. xviii + 342 pp.   [4, 236, 325, 346]

— 1954. *Genetic Homeostasis*. Edinburgh: Oliver & Boyd. vii + 134 pp.
[44, 202, 213, 263, 270, 271, 288, 331, 339]

— 1958. *The Genetic Basis of Selection*. New York: John Wiley & Sons, Inc. xvi + 298 pp.   [202, 263]

LERNER, I. M., and CRUDEN, D. 1951. The heritability of egg weight: the advantages of mass selection and of early measurements. *Poult. Sci.*, **30**: 34–41.   [168]

LEVENE, H. 1953. Genetic equilibrium when more than one ecological niche is available. *Amer. Nat.*, **87**: 331–333.   [43]

LI, C. C. 1955a. *Population Genetics*. Chicago: University of Chicago Press; London: Cambridge University Press. xi + 366 pp.
[4, 15, 20, 22, 24]

— 1955b. The stability of an equilibrium and the average fitness of a population. *Amer. Nat.*, **89**: 281–296.   [43]

LIVESAY, E. A. 1930. An experimental study of hybrid vigor or heterosis in rats. *Genetics*, **15**: 17–54.   [271]

LUSH, J. L. 1945. *Animal Breeding Plans*. Ames: Iowa State College Press. 3rd edn. viii + 443 pp.   [4]

— 1947. Family merit and individual merit as bases for selection. PT. I, PT. II. *Amer. Nat.*, **81**: 241–261; 362–379.   [236, 237]

— 1950. Genetics and animal breeding. *Genetics in the Twentieth Century*, ed. L. C. Dunn. New York: Macmillan Co. Pp. 493–525.   [200]

LUSH, J. L., and MOLLN, A. E. 1942. Litter size and weight as permanent characteristics of sows. *Tech. Bull. U.S. Dep. Agric.*, No. 836: 40 pp.
[167]

MACARTHUR, J. W. 1949. Selection for small and large body size in the house mouse. *Genetics*, **34**: 194–209.   [216, 295, 335]

McLAREN, A., and MICHIE, D. 1954. Factors affecting vertebral variation in mice. 1. Variation within an inbred strain. *J. Embryol. exp. Morph.*, **2**: 149–160.   [273, 274]

— 1955. Factors affecting vertebral variation in mice. 2. Further evidence on intra-strain variation. *J. Embryol. exp. Morph.*, **3**: 366–375.
[303, 306]

— 1956a. Factors affecting vertebral variation in mice. 3. Maternal effects in reciprocal crosses. *J. Embryol. exp. Morph.*, **4**: 161–166.
[306]

— 1956b. Variability of response in experimental animals. *J. Genet.*, **54**: 440–455.   [271]

MALÉCOT, G. 1948. *Les Mathématiques de l'Hérédité*. Paris: Masson et Cie. vi + 63 pp.   [4, 61, 69, 75, 88]

MANGELSDORF, P. C. 1951. Hybrid corn: its genetic basis and its significance in human affairs. *Genetics in the Twentieth Century*, ed. L. C. Dunn. New York: Macmillan Co. Pp. 555–571.   [110, 277]

MATHER, K. 1949. *Biometrical Genetics*. London: Methuen & Co., Ltd. ix + 162 pp.   [4, 106, 277, 340, 346]

1953a. Genetical control of stability in development. *Heredity*, 7: 297–
336.   [270]

1953b. The genetical structure of populations. *Symp. Soc. exp. Biol.* 7:
66–95.   [340]

1955a. Polymorphism as an outcome of disruptive selection. *Evolution*,
9: 52–61.   [43]

1955b. The genetical basis of heterosis. *Proc. roy. Soc., B.*, **144**: 143–
150.   [288]

MERRELL, D. J. 1953. Selective mating as a cause of gene frequency changes
in laboratory populations of *Drosophila melanogaster*. *Evolution*, 7:
287–296.   [34]

MORLEY, F. H. W. 1951. Selection for economic characters in Australian
Merino sheep. (1) Estimates of phenotypic and genetic parameters.
Sci. Bull. Dep. Agric. N.S.W., No. 73: 45 pp.   [144]

1954. Selection for economic characters in Australian Merino sheep.
IV. The effect of inbreeding. *Aust. J. agric. Res.*, **5**: 305–316.
                                                                        [249]

1955. Selection for economic characters in Australian Merino sheep.
V. Further estimates of phenotypic and genetic parameters. *Aust.
J. agric. Res.*, **6**: 77–90.   [168, 316]

MOURANT, A. E. 1954. *The Distribution of the Human Blood Groups*. Oxford:
Blackwell. xxi + 428 pp.   [5]

MULLER, H. J., and OSTER, I. I. 1957. Principles of back mutation as observed in *Drosophila* and other organisms. *Advances in Radio-
biology*. *Proc. 5th int. Conf. Radiobiol.* [*Stockh.*], 1956. Edinburgh:
Oliver & Boyd. Pp. 407–415.   [26]

NEWMAN, H. H., FREEMAN, F. N., and HOLZINGER, K. J. 1937. *Twins: a
Study of Heredity and Environment*. Chicago: University of Chicago
Press. xvi + 369 pp.   [185]

OSBORNE, R. 1957a. The use of sire and dam family averages in increasing
the efficiency of selective breeding under a hierarchical mating
system. *Heredity*, **11**: 93–116.   [241]

1957b. Correction for regression on a secondary trait as a method of
increasing the efficiency of selective breeding. *Aust. J. biol. Sci.*,
10: 365–366.   [328]

OSBORNE, R., and PATERSON, W. S. B. 1952. On the sampling variance of
heritability estimates derived from variance analyses. *Proc. roy.
Soc. Edinb., B.*, **64**: 456–461.   [183]

PAXMAN, G. J. 1957. A study of spontaneous mutation in *Drosophila
melanogaster*. *Genetica*, **29**: 39–57.   [342]

PEARSON, K., and LEE, A. 1903. On the laws of inheritance in man. I. Inheritance of physical characters. *Biometrika*, **2**: 357–462.   [163]

PENROSE, L. S. 1949. *The Biology of Mental Defect*. London: Sidgwick &
Jackson. xiv + 285 pp.   [163, 164]

1954. Some recent trends in human genetics. *Proc. 9th int. Congr.
Genet.* [*Bellagio (Como)*], 1953, PT. I (Suppl. to *Caryologia*, **6**): 521–
530.   [141]

PLUM, M. 1954. Computation of inbreeding and relationship coefficients. *J. Hered.*, **45**: 92–94. [89]

POWERS, L. 1950. Determining scales and the use of transformations in studies on weight per locule of tomato fruit. *Biometrics*, **6**: 145–163. [300]

—— 1952. Gene recombination and heterosis. *Heterosis*, ed. J. W. Gowen. Ames: Iowa State College Press. Pp. 298–319. [260]

RACE, R. R., and SANGER, R. 1954. *Blood Groups in Man.* Oxford: Blackwell. 2nd edn. xvi + 400 pp. [12, 16]

RASMUSON, M. 1952. Variation in bristle number of *Drosophila melanogaster*. *Acta zool.* [*Stockh.*], **33**: 277–307. [265]

—— 1956. Recurrent reciprocal selection. Results of three model experiments on *Drosophila* for improvement of quantitative characters. *Hereditas* [*Lund*], **42**: 397–414. [286]

REEVE, E. C. R. 1955*a*. Inbreeding with homozygotes at a disadvantage. *Ann. hum. Genet.* [*Lond.*], **19**: 332–346. [101]

—— 1955*b*. (Contribution to discussion.) *Cold Spr. Harb. Symp. quant. Biol.*, **20**: 76–78. [170]

—— 1955*c*. The variance of the genetic correlation coefficient. *Biometrics*, **11**: 357–374. [171, 318]

REEVE, E. C. R., and ROBERTSON, F. W. 1953. Studies in quantitative inheritance. II. Analysis of a strain of *Drosophila melanogaster* selected for long wings. *J. Genet.*, **51**: 276–316. [171, 316, 319]

—— 1954. Studies in quantitative inheritance. VI. Sternite chaeta number in *Drosophila*: a metameric quantitative character. *Z. indukt. Abstamm.-u. Vererblehre*, **86**: 269–288. [140, 145, 316]

RENDEL, J. M. 1954. The use of regressions to increase heritability. *Aust. J. biol. Sci.*, **7**: 368–378. [328]

RENDEL, J. M., ROBERTSON, A., ASKER, A. A., KHISHIN, S. S., and RAGAB, M. T. 1957. The inheritance of milk production characteristics. *J. agric. Sci.*, **48**: 426–432. [149]

ROBERTS, J. A. FRASER. 1957. Blood groups and susceptibility to disease: a review. *Brit. J. prev. Soc. Med.*, **11**: 107–125. [44]

ROBERTSON, A. 1952. The effect of inbreeding on the variation due to recessive genes. *Genetics*, **37**: 189–207. [268, 269]

—— 1954. Inbreeding and performance in British Friesian cattle. *Proc. Brit. Soc. Anim. Prod.*, **1954**: 87–92. [249]

—— 1955*a*. Prediction equations in quantitative genetics. *Biometrics*, **11**: 95–98. [236]

—— 1955*b*. Selection in animals: synthesis. *Cold Spr. Harb. Symp. quant. Biol.*, **20**: 225–229. [333, 337]

—— 1956. The effect of selection against extreme deviants based on deviation or on homozygosis. *J. Genet.*, **54**: 236–248. [338]

—— 1957*a*. Genetics and the improvement of dairy cattle. *Agric. Rev.* [*Lond.*], **2** (8): 10–21. [167]

—— 1957*b*. Optimum group size in progeny testing and family selection. *Biometrics*, **13**: 442–450. [243]

1959*a*. Experimental design in the evaluation of genetic parameters. *Biometrics*, **15**: 219–226.   [178, 182, 183]

1959*b*. The sampling variance of the genetic correlation coefficient. *Biometrics*, **15**: 469–485.   [318, 323]

ROBERTSON, A., and LERNER, I. M. 1949. The heritability of all-or-none traits: viability of poultry. *Genetics*, 34: 395–411.   [168, 303]

ROBERTSON, F. W. 1955. Selection response and the properties of genetic variation. *Cold Spr. Harb. Symp. quant. Biol.*, **20**: 166–177.
                                                                        [211, 212, 216]

1957*a*. Studies in quantitative inheritance. X. Genetic variation of ovary size in *Drosophila*. *J. Genet.*, **55**: 410–427.   [140, 144, 145, 168]

1957*b*. Studies in quantitative inheritance. XI. Genetic and environmental correlation between body size and egg production in *Drosophila melanogaster*. *J. Genet.*, **55**: 428–443.   [131, 140, 168]

ROBERTSON, F. W., and REEVE, E. C. R. 1952*a*. Studies in quantitative inheritance. I. The effects of selection of wing and thorax length in *Drosophila melanogaster*. *J. Genet.*, **50**: 414–448.   [223]

1952*b*. Heterozygosity, environmental variation and heterosis. *Nature* [*Lond.*], **170**: 296.   [270, 271]

ROBINSON, H. F., and COMSTOCK, R. E. 1955. Analysis of genetic variability in corn with reference to probable effects of selection. *Cold Spr. Harb. Symp. quant. Biol.*, **20**: 127–135.   [140, 284, 290]

ROBINSON, H. F., COMSTOCK, R. E., KHALIL, A., and HARVEY, P. H. 1956. Dominance versus over-dominance in heterosis: evidence from crosses between open-pollinated varieties of maize. *Amer. Nat.*, **90**: 127–131.   [290]

ROBSON, E. B. 1955. Birth weight in cousins. *Ann. hum. Genet.* [*Lond.*], **19**: 262–268.   [141, 163, 185]

RUSSELL, E. S. 1949. A quantitative histological study of the pigment found in the coat-color mutants of the house mouse. IV. The nature of the effects of genic substitution in five major allelic series. *Genetics*, **34**: 146–166.   [116, 126]

SCHÄFER, W. 1937. Über die Zunahme der Isozygotie (Gleicherbarkeit) bei fortgesetzter Bruder-Schwester-Inzucht. *Z. indukt. Abstamm.- u. Vererblehre*, **72**: 50–78.   [91, 97]

SEARLE, A. G. 1949. Gene frequencies in London's cats. *J. Genet.*, **49**: 214–220.   [18]

SHEPPARD, P. M. 1958. *Natural Selection and Heredity*. London: Hutchinson & Co. (Publishers) Ltd. 212 pp.   [44]

SHOFFNER, R. N. 1948. The reaction of the fowl to inbreeding. *Poult. Sci.*, **27**: 448–452.   [249]

SIERTS-ROTH, U. 1953. Geburts- und Aufzuchtgewichte von Rassehunden. *Z. Hundeforsch.*, **20**: 1–122.   [216]

ŚLIŻYŃSKI, B. M. 1955. Chiasmata in the male mouse. *J. Genet.*, **53**: 597–605.   [99]

SMITH, H. FAIRFIELD. 1936. A discriminant function for plant selection. *Ann. Eugen.* [*Lond.*], **7**: 240–250.   [325]

WILLI

WRIGH

SMITH, H. H. 1952. Fixing transgressive vigor in *Nicotiana rustica*. *Heterosis*, ed. J. W. Gowen. Ames: Iowa State College Press. Pp. 161–174. [260]

SNEDECOR, G. W. 1956. *Statistical Methods*. Ames: Iowa State College Press. 5th edn. xiii + 534 pp. [144, 173]

SPRAGUE, G. F. 1952. Early testing and recurrent selection. *Heterosis*, ed. J. W. Gowen. Ames: Iowa State College Press. Pp. 400–417. [283]

STERN, C. 1943. The Hardy-Weinberg law. *Science*, 97: 137–138. [9]
1949. *Principles of Human Genetics*. San Francisco: W. H. Freeman & Co. xi + 617 pp. [13, 183]

TANTAWY, A. O., and REEVE, E. C. R. 1956. Studies in quantitative inheritance. IX. The effects of inbreeding at different rates in *Drosophila melanogaster*. *Z. indukt. Abstamm.-u. Vererblehre*, 87: 648–667.
[249, 268, 290]

WADDINGTON, C. H. 1942. Canalisation of development and the inheritance of acquired characters. *Nature [Lond.]*, 150: 563. [310]
1953. Genetic assimilation of an acquired character. *Evolution*, 7: 118–126. [310, 311]
1957. *The Strategy of the Genes*. London: George Allen & Unwin, Ltd. ix + 262 pp. [43, 146, 271, 311, 340, 341, 342]

WALLACE, B. 1958. The comparison of observed and calculated zygotic distributions. *Evolution*, 12: 113–115. [12]

WALLACE, B., and VETUKHIV, M. 1955. Adaptive organization of the gene pools of *Drosophila* populations. *Cold Spr. Harb. Symp. quant. Biol.*, 20: 303–309. [262]

WARREN, E. P., and BOGART, R. 1952. *Effect of selection for age at time of puberty on reproductive performance in the rat*. Sta. Tech. Bull. Ore. agric. Exp. Sta., No. 25: 27 pp. [168]

WARWICK, B. L. 1932. *Probability tables for Mendelian ratios with small numbers*. Bull. Tex. agric. Exp. Sta., No. 463: 28 pp. [105]

WARWICK, E. J., and LEWIS, W. L. 1954. Increase in frequency of a deleterious recessive gene in mice. *J. Hered.*, 45: 143–145. [113]

WEINBERG, W. 1908. Über den Nachweis der Vererbung beim Menschen. *Jh. Ver. vaterl. Naturk. Württemb.*, 64: 369–382. [9]

WEIR, J. A. 1955. Male influence on sex ratio of offspring in high and low blood-pH lines of mice. *J. Hered.*, 46: 277–283. [321]

WEIR, J. A., and CLARK, R. D. 1955. Production of high and low blood-pH lines of mice by selection with inbreeding. *J. Hered.*, 46: 125–132. [321]

WHATLEY, J. A. 1942. Influence of heredity and other factors on 180-day weight in Poland China swine. *J. agric. Res.*, 65: 249–264. [167]

WILLIAMS, E. J. 1954. *The estimation of components of variability*. Tech. Pap. Div. math. Statist. C.S.I.R.O. Aust., No. 1: 22 pp. [173]

WRIGHT, S. 1921. Systems of mating. *Genetics*, 6: 111–178. [2, 4, 22, 90, 165]

1922. Coefficients of inbreeding and relationship. *Amer. Nat.*, **56**: 330–338.   [22, 87, 88]

1931. Evolution in Mendelian populations. *Genetics*, **16**: 97–159.
                                                              [4, 53, 54, 69, 75, 92]

1933. Inbreeding and homozygosis. *Proc. nat. Acad. Sci.* [*Wash.*], **19**: 411–420.   [90, 92]

1934. The method of path coefficients. *Ann. math. Statist.*, **5**: 161–215.
                                                                              [90]

1935*a*. The analysis of variance and the correlations between relatives with respect to deviations from an optimum. *J. Genet.*, **30**: 243–256.
                                                                             [338]

1935*b*. Evolution in populations in approximate equilibrium. *J. Genet.*, **30**: 257–266.   [338]

1939. Statistical genetics in relation to evolution. *Actualités scientifiques et industrielles*, **802**. Paris: Hermann et Cie. 63 pp.   [70]

1940. Breeding structure of populations in relation to speciation. *Amer. Nat.*, **74**: 232–248.   [71, 77]

1942. Statistical genetics and evolution. *Bull. Amer. math. Soc.*, **48**: 223–246.   [75, 79]

1943. Isolation by distance. *Genetics*, **28**: 114–138.   [77]

1946. Isolation by distance under different systems of mating. *Genetics*, **31**: 39–59.   [77]

1948. On the roles of directed and random changes in gene frequency in the genetics of populations. *Evolution*, **2**: 279–294.   [75]

1951. The genetical structure of populations. *Ann. Eugen.* [*Lond.*], **15**: 323–354.   [75, 76, 77]

1952*a*. The theoretical variance within and among subdivisions of a population that is in a steady state. *Genetics*, **37**: 312–321.   [57]

1952*b*. The genetics of quantitative variability. *Quantitative Inheritance*, edd. E. C. R. Reeve and C. H. Waddington. London: H.M.S.O. Pp. 5–41.   [219, 292, 297, 341]

1954. The interpretation of multivariate systems. *Statistics and Mathematics in Biology*, edd. O. Kempthorne, T. A. Bancroft, J. W. Gowen and J. L. Lush. Ames: Iowa State College Press. Pp. 11–33.
                                                                              [90]

1956. Modes of selection. *Amer. Nat.*, **90**: 5–24.   [263]

WRIGHT, S., and KERR, W. E. 1954. Experimental studies of the distribution of gene frequencies in very small populations of *Drosophila melanogaster*. II. Bar. *Evolution*, **8**: 225–240.   [74, 80, 81]

WRIGHT, S., and McPHEE, H. C. 1925. An approximate method of calculating coefficients of inbreeding and relationship from livestock pedigrees. *J. agric. Res.*, **31**: 377–383.   [87]

YOON, C. H. 1955. Homeostasis associated with heterozygosity in the genetics of time of vaginal opening in the house mouse. *Genetics*, **40**: 297–309.   [271]

ZELENY, C. 1922. The effect of selection for eye facet number in the white bar-eye race of *Drosophila melanogaster*. *Genetics*, **7**: 1–115. [107]

SUBJECT INDEX

Induce

chemics mutagens

(donization) Radiation